Current Trends in Eye Tracking Research

AF895894

Mike Horsley · Matt Eliot
Bruce Allen Knight · Ronan Reilly
Editors

Current Trends in Eye Tracking Research

Springer

Editors
Mike Horsley
Central Queensland University
Noosaville
Australia

Matt Eliot
Central Queensland University
Noosaville
Australia

Bruce Allen Knight
Central Queensland University
Noosaville
Australia

Ronan Reilly
National University of Ireland
Maynooth
Ireland

ISBN 978-3-319-02867-5 ISBN 978-3-319-02868-2 (e-Book)
DOI 10.1007/978-3-319-02868-2
Springer Cham Heidelberg New York Dordrecht London

Library of Congress Control Number: 2013956127

© Springer International Publishing Switzerland 2014
This work is subject to copyright. All rights are reserved by the Publisher, whether the whole or part of the material is concerned, specifically the rights of translation, reprinting, reuse of illustrations, recitation, broadcasting, reproduction on microfilms or in any other physical way, and transmission or information storage and retrieval, electronic adaptation, computer software, or by similar or dissimilar methodology now known or hereafter developed. Exempted from this legal reservation are brief excerpts in connection with reviews or scholarly analysis or material supplied specifically for the purpose of being entered and executed on a computer system, for exclusive use by the purchaser of the work. Duplication of this publication or parts thereof is permitted only under the provisions of the Copyright Law of the Publisher's location, in its current version, and permission for use must always be obtained from Springer. Permissions for use may be obtained through RightsLink at the Copyright Clearance Center. Violations are liable to prosecution under the respective Copyright Law.
The use of general descriptive names, registered names, trademarks, service marks, etc. in this publication does not imply, even in the absence of a specific statement, that such names are exempt from the relevant protective laws and regulations and therefore free for general use.
While the advice and information in this book are believed to be true and accurate at the date of publication, neither the authors nor the editors nor the publisher can accept any legal responsibility for any errors or omissions that may be made. The publisher makes no warranty, express or implied, with respect to the material contained herein.

Printed on acid-free paper

Springer is part of Springer Science+Business Media (www.springer.com)

Introduction

Eye tracking research and research methodologies are becomingly increasingly common in many disciplines from psychology and marketing to education and learning. This is because eye tracking research and research methodologies offer new ways of collecting data, framing research questions, and thinking about how we view, see, and experience the world. Researchers are also making new findings about the way that the visual system works and the way it interacts with attention, cognition, and behaviour.

As a result, research based on eye tracking research methods is increasing in every discipline. New studies using eye tracking technologies are continually being published and new applications of this innovative way of conducting research are being shared by researchers from every continent and country. Analysis of research using eye tracking methods is growing exponentially.

Current Trends in Eye Tracking Research presents a range of new research studies using eye tracking research and research methods from a wide variety of disciplines. The research studies have been chosen to chronicle the wide applications and uses of eye tracking research.

Current Trends in Eye Tracking Research is comprised of new and innovative studies using eye tracking research and research methods and showcases innovative ways of applying eye tracking technologies to interesting research problems. The book collects the research of over 55 researchers and academics currently using the eye tracking research and introduces the work of a number of eye tracking research laboratories and their key staff and research interests.

Current Trends in Eye Tracking Research is designed to explore a broad range of applications of this emerging and evolving research technology and to open the research space for wider sharing of new research methods and research questions. The book incorporates a number of new studies and introduces a number of new researchers to the practitioners of eye tracking research.

Current Trends in Eye Tracking Research also focuses on lessons learned in conducting eye movement research across multiple institutions, settings, and disciplines and innovative uses of existing technology as well as pioneering implementation of new technology in a range of research contexts and disciplines, key challenges, and important discoveries in moving from raw data to findings and challenges and opportunities related to situating individual research efforts in a larger research context.

Current Trends in Eye Tracking Research is divided into four key sections. Each section provides a central theme that integrates the many chapters in that section.

Part I is titled *Eye Tracking and the Visual System* and is concerned with research on the operation of the human visual system. The chapters in this section overview eye tracking and the human visual system research, and provide a series of chapters that examine how to explain the operation of the human visual system and fundamental research on the use of eye tracking to deepen and strengthen our understanding of the complexity of visual processes.

Part II is titled *Aligning Eye Tracking and EEG Data* and is concerned with research that reports on the alignment of EGG and eye tracking data. The chapters in this section overview fundamental research finding on how to link eye tracking and EEG data. The chapters in this section also address some critical research questions in integrating eye tracking data with other forms of data. The four chapters also overview current approaches to research on this alignment process.

Part III is titled *Eye Tracking and Marketing and Social Applications* and is concerned with eye tracking based research in a range of social science and marketing disciplines. Each chapter provides a different application from a different discipline—from marketing to aging, from mental illness to evaluating forgeries to understanding what people see when they read financial reports. Each chapter provides a novel application of eye tracking research methodology in the social sciences.

Part IV is titled *Eye Tracking and Education* and is concerned with research on learning using eye tracking methodologies. The five chapters focus on fundamental research problems in learning such as reading comprehension and the visual mechanics of comprehension, learning to read complex visual displays, and the development of student self-regulation skills. The section also explores the use of think aloud research protocols for multilingual learners.

Professor Mike Horsley
Director, Learning and Teaching
Education Research Centre
Central Queensland University, Australia

Acknowledgements

The editors would like to make a number of acknowledgements. The editors wish to acknowledge their families and friends, who have supported the project through many days and nights of research and writing. The editors also owe a debt of gratitude to the many research students who have stimulated their ideas and contributed to new ways of thinking about eye tracking research.

Also a special mention of gratitude is required for Vikki Walters from Central Queensland University. She contributed tirelessly to the formatting and initial editing of all the chapters.

The editors would also like to make a special mention of the all the chapter authors who responded in the most timely ways for redrafting and completing the many chapters in the Current Trends in Eye Tracking Research. Finally, the editors would like to acknowledge their publishers and the support of Bernadette Ohmer and Marianna Pascale from Springer.

Contents

Part II Aligning Eye Tracking and EEG Data

Part III Eye Tracking and Marketing and Social Applications

Contributors

Wan Fatimah Wan Ahmad Universiti Teknologi PETRONAS, Malaysia

Thierry Baccino Lutin Userlab, France

Stefanie I. Becker University of Queensland, Australia.

T. R. Beelders University of the Free State, South Africa

Melanie Birks James Cook University, Australia

Stephanie Alley Central Queensland University, Australia

P. J. Blignaut University of the Free State, South Africa

Alison Bowling Southern Cross University, Australia

James Breeze Objective Eye Tracking Pte Ltd, Australia

Donnel A. Briely University of Sydney, Australia

Marc Broadbent University of the Sunshine Coast, Australia

Nigel Chen University of Sydney, Australia

Michael Dambacher University of Konstanz, Germany

Richard Dewhurst Lund University, Sweden

Olaf Dimigen Humboldt University, Germany

Véronique Drai-Zerbib Lutin Userlab, France

Anja Draper Southern Cross University, Australia

A. G. Dyer RMIT University, Australia

Matt Eliot Central Queensland University, Australia

Bernard Fertil LSIS-UMR CNRS, France

Bryan Found Victoria Police, Australia

Amy L. Griffin University of New South Wales, Australia

Lyn Grigg University of New South Wales, Australia

Tracy Harwood Institute of Creative Technologies and Retail Lab, De Montfort University, UK

Kenneth Holmqvist Lund University, Sweden

Mike Horsley Central Queensland University, Australia

Martin Jones Retail Lab, De Montfort University, UK

Clare Kirtley University of Dundee, United Kingdom

Reinhold Kliegl University of Potsdam, Germany

Bruce Allen Knight Central Queensland University, Australia

En Li Central Queensland University, Australia

Yannick Lufimpu-Luviya LSIS-UMR CNRS, France

Ross G. Macdonald University of Dundee, United Kingdom

N. Mennie University of Nottingham, Malaysia

Djamel Merad LSIS-UMR CNRS, France

M. L. Merlino Kentucky State University, USA

Katy M. A. Mitchell University of Dundee, United Kingdom

N. Mohamad-Fadzil Universiti Kebangsaan, Malaysia

Z. Mohammed Universiti Kebangsaan, Malaysia

H. A. Mutalib Universiti Kebangsaan, Malaysia

Marcus Nyström Lund University, Sweden

Avni Pepe La Trobe University, Australia

Nayadin Persaud Central Queensland University, Australia

N. A. Razali Universiti Kebangsaan, Malaysia

Ronan Reilly National University of Ireland, Maynooth, Ireland

D. Rogers La Trobe University, Australia

N. H. Saliman Universiti Kebangsaan, Malaysia

Steven W. Savage University of Dundee, United Kingdom

M. M. Shahimin Universiti Kebangsaan, Malaysia

Jodi Sita La Trobe University, Australia

Ashok Sivaji MIMOS Berhad, Malaysia

Werner Sommer Humboldt University, Germany

Benjamin W. Tatler University of Dundee, United Kingdom

Corneel Vandelanotte Central Queensland University, Australia

Part I
Eye Tracking and the Visual System

The Active Eye: Perspectives on Eye Movement Research

Benjamin W. Tatler, Clare Kirtley, Ross G. Macdonald, Katy M. A. Mitchell and Steven W. Savage

Many of the behaviours that humans engage in require visual information for their successful completion. In order to acquire this visual information, we point our high-resolution foveae at those locations from which information is required. The foveae are relocated to new locations around three times every second. Eye movements, therefore, offer crucial insights into understanding human behaviour for two reasons. First, the locations selected for fixation provide us with insights into the changing moment-to-moment information requirements for the behaviours we engage in. Second, despite the fact that our eyes move, on average, three or four times per second, we are unaware of this and most of the time we are not conscious of where we are pointing our eyes. Thus, eye movements provide an ideal and powerful objective measure of ongoing cognitive processes and information requirements during behaviour. The utility of eye movements for understanding aspects of human behaviour is now recognised in a wide diversity of research disciplines. Indeed, the prevalence, diversity and utility of eye movements as research tools are evident from the contributions to be found in this volume.

In this brief overview, we take a glimpse at some of the emerging areas of study in eye movement research. To do so comprehensively and in a manner that reflects the impressive breadth of work contained in this volume would be a task that is both beyond the expertise of the authors and beyond the length of the chapter that we have been asked to write. Instead, we choose to introduce some emerging areas (with a clear bias towards our own research interests) that we feel will play an increasingly important role in shaping the direction that eye movement research will take over the coming years. A number of articles have reviewed eye movement research from particular perspectives and we refer the reader to several key reviews of eye movement research. Kowler (2011) provides a review of a wide variety of findings in eye movement research over the last 25 years or so. For a review of the link between eye movements and perception, see Schutz et al. (2011). Eckstein (2011) discusses contemporary and historical views on visual search and the roles that eye movements play in this process. While slightly earlier than the other reviews,

B. W. Tatler (✉) · C. Kirtley · R. G. Macdonald · K. M. A. Mitchell · S. W. Savage
University of Dundee, Dundee, UK
e-mail: b.w.tatler@activevisionlab.org

M. Horsley et al. (eds.), *Current Trends in Eye Tracking Research*,
DOI 10.1007/978-3-319-02868-2_1, © Springer International Publishing Switzerland 2014

Rayner (1998) offers an important overview of eye movements in reading. In this chapter, we focus upon the link between eye movements, perception and action.

1 Perception in Action

When we perceive our environment, we are acting in order to gain information that will help us perform the tasks in which we engage. In this way, perception is not simply the passive reception of information from our surroundings, but is an active part of how we operate in the world. This view is increasingly prominent in cognitive psychology (e.g. Hommel et al. 2001; Bridgeman and Tseng 2011). Indeed, Hommel et al. (2001) suggested that perception and action are 'functionally equivalent', with both processes working to allow us to build representations of the world around us. Perception and action processes appear to be linked in a bidirectional manner, so that each is able to affect the other: While perception informs the performance of action, action influences perceptual processes.

With this more active role for perception proposed, the question then is how to measure it. This is perhaps more difficult; as Bridgeman and Tseng (2011) state: Most effectors, such as the hands, double as tools for both action and perception. This is where eye movements become an invaluable tool: Eyes select and sample visual information and, thus, provide an online measure of perception, yet do not act directly upon the environment. Eye movements are an important means of investigating perception and action because they are perception in action, directed by the task to examine the world and allow us to complete the tasks set for us.

The importance of eye movements for coordinating perception and action can be seen clearly in the many studies that have made use of them. The eyes have two crucial functions: first, to gather information about the world and, second, to provide feedback during tasks, for example, when we manipulate an object. Using eye movements, these processes can be measured online as tasks are performed in both laboratory and real-world environments. For example, in the laboratory, Ballard et al. (1992) used a block-copying task in which participants moved a series of coloured squares from one location to a target area and arranged them to match a model depicting an arrangement of blocks that they had to recreate. The eye movements of the participants as they did this were shown to link strongly to the actions they were carrying out. The eyes followed a clear pattern of checking the model, preceding the hands to the blocks for the pick-up, then checking the model once more before placing the block in its correct position.

Ultimately, if we wish to understand the link between perception and action, we must do so in the context of natural behaviours conducted in real world environments. Mobile eye-tracking devices permit eye movement recordings to be made in untethered, real-world activities. This technological advancement has not only allowed researchers to study eye movements in the context of natural action but has also identified key insights into the relationship between vision and action that were not previously recognised. Mobile systems were developed in the 1950s by

Norman Mackworth and used in real environments in the 1960s (e.g. Mackworth and Thomas 1962; Thomas 1968). These devices were cumbersome and it was not until the 1990s that less obtrusive and more versatile mobile eye trackers were developed (Ballard et al. 1995; Land 1992). Using such devices, the tight link between vision and action is strikingly clear in real-world activities. Land et al. (1999) and Hayhoe (2000) measured participants' eye movements as they went through the stages of making a cup of tea or preparing a sandwich. Again, the findings demonstrated how vision acts to inform our behaviour: Throughout the constantly changing demands of the task, the participant's eyes precede the actions, fixating the required objects for the next step in the process. Furthermore, Hayhoe (2000) showed that when making a sandwich, the action intention could influence the deployment of attention. Participants were seated in front of either a non-cluttered tabletop, containing only the items needed for the sandwich-making task, or a busier tabletop, containing irrelevant objects along with the important ones. While these irrelevant objects were fixated, the greatest percentage of fixations came in the viewing period before the task began. Once the participants had started, task-irrelevant objects were rarely fixated: Almost all fixations were made to task-relevant items.

These examples illustrate the intimate link between vision and action and the manner in which eyes are deployed on a moment-to-moment basis to gather information and provide feedback for actions. The bidirectional nature of the perception–action coupling is evident in tasks where perceptual decisions are made in the presence of action. Indeed, before an action has begun, the intention to carry out an action influences how participants view a scene, even when the intention is created by a seemingly minor manipulation such as the performance of a particular grip type. For example, Bekkering and Neggers (2002) asked participants to find targets based on colour or orientation, in order to grasp or point at them. For orientation-defined targets, when participants searched to grasp the object, they made fewer incorrect saccades to the distracter objects compared to the situation when targets were defined by colour. This difference between colour- and orientation-based search was absent when participants were searching only to point to the object rather than grasp. The preparation of the grasp led to enhanced processing of the relevant feature for the action, in this case the objects' orientation, and, thus, detection of targets defined by that feature was enhanced. Similarly, Fagioli et al. (2007) asked participants to prepare different types of gestures, such as pointing or grasping. Before they could carry out these prepared actions, participants were given a detection task, which required them to find the odd one out in a set of objects. This target was defined by either its location or its orientation. Preparing a pointing gesture resulted in participants spotting the location oddity sooner, while the orientation oddity was spotted soonest when a grasping gesture was prepared. Thus, even when the action prepared did not directly relate to the following task, the enhanced processing of relevant dimensions was continued. Symes et al. (2008) used this action–preparation paradigm in a different task setting to look at change detection. Here, power and precision grip types were formed by participants during change blindness trials, and it was demonstrated that change detection improved for objects whose size matched the grip type held by the participant.

In all three studies that have used this paradigm, the effect is clear: By forming an intention to act, sometimes not even requiring the actual action posture itself, the perception of the environment is changed. By forming an intention to grip, information that informs gripping, like the orientation of an object or its size, becomes more relevant and prioritised in the examination of the scene. Our perceptions are influenced by our intentions to act, and, thus, perception is used here to gain the information we know we may require.

Eye movements are an invaluable measure in a paradigm such as this. Not all the above studies used eye movements as a measure: They are used most in the Bekkering and Neggers (2002) study, but it is clear how eye movements can add to this kind of research. They give us the ability to see how the influence of the intention to act unfolds across the task and to see what measures are most affected—saccade time, fixation duration and scan path, amongst many others. As in the studies by Land et al. (1999) and Hayhoe (2000), eye movements give us a window onto how perception operates across the course of a task, from the first intention to act and through the process of carrying out the task itself.

The relationship between perception and action means not only that eye movements offer a crucial tool for understanding this relationship but also that we must be cautious when studying eye movements and perception in the absence of action. It is becoming increasingly clear that any exploration of visual perception and eye movements should consider the possible influence of action. For example, if we wish to understand memory representations, it is important to consider these in the context of real environments (Tatler and Land 2011; Tatler and Tatler 2013) and natural behaviours (Tatler et al., 2013). Similarly, any understanding of the factors that underlie decisions about when and where to move the eyes must consider these decisions in the context of natural behaviours (Tatler et al. 2011; Tatler this volume). Of course, this is not to say that all eye movement research should be conducted in real-world settings using mobile eye trackers. Many of the behaviours we engage in involve being seated in front of a display screen of some sort: for example, working at a computer or using a handheld computer device. However, even in these situations, an understanding of perception in the context of action is important. When using the Internet, we do not passively watch but actively interact with the viewed content—scrolling, clicking and entering text as needed. Similarly, computing devices are increasingly using touch and gesture interfaces. The bidirectional relationship between perception and action, therefore, necessitates that these interactive situations are studied in a manner that is relevant to the interactions being undertaken.

2 Social Interaction

As we increasingly move towards the study of eye movements and perception in ecologically valid situations, it becomes clearer that not only might it be inappropriate to study vision in isolation from action in many circumstances but it might also be inappropriate to study individuals behaving in isolation from other individuals. Humans are highly social beings and many of the behaviours we engage in are carried out in the presence of, in collaboration with or in competition with others.

We have a strong tendency from an early age to attend to the same locations that others are attending to. The intimate link between eye direction and our behavioural goals and intentions means that eyes provide a strong cue to understand where another individual is attending. Human eyes appear uniquely well equipped for signalling eye direction to others: We have whiter and more exposed scleras than other great apes (Kobayashi and Kohshima 1997), and the high contrast between the sclera and iris in human eyes provides easily detectable directional signals. Indeed, we are extremely good at detecting the direction of another person's gaze (Symons et al. 2004). Not only are we able to detect where someone else is looking but we are also able to use this information to orient our own eyes to the same locations in space. This tendency to follow the gaze direction of another individual can be seen from as early as infancy (e.g. Senju and Csibra 2008), and it has been suggested that it leads to a shared mental state that is central to the development of 'Theory of Mind' (Baron-Cohen 1995).

How an individual's gaze direction cues the gaze direction of an observer has been the subject of much of the eye movement research on social attention. Laboratory-based experiments using Posner (1980)-type paradigms to investigate the attentional effects of gaze cues have mostly found that participants' eyes reflexively orient to gazed locations (Ricciardelli et al. 2002; Tipples 2002; Galfano et al. 2012). Studies using more complex scenes appear to support these findings; when viewing images containing people, observers have a strong tendency to fixate on the eyes of individuals (Birmingham et al. 2009) or the objects that they are looking at (Castelhano et al. 2007). However, recent studies using real-world settings suggest that this tendency might be critically modulated by the social factors during natural interactions.

Laidlaw et al. (2011) recorded participants sitting in a waiting room and found that they were more likely to look at a confederate displayed on a video monitor than the same confederate present in the waiting room. Similar results for gaze following rather than seeking were found by Gallup et al. (2012). They observed people walking past an attractive item in a hallway and found that people were more likely to look in the same direction as somebody walking in front of them than somebody walking towards them. The results of these studies were explained by their respective authors as being due to participants trying to avoid potential interactions with strangers. It seems that the mere potential for a social interaction changes the way in which we seek and follow gaze cues. These findings highlight the limitations of using laboratory-based paradigms to investigate natural gaze-cueing behaviour.

Freeth et al. (2013) investigated the effect of the presence of a speaker on a listener's gaze behaviour when a social interaction was actually taking place. Participants answered questions from an experimenter who was either physically present or on a video monitor. There was no significant difference found across conditions in terms of the amount of time participants spent looking at the face of the speaker. However, the presence of eye contact caused participants to look at the speaker's face for longer in the real-world condition only. This shows that the effect of a speaker's gaze behaviour on the eye movements of a listener is dependent on the speaker being present.

If we are to understand the use of gaze cues in social interactions, then, it is important to remember that in most natural situations, gaze cues are not employed in isolation: Typically, these occur as part of an interaction and are accompanied by other communicative signals like gestures and spoken language. In recent years, research carried out in more ecologically valid environments (including real-world paradigms) has not only challenged the idea of reflexive gaze following but has also been able to consider important aspects of natural gaze cue utilisation not considered in Posner-type tasks. In particular, there has been emerging interest in studying the role of gaze cues in natural communication and the effects of social factors on gaze seeking and following.

In natural communication, gaze cues are usually used alongside spoken language. Therefore, understanding the interaction between these cues and spoken language is vital for understanding how gaze cues are naturally utilised. Hanna and Brennan (2007) used a real-world communicative paradigm to investigate natural dialogue and gaze cues. They found that listeners in a block-identification task would use the gaze cues of a speaker to find a target block before the point of verbal disambiguation, showing that gaze cues are used to aid and speed up communication during a collaborative task. In an experiment with more controlled language stimuli (Nappa et al. 2009), young children were found to use the object-directed gaze cues of an adult (presented on a screen) to interpret the meaning of made-up verbs used in spoken sentences. A similar study by Stuadte and Crocker (2011) used an adult population and showed participants videos of a robot describing the spatial and featural relations between a series of visible items, whilst providing gaze cues. The robot made incorrect statements about the relations between the items that had the potential to be corrected in two different ways. The experimenters found that participants would correct in the way that used the gazed item as the object of the sentence, suggesting that they were inferring meaning from the robot's gaze. These results collectively show that, when used alongside language, gaze cues are used to solve ambiguities in spoken language and aid in the understanding of another's intentions.

Other research on gaze cues and spoken language has focused on how changing language can affect the utilisation of gaze cues. In a task in which gaze cues were inessential for its successful completion (Knoeferle and Kreysa 2012), participants followed gaze cues more often when hearing a German sentence in the common subject–verb–object (SVO) structure than the less common (but still grammatically legal) object–verb–subject (OVS) structure. The authors suggested this finding was due to the extra difficulty in processing the OVS sentences leaving fewer processing resources for gaze cue utilisation. Macdonald and Tatler (2013) investigated the effect of changing language specificity on the use of gaze cues using a real-world communicative task, involving a one-to-one interaction between an instructor (experimenter) and participant. The instructor manipulated his use of gaze as well as the specificity of his instructions in a simple block-building task. Participants were found to only seek and follow gaze cues when the language was ambiguous, suggesting that gaze cues are used flexibly, depending on other information that is available. It, therefore, appears that the difficulty and specificity of language affects the utilisation of gaze cues during communication.

The above results show the value of using real-world paradigms when investigating gaze utilisation, as the effects of language and social context on gaze behaviour can be taken into account. Gaze cues have been shown to support and disambiguate spoken language as well as provide insight into a speaker's intentions. Our gaze-seeking and -following behaviour has been shown to be sensitive to potential social interactions and our social perceptions of those with whom we interact. The benefits of using more ecologically valid paradigms across different areas of social cognition and social neuroscience are the subject of a number of recent review articles (Risko et al. 2012; Skarratt et al. 2012; Przyrembel et al. 2012), and with technological advances providing more opportunities (see Clark and Gergle 2011 for discussion), the trend for investigating social interactions in the real world is likely to continue.

3 Magic and Misdirection

While we are still very much discovering the role of the eyes in natural social interaction, magicians (and other experts in misdirection) seem to have possessed mastery of this situation for centuries (e.g. see Kuhn and Martinez 2011; Lamont and Wiseman 1999). Misdirection, in the broadest sense, is the means by which a magician diverts the audience's attention from the mechanics of a trick, for example, the palming of a coin or pocketing of a card (Kuhn and Tatler 2011). More specifically, misdirection is an umbrella term for a number of different behaviours, including gesture, speech, posture and gaze cues. A magician must include some or all of these aspects at once for misdirection to be successful. As yet, our understanding of the cumulative effect that these behaviours have on an audience's attention is incomplete, and Kuhn et al. (2008), amongst others, have argued for more research in this area because of the rich insights it can potentially offer about psychological processes. Much can be learnt about visual perception and cognition from studying the conditions in which we fail to perceive or understand events, or in which we can be made to believe that we have seen something that did not occur (Kuhn and Martinez 2011). Magic, therefore, offers a medium in which we can study psychological processes in an ecologically valid, real-world situation, but still manipulate the nature of cues used by the magician in order to misdirect the observer.

Kuhn and Tatler (2005) were the first researchers to examine an observer's eye movements as they watched a magic trick. They developed a trick in which a magician (Gustav Kuhn) made a cigarette and lighter disappear using a combination of two methods of misdirection—gaze cues and gesture—to conceal a simple drop of each object onto the magician's lap. The trick was unusual in that the drop of the cigarette was performed in full view of the participant: The magician dropped the cigarette from a height at which it would be visible for about 120 ms as it dropped. This trick was performed live in front of participants, in a one-to-one interaction with the magician; half of the participants expected a magic trick, half did not. Surprisingly, only two out of the 20 participants noticed the drop on the first performance of the trick; however, when the performance was repeated, all noticed the drop. At the time that the magician dropped the cigarette from one hand, participants tended to be looking either at the magician's other hand or at his face. This was the case irrespective of whether the participant was expecting to see a magic trick or not and persisted even on the second trial when all participants perceived the drop. This led Kuhn and Tatler (2005) to conclude that prior information seemed to have no effect on strategic eye movements in this situation as they were similar across both groups and that a magician manipulates an observer's attention rather than their gaze because the eye movement behaviour was the same.

What aspects of the magician's performance resulted in the successful misdirection as the cigarette was dropped? Kuhn and Tatler (2005) ruled out the occurrence of blinks or eye movements, or the distance into peripheral vision of the dropping cigarette as possible reasons for the success of the magician's misdirection and speculated that it was the gaze direction of the magician that was crucial for the misdirection in this trick. Consistent with this possibility, the correlation between the gaze direction of the magician and the observer was highest at the misdirection events (the two object drops) during this performance (Tatler and Kuhn 2007) and, at these times, most participants were fixating the gaze target of the magician. However, these correspondences alone are not sufficient to claim that it was the magician's gaze that was central to this misdirection because the misdirecting gaze was accompanied by movement and sound cues at the magician's gaze target: The magician not only looked at the other hand when dropping an object but also waved it and clicked his fingers. In order to tease apart gaze cues from these other potential cues for misdirection, Kuhn et al. (2009) used a modified version of the trick using only a single drop (of a cigarette lighter). Crucially, two versions of the performance were filmed: In each case, the misdirection cues from the non-dropping hand (movement, etc.) were the same at the time of the drop, but in one video the normal misdirecting gaze was given by the magician, whereas in the other the magician maintained fixation on the hand from which the lighter was being dropped. The results showed that observers were significantly less likely to detect the drop when the gesture was supported by gaze cues away from the concealed event. Furthermore, it was shown that when observers watched the non-misdirected trick (where the gesture was not supported by gaze cues), they fixated significantly closer to the dropping lighter (Kuhn et al. 2009). These results demonstrate that the magician's gaze is a crucial cue for both where the observer looks during the performance and whether or not the magic is successful.

As a means of understanding the importance and use of gaze cues in interaction, misdirection is a powerful tool. First, we can manipulate the manner in which gaze cues are provided or supported by other cues and study the effects of these manipulations on observers' gaze behaviour and perception of the events. Second, we can use magic performances and an ecologically valid setting to understand more about the use and understanding of gaze cues in special populations (Kuhn et al. 2010). When watching an illusion, individuals with autistic spectrum disorders (ASDs) showed the same gaze behaviour as typically developing individuals but were slower to launch saccades to magician's face. These results challenge common notions that people on the autistic spectrum have general problems with social attention: Here, general behaviour was very much like that found in individuals without ASD, and the difference was rather subtle. While little has been done with special populations to date, magic offers a potentially valuable research tool for exploring aspects of social attention and wider visual cognition in these populations.

There is growing interest in the psychology of magic to explore a range of issues in cognitive psychology (Kuhn et al. 2008; Martinez-Conde and Macknik 2008; Macknik et al. 2010). Given the inherently visual nature of many of the striking magical performances and their reliance on illusion, misdirection and other magical acts that at least partly involve our visual sense, it seems likely that eye movement recordings will play a central role in this emerging field of research.

4 Distraction

In magical performances, we often fail to notice what should be an easily detectable visual event because we have been misdirected by a magician with mastery in controlling our attention. However, failing to detect what should be an obvious event is not restricted to situations in which we have been actively misdirected: All too commonly, we may miss external events and this can occur for a number of reasons and with a number of consequences. Failures to detect external visual events can be particularly problematic in some situations: For example, failing to detect a hazard when driving can have critical safety implications. In driving situations, a key factor that can result in failures to detect hazards is being distracted from the driving task in some way.

Researchers have quantified the variety of different causes for driver distraction into three major categories: visual, cognitive and physical. Although the ultimate outcome of these distractions is the same—an increase in crash risk—the underlying cognitive mechanisms are different (Regan et al. 2008; Anstey et al. 2004). Research has also shown that visual and cognitive task demands affect eye movements within driving situations in qualitatively different ways.

Visual distraction can be caused by a variety of different factors. The primary commonality, however, is the increase in the visual load, which is typically achieved by including an additional secondary visual task such as planning a route in a navigation system or by manipulating the visual information within the driving scene itself (Konstantopoulos et al. 2010). Interestingly, visual distraction appears to influence eye movement behaviour in a number of ways. Di Stasi et al. (2010) manipulated visual task demand by increasing traffic density. Results indicated that this increase in the visual content of the driving scene results in slower saccade peak velocities. Another measure found to have been effected by visual load is blink durations (Recarte et al. 2008; Veltman and Gaillard 1996). Research by Ahlstrom and Friedman-Berg (2006) has indicated a linear decrease in blink durations as a function of visual task demand. Benedetto et al. (2011) examined the effects of interacting with an in-vehicle information system (IVIS) on drivers' blink rates and blink durations during a simulated lane-changing task. Analyses indicated that as visual task demand was increased, blink durations significantly decreased. Results also indicated that blink rates were not significantly affected by visual task demand. It was argued that changes in eye movement metrics such as fixation number, fixation duration, saccade amplitude and gaze position were the result of gaze switching between primary and secondary visual tasks. However, as the observed pattern of gaze switching cannot account for the decrease in blink durations, this measure has been considered a reliable indicator of driver visual task demand (Benedetto et al. 2011).

Research has shown that cognitive task demand affects eye movement measures in a qualitatively different manner than visual task demand manipulations. Recently, results from a hazard perception study have indicated that saccade peak velocity was significantly increased as a result of increased cognitive task demand (Savage et al. 2013). Cognitive task demand influences the spread of fixations in driving situations: When cognitive load is high, there is an increased concentration of gaze towards the centre of the road (Recarte and Nunes 2003; Savage et al. 2013). Cognitive load also influences blinking behaviour in drivers: When cognitive load is high, people blink more often (Recarte et al. 2008) and for a longer duration (Savage et al. 2013).

The fact that increasing either visual or cognitive load results in changes in eye movement behaviour means that we might be able to exploit these characteristic eye movement changes to identify periods of distraction during driving (Groeger 2000; Velichkovsky et al. 2002). The safety implications of this are self-evident: If eye movements can be used as a diagnostic marker of distraction, then we can use these to detect periods of distraction and intervene, alerting the driver to the danger of their current state. Importantly, there is clear utility in being able to differentiate visual and cognitive distraction: Any intervention may need to be tailored to whether the current situation involves an unusually high visual load—which may be due to external events in the environment—or an unusually high cognitive load—which may be more likely due to distractions by conversation and contemplation of language. Intervening appropriately may be safety critical in some situations.

Importantly, not only do we find that visual and cognitive load appear to influence eye movement behaviour in driving situations but the above findings also suggest that the manner in which these two types of load impact eye movement behaviour may be rather different. In particular, saccade peak velocity was significantly reduced as a result of increased visual load (Di Stasi et al. 2010) but significantly increased as a result of increased cognitive load (Savage et al. 2013). Similarly, increases in visual task demand have been shown to result in significantly shorter blinks (Ahlstrom and Friedman-Berg 2006), whereas blink durations increase in situations of high cognitive load (Savage et al. 2013).

As eye movement metrics are affected in qualitatively and quantitatively different ways by both cognitive and visual demand manipulations, eye movements offer a potentially powerful diagnostic tool with which to examine the interaction of the different attention networks as well as assess the driver's current mental state. The use of eye movement measures as diagnostic markers for mental state in driving is an emerging area with important practical implications. At present, there is a need to continue to identify those aspects of eye movement characteristics that will provide robust and specific markers of particular mental states before these can be applied directly to in-vehicle interventions. This research effort somewhat mirrors research effort in the potential use of eye movements as diagnostic markers of disease in clinical settings: Many neurological and psychiatric conditions are associated with atypical eye movement behaviours (Diefendorf and Dodge 1908; Lipton et al. 1983; Trillenberg et al. 2004). While, for some conditions, it is now possible to distinguish affected individuals from healthy controls with an impressive degree of accuracy (Benson et al. 2012), a remaining challenge in this field is to identify oculomotor markers that are specifically diagnostic of particular disorders.

5 Conclusion

Eye movements provide powerful research tools for those interested in a wide variety of aspects of human cognition and behaviour. The selective nature of viewing—high acuity sampling is restricted in both space and time—means that the locations selected for scrutiny by high acuity vision reveal much about the moment-to-moment demands of ongoing cognition and action. It is, therefore, unsurprising that the use of eye tracking as a behavioural measure is now very widespread and encompasses a diversity of research disciplines. Indeed, the diversity of applications of eye tracking is reflected in the contributions to this volume. Two key aspects of eye movement behaviour are becoming increasingly clear that straddle the different research interests for which eye tracking is employed. First, the intimate link between vision and action means that visual perception and cognition should be studied in the presence of the actions that we are interested in characterising. Second, the intimate link between eye movements and ongoing cognition means that eye movements offer important potential diagnostic markers of mental state. In our continuing efforts to produce ecologically valid accounts of human behaviour in a variety of situations, eye movements are likely to assume an increasingly pivotal role in shaping our understanding of perception, cognition and action.

References

Ahlstrom, U., & Friedman-Berg, F. J. (2006). Using eye movement activity as a correlate of cognitive workload. *International Journal of Industrial Ergonomics, 36,* 623–636.

Anstey, K. J., Wood, J., Lord, S., & Walker, J. G. (2004). Cognitive, sensory and physical factors enabling driving safety in older adults. *Clinical Psychology review, 25,* 45–65.

Ballard, D. H., Hayhoe, M. M., Li, F., & Whitehead, S. D. (1992). Hand-eye coordination during sequential tasks. *Philosophical Transactions of The Royal Society B, 337,* 331–339.

Ballard, D. H., Hayhoe, M. M., & Pelz, J. B. (1995). Memory representations in natural tasks. *Journal Of Cognitive Neuroscience, 7*(1), 66–80.
Baron-Cohen, S. (1995). *Mindblindness: An essay on autism and theory of mind.* Cambridge: MIT Press.
Bekkering, H., & Neggers, S. F. W. (2002). Visual search is modulated by action intentions. *Psychological Science, 13*(4), 370–374.
Benedetto, S., Pedrotti, M., Minin, L., Baccino, T., Re, A., & Montanari, R. (2011). Driver workload and blink duration. *Transportation Research: Part F, 14,* 199–208.
Benson, P. J., Beedie, S. A., Shephard, E., Giegling, I., Rujescu, D., & Clair, D. S. (2012). Simple viewing tests can detect eye movement abnormalities that distinguish schizophrenia cases from controls with exceptional accuracy. *Biological psychiatry, 72*(9), 716–724. doi:10.1016/j.biopsych.2012.04.019..
Birmingham, E., Bischof, W. F., & Kingstone, A. (2009). Get real! resolving the debate about equivalent social stimuli. *Visual Cognition, 17*(6–7), 904–924.
Bridgeman, B., & Tseng, P. (2011). Embodied cognition and the perception-action link. *Physics of life reviews, 8*(1), 73–85.
Castelhano, M. S., Wieth, M., & Henderson, J. M. (2007). I see what you see: Eye movements in real-world scenes are affected by perceived direction of gaze. *Lecture Notes in Computer Science, 4840,* 251–262.
Clark, A. T., & Gergle, D. (2011). Mobile dual eye-tracking methods: challenges and opportunities. Paper presented at DUET 2011: Duel Eye Tracking workshop at ECSCW 2011
Di Stasi, L. L., Renner, R., Staehr, P., Helmert, J. R., Velichkovsky, B. M., Cañas, J. J., Catena, A., & Pannasch, S. (2010). Saccadic peak velocity is sensitive to variations in mental workload in complex environments. *Aviation, Space, and Environmental Medicine, 81,* 413–417.
Diefendorf, A. R., & Dodge, R. (1908). An experimental study of the ocular reactions of the insane from photographic records. *Brain. A Journal of Neurology, 31,* 451–489.
Eckstein, M. P. (2011). Visual search: a retrospective. *Journal of Vision, 11*(5):14, 1–36.
Fagioli, S., Hommel, B., & Schubotz, R. I. (2007). Intentional control of attention: Action planning primes action-related stimulus dimensions. *Psychological research, 71*(1), 22–29.
Freeth, M., Foulsham, T., & Kingstone, A. (2013). What affects social attention? Social presence, eye contact and autistic traits. *PLoS ONE, 8*(1), e53286.
Galfano, G., Dalmaso, M., Marzoli, D., Pavan, G., Coricelli, C., & Castelli, L. (2012). Eye gaze cannot be ignored (but neither can arrows). *Quarterly Journal of Experimental Psychology, 65,* 1895–1910.
Gallup, A. C., Chong, A., & Couzin, I. D. (2012). The directional flow of visual information transfer between pedestrians. *Biology Letters, 8*(4), 520–522.
Groeger, J. A. (2000). *Understanding driving.* Hove: Psychology Press.
Hanna, J. E., & Brennan, S. E. (2007). Speakers' eye gaze disambiguates referring expressions early during face-to-face conversation. *Journal of Memory and Language, 57,* 596–615.
Hayhoe, M. M. (2000). Vision using routines: A functional account of vision. *Visual Cognition, 7,* 43–64.
Hommel, B., Müsseler, J., Aschersleben, G., & Prinz, W. (2001). The theory of event coding (TEC): a framework for perception and action planning. *The behavioural and brain sciences, 24*(5), 849–78; discussion 878–937
Knoeferle, P., & Kreysa, H. (2012). Can speaker gaze modulate syntactic structuring and thematic role assignment during spoken sentence comprehension? *Frontiers in Psychology, 3,* 538.
Kobayashi, H., & Kohshima, S. (1997). Unique morphology of the human eye. *Nature, 387,* 767–768.
Konstantopoulos, P., Chapman, P., & Crundall, D. (2010). Driver's visual attention as a function of driving experience and visibility. Using a driving simulator to explore drivers' eye movements in day, night and rain driving. *Accident Analysis and Prevention, 42,* 827–834.
Kowler, E. (2011). Eye movements: The past 25 years. *Vision Research, 51*(13): 1457-83.
Kuhn, G., & Tatler, B. W. (2005). Magic and fixation: Now you don't see it, now you do. *Perception, 34,* 1155–1161.

Kuhn, G., & Martinez, L. M. (2011). Misdirection–past, present, and the future. *Frontiers in Human Neuroscience, 5,* 172.

Kuhn, G., & Tatler, B. W. (2011). Misdirected by the gap: The relationship between inattentional blindness and attentional misdirection. *Consciousness and Cognition, 20*(2), 432–436. doi:10.1016/j.concog.2010.09.013.

Kuhn, G., Amlani, A. A., & Rensink, R. A. (2008). Towards a science of magic. *Trends in Cognitive Sciences, 12*(9), 349–354. doi:10.1016/j.tics.2008.05.008.

Kuhn, G., Kourkoulou, A., & Leekam, S. R. (2010). How magic changes our expectations about autism. *Psychological Science, 21*(10), 1487–1493. doi:10.1177/0956797610383435..

Kuhn, G., Tatler, B. W., & Cole, G. (2009). You look where I look! Effects of gaze cues on overt and covert attention in misdirection. *Visual Cognition, 17,* 925–944.

Laidlaw, K. E. W., Foulsham, T., Kuhn, G., & Kingstone, A. (2011). Potential social interactions are important to social attention. *Proceedings of the National Academy of Sciences, 108,* 5548–5553

Lamont, P., & Wiseman, R. (1999). *Magic in theory.* Hartfield: Hermetic Press.

Land, M. F. (1992). Predictable eye-head coordination during driving. *Nature, 359*(6393), 318–320.

Land, M. F., Mennie, N., & Rusted, J. (1999). The roles of vision and eye movements in the control of activities of daily living. *Perception, 28,* 1311–1328.

Lipton, R. B., Levy, D. L., Holzman, P. S., & Levin, S. (1983). Eye movement dysfunctions in psychiatric patients. *Schizophrenia Bulletin, 9*(1), 13–32.

Macdonald, R. G., & Tatler, B. W. (2013). Do as eye say: Gaze cueing and language in a real-world social interaction. *Journal of Vision, 13*(4):6, 1–12.

Macknik, S. L., Martinez-Conde, S., & Blakeslee, S. (2010). *Sleights of Mind: What the Neuroscience of Magic Reveals about Our Everyday Deceptions.* New York: Henry Holt and Company

Mackworth, N. H., & Thomas, E. L. (1962). Head-mounted eye-movement camera. J. Opt. Soc. *America, 52,* 713–716.

Martinez-Conde, S., & Macknik, S. L. (2008). Magic and the brain. *Scientific American, 299*(6), 72–79.

Nappa, R., Wessel, A., McEldoon, K. L., Gleitman, L. R., & Trueswell, J. C. (2009). Use of speaker's gaze and syntax in verb learning. *Language learning and Development (Cambridge, England), 5,* 203–234.

Posner, M. I. (1980). Orienting of attention. *The Quarterly Journal of Experimental Psychology, 32,* 3–25.

Przyrembel, M., Smallwood, J., Pauen, M., & Singer, T. (2012). Illuminating the dark matter of social neuroscience: Considering the problem of social interaction from philosophical, psychological, and neuroscientific perspectives. *Frontiers in Human Neuroscience, 6,* 190.

Rayner, K. (1998). Eye movements in reading and information processing: 20 years of research. *Psychological bulletin, 124*(3), 372–422.

Recarte, M. A., & Nunes, L. M. (2003). Mental workload while driving: Effects on visual search, discrimination and decision making. *Journal of Experimental Psychology: Applied, 9,* 119–137.

Recarte, M. A., Pérez, E., Conchillo, A., & Nunes, L. M. (2008). Mental workload and visual impairment: Differences between pupil, blink, and subjective rating. *The Spanish Journal of Psychology, 11,* 374–385.

Regan, M. A., Lee, J. D., & Young, K. L. (2008). *Driver distraction: Theory, effects and mitigation.* Boca Raton, FL: CRC Press.

Ricciardelli, P., Bricolo, E., Aglioti, S. M., & Chelazzi, L. (2002). My eyes want to look where your eyes are looking: Exploring the tendency to imitate another individual's gaze. *Neuroreport, 13*(17), 2259–2264.

Risko, E. F., Laidlaw, K., Freeth, M., Foulsham, T., & Kingstone, A. (2012). Social attention with real versus reel stimuli: Toward an empirical approach to concerns about ecological validity. *Frontiers in Human Neuroscience, 6,* 143.

Savage, S. W., Potter, D. D., & Tatler, B. W. (2013). Does preoccupation impair hazard perception? A simultaneous EEG and Eye Tracking study. *Transportation Research Part F, 17,* 52–62.
Schütz, A. C., Braun, D. I., & Gegenfurtner, K. R. (2011). Eye movements and perception: A selective review. *Journal of Vision, 11*(5), 1–30.
Senju, A., & Csibra, G. (2008). Gaze following in human infants depends on communicative signals. *Current Biology, 18,* 668–671.
Skarratt, P. A., Cole, G. G., & Kuhn, G. (2012). Visual cognition during real social interaction. *Frontiers in Human Neuroscience, 6,* 196.
Staudte, M., & Crocker, M. W. (2011). Investigating joint attention mechanisms through spoken human-robot interaction. *Cognition, 120,* 268–291.
Symes, E., Tucker, M., Ellis, R., Vainio, L., & Ottoboni, G. (2008). Grasp preparation improves change detection for congruent objects. *Journal of experimental psychology. Human perception and performance, 34*(4), 854–871.
Symons, L. A., Lee, K., Cedrone, C. C., & Nisimura, M. (2004). What are you looking at? Acuity for triadic eye gaze. *Journal of General Psychology, 131,* 451–469.
Tatler, B. W., & Kuhn, G. (2007). Don't look now: The magic of misdirection. *Eye Movements: A window on mind and brain*, 697–714
Tatler, B. W., & Land, M. F. (2011). Vision and the representation of the surroundings in spatial memory. *Philosophical Transactions of the Royal Society B-Biological Sciences, 366*(1564), 596–610. doi:10.1098/rstb.2010.0188.
Tatler, B. W., & Tatler, S. L. (2013). The influence of instructions on object memory in a real-world setting. *Journal of Vision, 13*(2):5, 1–13.
Tatler, B. W., Hayhoe, M. M., Land, M. F., & Ballard, D. H. (2011). Eye guidance in natural vision: Reinterpreting salience. *Journal of Vision, 11(5), 5,* 1–23. doi:10.1167/11.5.5.
Tatler, B. W., Hirose, Y., Finnegan, S. K., Pievilainen, R., Kirtley, C., & Kennedy, A. (2013). Priorities for selection and representation in natural tasks. Philosophical Transactions of the Royal Society B-Biological Sciences, 368(1628), 20130066. doi:10.1016/j.actpsy.2004.01.004
Thomas, E. L. (1968). Movements of the eye. *Scientific American, 219*(2), 88–95.
Tipples, J. (2002). Eye gaze is not unique: Automatic orienting in response to uninformative arrows. *Psychonomic Bulletin & Review, 9,* 314–318.
Trillenberg, P., Lencer, R., & Heide, W. (2004). Eye movements and psychiatric disease. *Current Opinion in Neurology, 17*(1), 43–47.
Velichkovsky, B. M., Rothert, A., Kopf, M., Dornhoefer, S. M., & Joos, M. (2002). Towards an express-diagnostics for level of processing and hazard perception. *Transportation Research. Part F, 5,* 145–156.
Veltman, J. A., & Gaillard, A. W. K. (1996). Physiological indices of workload in a simulated flight task. *Biological Psychology, 42,* 323–342.

Eye Movements from Laboratory to Life

Benjamin W. Tatler

The manner in which we sample visual information from the world is constrained by the spatial and temporal sampling limits of the human eye. High acuity vision is restricted to the small central foveal region of the retina, which is limited to just a few degrees of visual angle in extent. Moreover, visual sampling is effectively limited to when the retinal image is relatively stabilised for periods of fixation (Erdmann and Dodge 1898), which last on average around 200–400 ms when viewing text, scenes or real environments (Land and Tatler 2009; Rayner 1998). It is clear from these severe spatiotemporal constraints on visual sampling that high acuity vision is a scarce resource and, like any scarce resource, it must be distributed carefully and appropriately for the current situation.

The selection priorities that underlie decisions about where to direct the eyes have interested researchers since eye movement research was in its infancy. While stimulus properties were shown to influence fixation behaviour (McAllister 1905), it was soon recognised that the relationship between the form of the patterns viewed and the eye movements of the observer was not as close as early researchers had expected (Stratton 1906). Moreover, the great variation in fixation patterns between individuals (McAllister 1905) made it clear that factors other than stimulus properties were likely to be involved in allocating foveal vision.

In light of evidence gathered from observers viewing the Müller-Lyer illusion (Judd 1905), Poggendorff illusion (Cameron and Steele 1905) and Zöllner illusion (Judd and Courten 1905), Judd came to the conclusion that "the actual movements executed are in no small sense responses to the verbal stimuli which the subject receives in the form of general directions. The subject reacts to the demands imposed upon him by the general situation… The whole motive for movement is therefore not to be sought in the figures themselves" (Judd, 1905, p. 216–217).

The relative importance of external factors relating to the stimulus properties and internal factors relating to goals of the observer became a prominent theme in eye movement research and continues to underlie many aspects of contemporary eye movement research. While early research in this domain used simple patterns and

B. W. Tatler (✉)
University of Dundee, Dundee, UK
e-mail: b.w.tatler@activevisionlab.org

M. Horsley et al. (eds.), *Current Trends in Eye Tracking Research*,
DOI 10.1007/978-3-319-02868-2_2, © Springer International Publishing Switzerland 2014

line illusions (due to technological limitations in display and recording devices), more recent research has considered how we view complex scenes in an attempt to produce an ecologically valid account of eye guidance.

1 Eye Guidance in Scene Viewing

When viewing complex scenes, fixations are allocated preferentially to certain locations, while other locations receive little or no scrutiny by foveal vision (Buswell 1935). Moreover, the regions selected for fixations are similar between individuals: different people select similar locations in scenes to allocate foveal vision to (Buswell 1935; Yarbus 1967). Such similarity in fixation behaviour implies common underlying selection priorities across observers. Buswell (1935) recognised that these common selection priorities are likely to reflect a combination of common guidance by low-level information in scenes and by high-level strategic factors. However, what external factors are involved in prioritising locations for fixation and the manner in which low- and high-level sources of information combine to produce fixation behaviour were not clear. Since Buswell's seminal work, a considerable body of evidence has been accumulated regarding these issues and there now exist computational models of scene viewing that propose particular low-level features as prominent in fixation allocation, and specific ways in which high-level sources of information may be combined with low-level image properties in order to decide where to fixate.

1.1 Low-Level Factors in Eye Guidance

From the extensive literature on how humans search arrays of targets, it is clear that basic visual features can guide attention (Wolfe 1998) and models based solely on low-level features can offer effective accounts of search behaviour (Treisman and Gelade 1980; Wolfe 2007). Koch and Ullman (1985) proposed an extension of these feature-based accounts of visual search to more complex scenes, and this was later implemented as a computational model (Itti and Koch 2000; Itti et al. 1998). In this model, low-level features are extracted in parallel across the viewed scene using a set of biologically plausible filters. Individual feature maps are combined across features and spatial scales via local competition in order to produce a single overall visual conspicuity map referred to as a salience map (see Fig. 1). In this account, attention is allocated to the location in the scenes that corresponds to the most salient location in the salience map. Once attended, the corresponding location in the salience map receives transient local inhibition, and attention is relocated to the next most salient location. Thus, attention is allocated serially to locations in the scene in order of most to least conspicuous in the salience map.

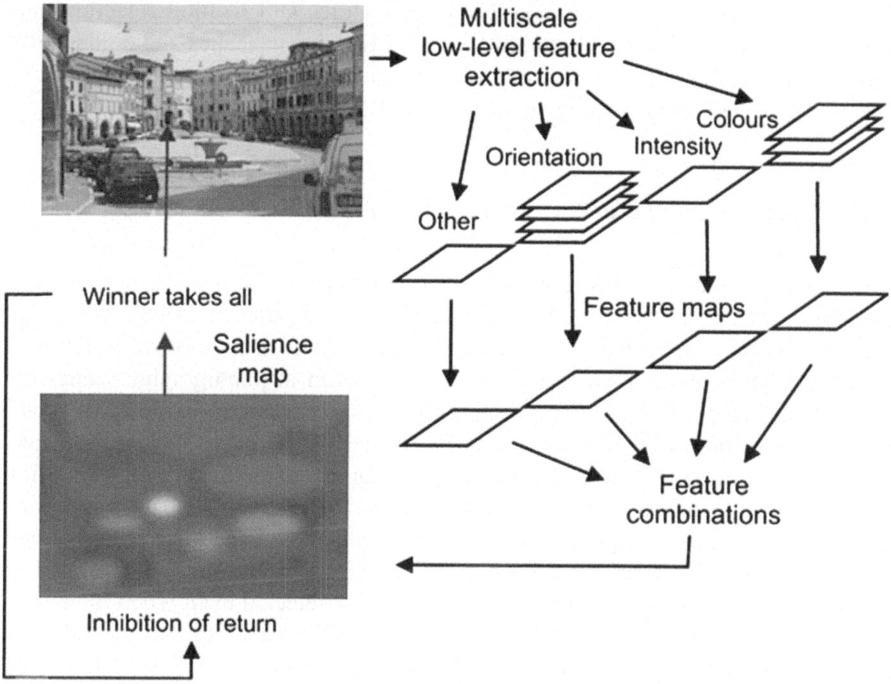

Fig. 1 Schematic of Itti and Koch's (2000) salience model, redrawn for Land and Tatler (2009)

The salience model replicates human search behaviour well when searching for feature singletons or conjunctions of two features (Itti and Koch 2000), and the extent to which it can explain attention allocation in more complex scenes has been the topic of a large volume of research. Most evaluations of the explanatory power of the salience model (and other similar models based on low-level feature-based attention allocation) use one of two approaches: measuring local image statistics at fixated locations (e.g. Reinagel and Zador 1999) or using the model to predict locations that should be fixated and seeing what proportion of human fixations fall within these predicted locations (e.g. Torralba et al. 2006). Both approaches seem to support a role for low-level information in fixation selection. Fixated locations have higher salience than control locations (e.g. Parkhurst et al. 2002), and more fixations are made within locations predicted by salience models than would be expected by chance (e.g. Foulsham and Underwood 2008). However, despite these apparently supportive results, the explanatory power of purely low-level models is limited: The magnitude of featural differences between fixated and control locations or how likely fixations are to fall within regions predicted by the models is typically small (Einhauser et al. 2008; Nyström and Holmqvist 2008; Tatler et al. 2005), suggesting that these models can only count for a limited fraction of fixation behaviour. Moreover, these basic results that appear to support low-level models must be interpreted with caution. Correlations between low-level features and fixation selection may arise because of correlations between low-level features in scenes and higher-level scene content rather than because of a causal link between low-level properties and eye guidance (Henderson 2003; Henderson et al. 2007; Tatler 2007).

1.2 Higher-Level Factors in Eye Guidance

Low-level conspicuity tends to correlate with higher-level scene structure: Salient locations typically fall within objects in scenes (Elazary and Itti 2008). Moreover, the distribution of objects in a scene is a better account of fixation selection than salience. The locations that people select for fixation in photographic scenes are better described by the locations of objects in the scenes than by the peaks in a low-level salience map (Einhauser et al. 2008). Indeed, object-based descriptions may be a more appropriate level of scene description for understanding fixation selection than low-level feature descriptions (Nuthmann and Henderson 2010). It is possible that low-level visual conspicuity might offer a convenient heuristic for the brain to select locations that are likely to contain objects (Elazary and Itti 2008). However, semantically interesting locations are preferentially selected even when their low-level information is degraded: A blurred face will still attract fixations even though it has little signature in a salience map (Nyström and Holmqvist 2008). This result implies that even though low-level conspicuity tends to correlate with objects, it is not sufficient to explain why people select objects when viewing a scene.

In light of the shortcomings of purely low-level models of fixation selection, a number of models have been proposed that incorporate high-level factors. Navalpakkam and Itti (2005) suggested that higher-level knowledge might result in selective tuning of the various feature maps that make up the overall salience map. If the features of a target object are known, the corresponding channels in the salience map can be selectively weighted, and this should enhance the representation of the target object in the salience map. Other sources of knowledge about objects present potential candidates that may guide our search for them. Most objects are more likely to occur in some places than others—for example, clocks are more likely to be found on walls than on floors or ceilings. Torralba et al. (2006) suggested that these typical spatial associations between objects and scenes can be used to produce a contextual prior describing the likely location of an object in a scene. This contextual prior can then be used to modulate a low-level conspicuity map of the scene, producing a context-modulated salience map. Therefore, the suggestion is that, in general, gaze will be directed to locations of high salience that occur within the scene regions in which the target is expected to be found. Previous experience of objects can be used not only to form contextual priors describing where objects are likely to be found but also to produce "appearance priors" describing the likely appearance of a class of objects (Kanan et al. 2009). Again, if searching for a clock, we can use prior knowledge about the likely appearance of clocks to narrow down the search to clock-like objects in the scene irrespective of where they occur. Kanan et al. (2009) proposed a model in which the appearance prior is used to modulate a low-level salience map in much the same way as Torralba et al. (2006) proposed for their context modulation. As such, in Kanan et al.'s (2009) model, gaze selects locations of high salience that coincide with scene regions that share properties characteristic of the target object's class. Modulating salience maps using context priors or appearance priors improves the performance of the model (Kanan et al. 2009; Torralba et al. 2006), suggesting that decisions about where to look when viewing scenes are likely to involve these types of information. Indeed, if both context and appearance priors are used to modulate a salience map, the resultant model is able to predict the likely locations that humans will fixate with remarkably high accuracy (Ehinger et al. 2009).

Many current models incorporate higher-level factors as modifiers of a basic low-level salience map. However, others suggest alternative cores to their models. In Zelinsky's (2008) target acquisition model, visual information is not represented as simple feature maps but as higher-order derivatives that incorporate object knowledge. Similarly, in Wischnewski et al.'s (2010) model, selection involves static and dynamic proto-objects rather than first-order visual features. Nuthmann and Henderson (2010) propose an object-level description as the core component of deciding where to look. These models each offer good explanatory power for scene viewing and demonstrate that basic visual features need not be the language of priority maps for fixation selection.

1.3 Behavioural Goals in Eye Guidance

Since they first proposed the salience model, Itti and Koch (2000) recognised that it would always be limited by its inability to account for the influence of behavioural goals on fixation selection. The importance of behavioural goals and the profound effect they have upon where people look have been recognised since the earliest work on illusions and scene viewing. As we have seen, Judd (1905) came to the conclusion that the instructions given to participants had more of an effect on where people fixated than did the stimuli when they were viewing simple line illusions. Buswell (1935) extended this idea to complex scene viewing. He showed that fixation behaviour when viewing a photograph of the Tribune Tower in Chicago with no instructions was very different from fixation behaviour by the same individual when asked to look for a face at one of the windows in the tower (Fig. 2). Yarbus (1967) later provided what has now become a classic demonstration of the profound effect task instructions have on viewing behaviour. A single individual viewed Repin's *They did not expect him* seven times, each time with a different instruction prior to viewing. Fixation behaviour was markedly different each time, and the locations fixated corresponded to those that might be expected to provide information relevant to the task suggested by the instructions (Fig. 3). These demonstrations provide a profound and important challenge for any model of fixation behaviour. Empirical evaluations of the explanatory power of low-level feature

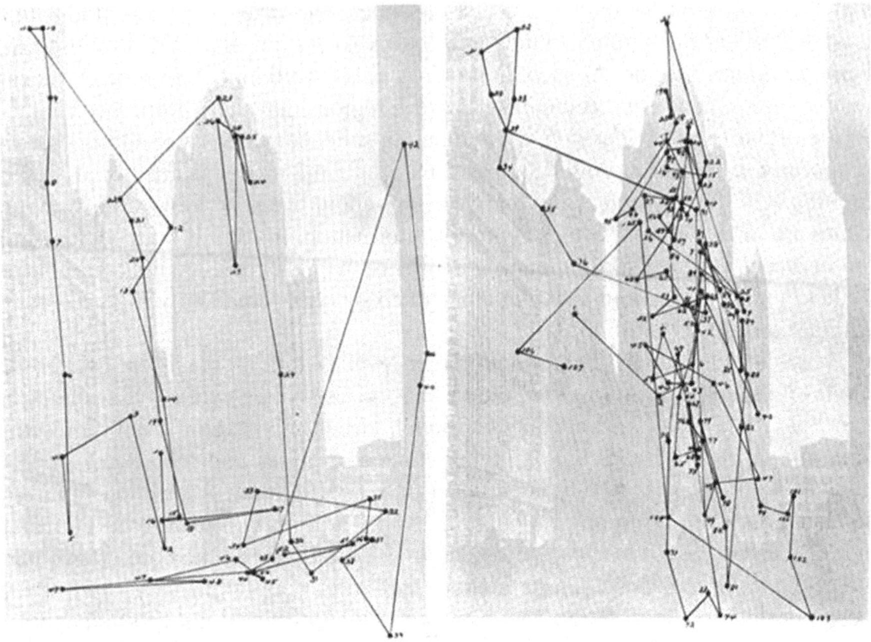

Fig. 2 *Left,* eye movements of an individual viewing the Chicago Tribune Tower with no specific instructions. *Right,* eye movements of the same individual when instructed to look for a face at a window in the tower. (Adapted from Buswell 1935)

salience during goal-directed looking tasks have shown that correlations between salience and selection are very low or absent when the observer is engaged in an explicit task such as search (Einhauser et al. 2008; Henderson et al. 2007; Underwood et al. 2006) or scene memorisation (Tatler et al. 2005). Where greater explanatory power has been found has been in cases where the task is not defined—the so-called free-viewing paradigm. In this task, participants are given no instructions other than to look at the images that they will be presented with. One motivation for employing this free-viewing paradigm is that it may be a way of isolating task-free visual processing, minimising the intrusion of higher-level task goals on fixation selection (Parkhurst et al. 2002). However, this paradigm is unlikely to produce task-free viewing in the manner hoped and is more likely to provide a situation where viewers select their own priorities for inspection (Tatler et al. 2005, 2011). It is also worth noting that even in such free-viewing situations, correlations between features and fixations are weak (Einhauser et al. 2008; Nyström and Holmqvist 2008; Tatler and Kuhn 2007).

1.4 Limits of the Screen

State-of-the-art models of scene viewing are able to make predictions that account for an impressive fraction of the locations fixated by human observers (Ehinger

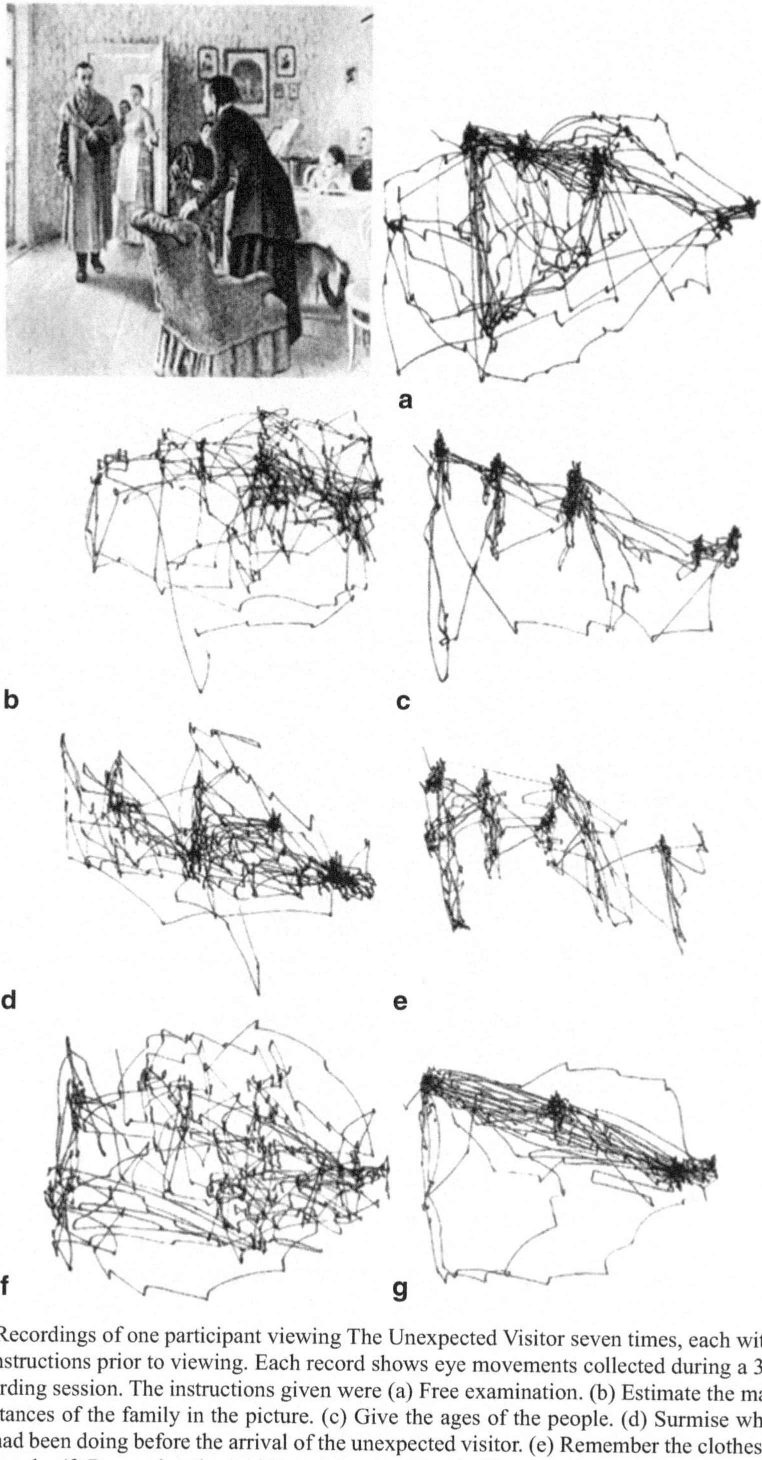

Fig. 3 Recordings of one participant viewing The Unexpected Visitor seven times, each with different instructions prior to viewing. Each record shows eye movements collected during a 3-minute recording session. The instructions given were (a) Free examination. (b) Estimate the material circumstances of the family in the picture. (c) Give the ages of the people. (d) Surmise what the family had been doing before the arrival of the unexpected visitor. (e) Remember the clothes worn by the people. (f) Remember the position of the people and objects in the room. (g) Estimate how long the unexpected visitor had been away from the family. (Illustration adapted from Yarbus, 1967, Figure 109, for Land and Tatler, 2009)

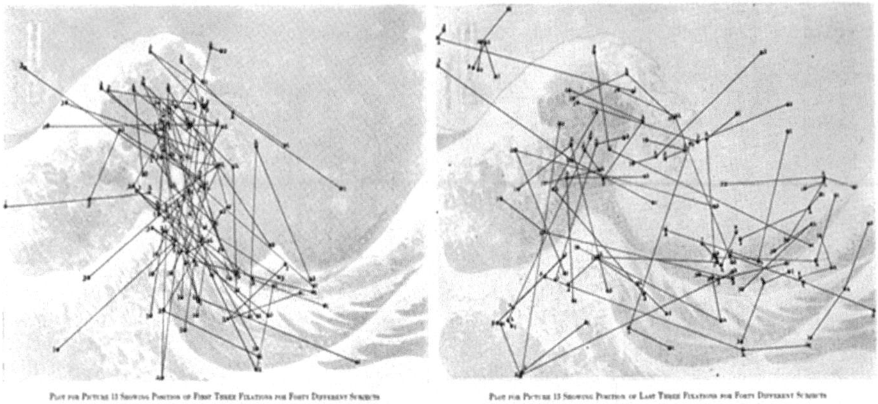

Fig. 4 Left, eye movements of 40 subjects during the first second of viewing *The Wave*. Right, eye movements of 40 subjects during the final second of viewing *The Wave*. From Buswell (1935).

et al. 2009). However, it is important to remember that the majority of evidence regarding the control of fixation selection in scene viewing comes from studies in which participants view static photographic (or photorealistic) images displayed on computer monitors. Static scenes are, of course, very different from real environments in many ways and it is important to ask the extent to which the principles of fixation selection identified in such studies generalise beyond the limits of the computer screen. There are at least four key aspects of static scene-viewing paradigms that must be considered. First, scenes typically appear with a sudden onset, are viewed for a few seconds and then disappear again. Second, the viewed scene is wholly contained within the frame of the monitor. Third, static scenes necessarily lack the dynamics of real environments. Fourth, the tasks that we engage in when viewing images on screens are rather unlike those that we engage in in more natural contexts.

Viewing behaviour is very different in the first second or two following scene onset than it is later on in the viewing period (Buswell 1935; Fig. 4). Locations selected for fixation are more similar across observers soon after scene onset than they are after several seconds of viewing (Buswell 1935; Tatler et al. 2005). Early consistency across participants followed by later divergence in fixation selection could imply that early fixations are more strictly under the control of low-level salience (Carmi and Itti 2006; Parkhurst et al. 2002) or alternatively that higher-level strategies for viewing are common soon after scene onset but later diverge (Tatler et al. 2005). Whatever the underlying reasons for these changes in viewing behaviour over time, the mere fact that viewing behaviour is very different soon after scene onset than it is later on raises concerns about the generalisability of findings from scene-viewing paradigms. It seems likely that the priorities for selection are rather different in the first second or two of viewing than they are for subsequent fixations. Given that sudden whole-scene onsets are not a feature of real-world environments, it may be that the factors that underlie saccade-targeting

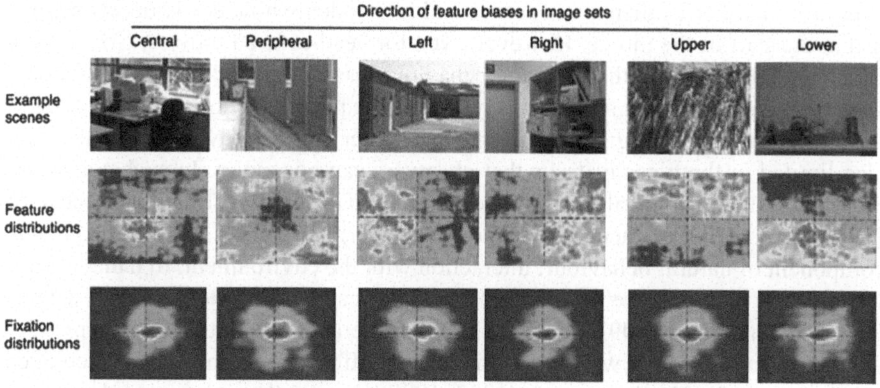

Fig. 5 The central fixation bias in fixation behaviour when viewing images on a computer monitor. Fixation distributions (*bottom row*) show a strong central tendency irrespective of the distribution of features in the images (*middle row*). (Redrawn from Tatler 2007)

decisions soon after scene onset do not reflect those that underlie natural saccade target selection. As such, this potentially limits the utility of models developed using these data.

When viewing scenes on a monitor, observers show a marked tendency to fixate the centre of the scene more frequently than the periphery (e.g. Parkhurst et al. 2002). Compositional biases arising from photographers' tendencies to put objects of interest near the centre of the viewfinder mean that images typically used in static scene-viewing paradigms often have centrally weighted low-level feature distributions. However, the distribution of low-level features in scenes is not sufficient to explain this tendency to preferentially fixate the centre of scenes (Tatler 2007). When viewing scenes with feature distributions that are not centrally biased, the tendency to fixate the centre of the scene persists, and indeed the overall distribution of fixation locations is not shifted by the distribution of features across the scene (Fig. 5). Not only is this result challenging for low-level salience models but also it raises a more serious concern for screen-based experiments: that these central fixation tendencies exist irrespective of the content of the scenes shown to the observers. There are a number of reasons that this tendency to look at the screen centre may be adaptive—it provides an optimal view of the whole scene, a good starting point for scene exploration and a location where objects of interest are expected given previous experience of photographs—but the factors that underlie these decisions to look at the screen centre are not strictly visual. As such, attempting to model these selections on the basis of the targeted visual information may be rather misleading.

Of course, static scenes necessarily lack the dynamics of real environments, but one potential solution here is to use dynamic moving images to overcome this shortcoming. By passively recording a movie of a scene from a single static viewpoint (Dorr et al. 2010) or recording a head-centred view of an environment (Cristino and Baddeley 2009), it is possible to produce dynamic scenes that have less pronounced compositional biases than static scenes and no sudden whole-scene onsets beyond that at the start of the movie. However, even for head-centred movies, Cristino and Baddeley (2009) found that viewing behaviour was dominated by scene structure, with fixations showing a spatial bias related to the perceived horizon in the scene.

Screen-based viewing paradigms—using either static or dynamic scenes—are also limited in the types of tasks that observers can engage in. In such situations, task manipulations typically involve responding to different instructions, such as to freely view, search or memorise scenes. However, these tasks lack a fundamental component of natural behaviour: interaction with the environment. In natural tasks, we typically employ gaze in a manner that is intricately linked to our motor actions (see Land and Tatler 2009). The lack of motor interaction with the scene in picture-viewing paradigms may well have fundamental effects upon how gaze is deployed (Steinman 2003). Epelboim et al. (1995, 1997) showed that many aspects of gaze coordination change in the presence of action, including the extent to which gaze shifts involve head as well as eye movements, the extent to which the eyes converge on the plane of action and the relationship between saccade amplitude and peak velocity. The limitation of using screen-based paradigms to study real-world behaviours was highlighted by Dicks et al. (2010) in a task that required goalkeepers to respond to either a real person running to kick a football or a life-sized video of the same action. Furthermore, the nature of the response was varied such that the goalkeepers responded verbally, moved a lever or moved their body to indicate how they would intercept the ball's flight. The locations fixated by the goalkeeper differed between real and video presentations and also with the type of response required. Importantly, viewing behaviour was different when observing a real person and responding with a whole body movement than in any other condition. This highlights the importance of studying visual selection in a natural task setting and suggests that any removal of naturalism can result in fixation behaviour that is unlike that produced in real behaviour.

2 Eye Guidance in Natural Tasks

From its evolutionary origins, a fundamental function of vision has been to provide information that allows the organism to effectively and appropriately carry out actions necessary for survival. Decisions about when and where to move the eyes in real-world situations are therefore likely to be intimately linked to the information demands of the current actions. Thus, it is appropriate to consider gaze not as an isolated system but as part of a broader network of vision, action and planning as we interact with the environment (Fig. 6). Thus, if we are to produce an ecologically valid account of the factors underlying fixation selection, we must consider whether models developed using laboratory-based paradigms can be extended to more natural settings.

To date, the computational models developed for scene-viewing paradigms have rarely been tested in the context of natural behaviour. One exception to this comes

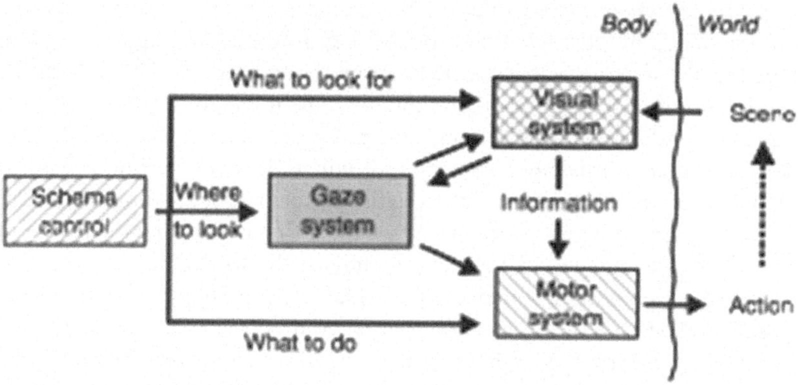

Fig. 6 Schematic illustration of interplay between gaze control, visual processing, motor action and schema planning in natural behaviour

from Rothkopf et al. (2007) who showed that in a virtual reality walking task, low-level salience was unable to account for fixation selection. Instead, fixations were made to task-relevant objects and locations in the environment irrespective of their low-level visual salience. While more state-of-the-art models incorporating higher level factors (Ehinger et al. 2009; Kanan et al. 2009; Torralba et al. 2006) have yet to be tested in natural settings, the fundamental failure of the pure salience model in a naturalistic setting raises concerns about the utility of these types of model, which retain visual conspicuity as their core. An alternative, and necessary, approach is to consider what principles for fixation selection can be identified from studies of eye movements during natural tasks and use these to specify the aspects of behaviour that any model of fixation selection in natural tasks must be able to account for.

Eye movements have been studied in a wide variety of real-world activities from everyday domestic tasks to driving, to ball sports (see Land and Tatler 2009). Across all of these tasks, it is clear that where we look is intimately linked to our actions. This simple and universal finding itself clearly demonstrates the fundamental influence that the active task requirements place on guiding eye movement behaviour. The intricate link between our behavioural goals and the allocation of overt visual attention is highlighted by the fact that when engaged in a natural task, we rarely fixate objects that are not relevant to our overall behavioural goals (Hayhoe et al. 2003; Land et al. 1999). In comparison, before beginning the task we are equally likely to fixate objects that will later be task relevant or irrelevant (Hayhoe et al. 2003). But the influence of natural behaviour on viewing is not simply to impose a preference to look at objects relevant to the overall goals of the behaviour. What is clear is that the eyes are directed to the locations that are relevant to the task on a moment-to-moment basis. That is, at each moment in time we look at the locations that convey information that allows us to act upon the environment in order to complete our current motor acts (Ballard et al. 1992; Hayhoe et al. 2003; Land et al. 1999; Land and Furneaux 1997; Patla and Vickers 1997; Pelz and Canosa 2001).

For example, when approaching a bend in the road, drivers fixate the tangent point of the bend, and this location provides key information required to compute the angle that the steering wheel should be turned (Land and Lee 1994). In table tennis (Land and Furneaux 1997) and cricket, we look at the point where the ball will bounce (Land and McLeod 2000), and this point offers crucial information about the likely subsequent trajectory that the ball will follow. These findings illustrate that spatial selection is intimately linked to the current target of manipulation. Thus, in order to understand where people look, we must first understand the nature of the behaviour they are engaged in and the structure of the task. Of course, this means that spatial selection will be somewhat parochial to the particular task that a person is engaged in. The type of information that is required to keep a car on the road is likely to be very different from that required to make a cup of tea. As such, the type of information that governs spatial selection by the eye is likely to be very different in different tasks.

While spatial selection is, in some ways, parochial to the task, temporal allocation of gaze is strikingly similar across many real activities. For many activities, gaze tends to be directed to an informative location around 0.5–1 s before the corresponding action. In tea making, the eyes fixate an object on average 0.5–1 s before the hands make contact with the object. In music reading (Furneaux and Land 1999) and speaking aloud (Buswell 1920), the eyes are typically 0.5–1 s ahead of key presses and speech respectively. During locomotion, the eyes fixate locations about 0.5–1 s ahead of the individual, and this is found when walking (Patla and Vickers 2003), driving at normal speed (Land and Lee 1994) or driving at high speed (Land and Tatler 2001). The correspondence in eye-action latency across such different tasks suggests that this temporal allocation of gaze is not only under strict control but also under common control in many real-world activities. As such, any account of gaze allocation in natural tasks must be able to explain this temporal coupling between vision and action in which gaze is allocated in anticipation of the upcoming action.

Of course, there are exceptions to the typical 0.5–1 s eye-action latency found in many natural tasks. In particular, in ball sports like cricket, squash and table tennis, there simply is not enough time to keep the eyes this far ahead of action. In these situations, anticipatory allocation of gaze is still seen albeit over rather different timescales to other tasks. In cricket (Land and McLeod 2000) and table tennis (Land and Furneaux 1997), gaze is directed to the point in space where the ball will bounce about 100 ms before the ball arrives. Similarly, in squash the eyes arrive at the front wall about 100 ms ahead of the ball (Hayhoe et al. 2011). If the ball bounces off a wall, gaze is allocated to a location that the ball will pass through shortly after it bounces off the wall with an average of 186 ms before the ball passes through this space (Hayhoe et al. 2011).

The examples described above illustrate that gaze is used to acquire information required for ongoing action and is allocated ahead of action. Correct spatiotemporal allocation of gaze is central to successful task performance in many situations. For example, in cricket both a skilled and an unskilled batsman were found to look at the same locations (the release of the ball and the bounce point), but the skilled

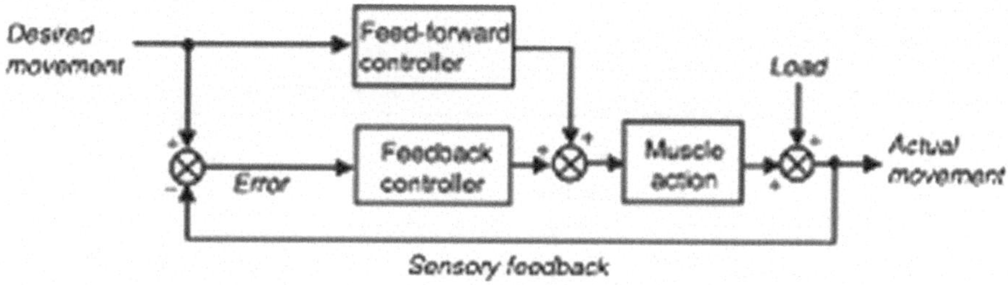

Fig. 7 Action control using feedforward and feedback loops. (From Land and Tatler 2009)

batsman looked at the bounce point about 100 ms before the ball arrived, whereas the unskilled batsman fixated this location at or slightly after the ball arrived at the bounce point (Land and McLeod 2000). Given the importance of appropriate spatiotemporal allocation of gaze in natural behaviours, what internal processes might underlie this visuomotor co-ordination in space and time? Anticipatory allocation of gaze ahead of ongoing action could be achieved if we allocate gaze on the basis of internal predictive models (Hayhoe et al. 2011; Land and Tatler 2009). The idea that the brain constructs internal predictive models of external events has been around for some time (e.g. Miall and Wolpert 1996; Wolpert et al. 1995; Zago et al. 2009). An elegant example of the importance of both feedback and prediction in visuomotor control was provided by Mehta and Schaal (2002). When balancing a 1-m pole on a table tennis bat, visual feedback alone was inadequate: If the tip of the pole was touched, disturbing the pole, the delay between visual sampling of this event and an appropriate motor response was slower (220 ms) than the maximum possible delay for normal balancing (160 ms). This suggests that to balance the pole effectively, visual feedback was too slow, and so task performance must be reliant on internal prediction. The use of forward models in this behaviour was underlined by the finding that participants were able to continue to balance the pole even when vision was removed for periods of up to 500–600 ms. Mehta and Schaal (2002) explained this behaviour as involving a Kalman filter where raw sensory feedback is compared to a copy of the motor command to the muscles in order to provide an optimised prediction of the consequences of action (Fig. 7). Such a scheme has the advantage of being able to use prediction alone in the absence of visual feedback and, thus, can tolerate brief interruptions to sensory feedback.

However, the scheme illustrated in Fig. 7 is unlikely to be sufficient for more complex tasks like the ball sports and everyday activities discussed earlier. In these situations, gaze acquires information about the future state of the world by looking at locations where action is about to occur: Objects are fixated 0.5–1 s before they are manipulated; the space where an object will be set down is fixated about half a second before the object is placed there; the spot where a ball will soon pass through is fixated 100–200 ms before the ball arrives. These anticipatory allocations of gaze certainly involve internal predictive models, but these models are not predictors in the sense described in Fig. 7. Rather, these models are mechanisms for providing

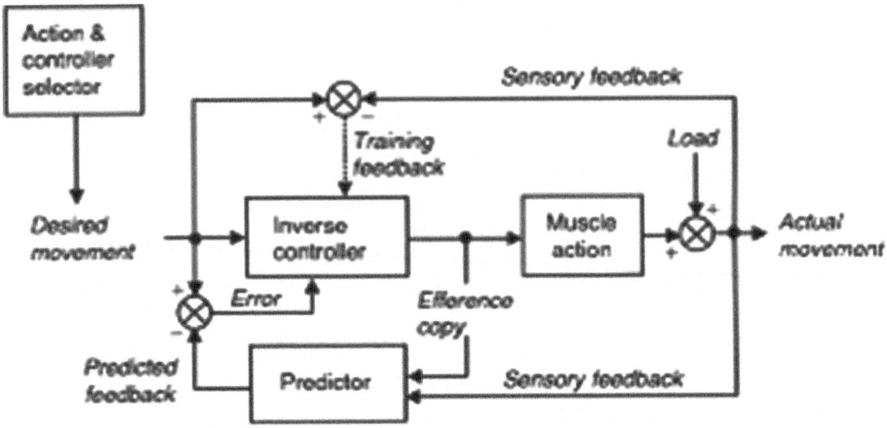

Fig. 8 Control of action using an inverse controller to refine task performance, together with a predictor in the feedback loop that provides delay-free feedback. (Modified from the "motor control system based on engineering principles" of Frith et al. 2000 for Land and Tatler 2009)

feedforward input to the motor controllers, which manage the relationship between the desired goal and the motor commands required to achieve that goal. The model illustrated in Fig. 8 depicts a situation that is suitable for understanding complex skilled behaviour. The inclusion of an inverse controller provides a mechanism for learning by transforming the desired sensory consequences of an action back into the motor commands that will produce those consequences. The mismatch between the desired and actual sensory consequences of the actions produced by the inverse controller provides the signal with which the controller can be improved. This model places learning at the heart of visuomotor co-ordination. Initially, for a novel visuomotor task, this system should operate essentially by trial and error, using feedback to improve performance. But after sufficient training, the controller can operate in an open-loop manner using the desired result as its input. Evidence in support of this scheme was provided by Sailer et al. (2005) who studied eye–hand co-ordination while learning a novel visuomotor task in which a manual control device was manipulated in order to move a cursor to targets on a computer monitor. Initially, the eyes lagged the movements of the cursor. In this phase, gaze was presumably deployed to provide feedback about the consequences of motor acts. However, after sufficient training, participants were able to perform the task well and gaze was deployed ahead of action, with the eyes leading the movements of the cursor by an average of about 0.4 s.

Not only can the scheme illustrated in Fig. 8 be used to explain visuomotor skill acquisition, but also it can provide a framework for online refinement of the internal models in the light of incoming sensory evidence. In cricket, a general model of how the ball will behave at the bounce point can be built up over years of experience, but the general model must be flexible enough to be adapted to the current pitch conditions for any given innings. The defensive play that batsmen typically engage in at the start of their innings presumably reflects this refinement of the general model based on sensory input for the current conditions (Land and McLeod 2000). Similar online adaptations of internal models based on current experience have been found when unexpected changes are made during ongoing behaviour.

Hayhoe et al. (2005) provide a nice example of how we are able to adapt our internal forward models to an unexpected change in the environment. Three people stood in a triangular formation and threw a tennis ball to each other. Like cricket, when receiving a ball, participants first fixated the release point of the ball before making an anticipatory saccade to the predicted bounce point, and then tracked the ball after its bounce. However, after several throws, one of the participants surreptitiously switched the tennis ball for a bouncier ball. When this happened, the usual oculomotor tracking of the ball broke down on the first trial with the new ball; instead, participants reverted to making a series of saccades. However, the flexibility of the internal predictors was demonstrated first by the fact that participants still caught this unexpected ball, and second by the adaptation in behaviour that followed over the next few trials with the new ball. Over the next six trials, arrival time at the bounce point advanced such that by the sixth throw with the new ball the participant was arriving at the bounce point some 100 ms earlier than on the first trial. Furthermore, the pursuit behaviour was rapidly reinstated, with pursuit accuracy for the new, bouncier ball about as good as it had been for the tennis ball by the third throw of the new ball. Thus, not only do the results demonstrate a reliance on forward models for task performance and the allocation of gaze, but they also demonstrate that these models can rapidly adapt to change in the environment.

When observers walk toward other people who they have encountered previously and who may attempt to collide with them, Jovancevic-Misic and Hayhoe (2009) showed that observers can use prior experience of these individuals to allocate gaze on the basis of the predicted threat the individual poses. Those people who the observers predicted were likely to collide with them were be looked at for longer than those who observers predicted were unlikely to collide with them, based on previous encounters. Moreover, if after several encounters, the behaviour of the oncoming individuals changed such that those who were previously of low collision threat were now trying to collide with the observer and vice versa, gaze allocation rapidly adapted to these changed roles over the next couple of encounters.

The model of visuomotor co-ordination outlined in Fig. 8 provides a framework for understanding spatiotemporal allocation of gaze for the actions required to serve ongoing behavioural goals. This model can be used to explain how gaze is allocated ahead of action in skilled behaviour and places emphasis on the importance of learning and online refinement of internal models. Learning in the proposed inverse controller can be achieved via simple reinforcement. Reward mechanisms therefore may play a crucial role not only in the development of these internal models but also in the moment-to-moment allocation of gaze. In support of this possibility, the eye movement circuitry is sensitive to reward (Montague and Hyman 2004; Schultz 2000) and, therefore, reward-based learning of gaze allocation is neurally plausible. Sprague and colleagues (e.g. Sprague et al., 2007) have begun to develop reward-based models of gaze behaviour in complex tasks. In a walking task that involves three concurrent sub-goals (avoid obstacles, collect "litter" and stay on the path), some reward value can be assigned to each sub-task. Gathering information for a sub-task is therefore rewarded. In this model, attention can only be allocated to one sub-task at a time, and uncertainty about non-attended sub-tasks increases over time. As uncertainty increases, so does the amount of information (i.e. the reduction in uncertainty) that will be gained by attending to that sub-task. The model allocates attention over time on the basis of the expected reward associated with attending to each sub-task and reducing uncertainty about that sub-task (Sprague et al. 2007). This model offers a proof of principle that gaze allocation in natural tasks can be explained using reward-based models.

Reward-based explanations of sensorimotor behaviour are emerging across a variety of experimental settings (e.g. Tassinari et al. 2006; Trommershäuser et al. 2008). Hand movements are optimised to maximise externally defined reward (e.g. Seydell et al. 2008; Trommershäuser et al. 2003). Saccadic eye movements show similar sensitivity to external monetary reward (Stritzke et al. 2009) and are consistent with an ideal Bayesian observer that incorporates stimulus detectability and reward (Navalpakkam and Itti 2010). It seems likely therefore that reward-based underpinnings to saccadic decisions may become increasingly important to our understanding of eye movements in laboratory and real environments. Moreover, reward-based models of fixation selection provide a promising new direction for research and language for describing the priority maps that are likely to underlie decisions about when and where to move the eyes.

References

Ballard, D. H., Hayhoe, M. M., Li, F., & Whitehead, S. D. (1992). Hand-eye coordination during sequential tasks. *Philosophical Transactions of the Royal Society of London Series B-Biological Sciences, 337*(1281), 331–338.

Buswell, G. T. (1920). *An experimental study of the eye-voice span in reading*. Chicago: Chicago University Press.

Buswell, G. T. (1935). *How people look at pictures: A Study of the Psychology of Perception in Art*. Chicago: University of Chicago Press.

Cameron, E. H., & Steele, W. M. (1905). The poggendorff illusion. *Psychological Monographs, 7*(1), 83–111.

Carmi, R., & Itti, L. (2006). Visual causes versus correlates of attentional selection in dynamic scenes. *Vision Research, 46*(26), 4333–4345. doi:10.1016/j.visres.2006.08.019.

Cristino, F., & Baddeley, R. J. (2009). The nature of the visual representations involved in eye movements when walking down the street. *Visual Cognition, 17*(6–7), 880–903.

Dicks, M., Button, C., & Davids, K. (2010). Examination of gaze behaviors under in situ and video simulation task constraints reveals differences in information pickup for perception and action. *Attention, Perception, & Psychophysics, 72*(3), 706–720. doi:10.3758/APP.72.3.706.

Dorr, M., Martinetz, T., Gegenfurtner, K. R., & Barth, E. (2010). Variability of eye movements when viewing dynamic natural scenes. *Journal of Vision, 10*(10), *28*, 1–17. doi:10.1167/10.10.28.

Ehinger, K. A., Hidalgo-Sotelo, B., Torralba, A., & Oliva, A. (2009). Modeling Search for People in 900 Scenes: A combined source model of eye guidance. *Visual Cognition, 17*(6–7), 945-978. doi:10.1080/13506280902834720.

Einhauser, W., Spain, M., & Perona, P. (2008). Objects predict fixations better than early saliency. *Journal of Vision*, *8*(14):18, 1–26, http://www.journalofvision.org/ content/8/14/18, doi: 10.1167/8.14.18.

Elazary, L., & Itti, L. (2008). Interesting objects are visually salient. *Journal of Vision*, *8*(3), 3.1–15. doi:10.1167/8.3.3.

Epelboim, J. L., Steinman, R. M., Kowler, E., Edwards, M., Pizlo, Z., Erkelens, C. J., & Collewijn, H. (1995). The function of visual search and memory in sequential looking tasks. *Vision Research*, *35*(23–24), 3401–3422.

Epelboim, J. L., Steinman, R. M., Kowler, E., Pizlo, Z., Erkelens, C. J., & Collewijn, H. (1997). Gaze-shift dynamics in two kinds of sequential looking tasks. *Vision Research*, *37*(18), 2597–2607.

Erdmann, B., & Dodge, R. (1898). *Psychologische Untersuchungen über das Lesen auf experimenteller Grundlage*. Halle: Niemeyer.

Foulsham, T., & Underwood, G. (2008). What can saliency models predict about eye movements? Spatial and sequential aspects of fixations during encoding and recognition. *Journal of Vision*, *8*(2), 6.1–17. doi:10.1167/8.2.6.

Frith, C.D., Blakemore, S-J., & Wolpert, D.M. (2000). Abnormalities in the awareness of and control of action. *Philosophical Transactions of the Royal Society of London B* 355, 1771–1788.

Furneaux, S., & Land, M. F. (1999). The effects of skill on the eye-hand span during musical sight-reading. *Proceedings of the Royal Society of London Series B-Biological Sciences*, *266*(1436), 2435–2440.

Hayhoe, M. M., McKinney, T., Chajka, K., & Pelz, J. B. (2011). Predictive eye movements in natural vision. *Experimental Brain Research*, *217*(1), 125–136. doi:10.1007/s00221-011-2979-2.

Hayhoe, M., Mennie, N., Sullivan, B., & Gorgos, K. (2005) The role of internal models and prediction in catching balls. In: Proceedings of AAAI fall symposium series.

Hayhoe, M. M., Shrivastava, A., Mruczek, R., & Pelz, J. B. (2003). Visual memory and motor planning in a natural task. *Journal of Vision*, *3*(1), 49–63. doi:10:1167/3.1.6.

Henderson, J. M. (2003). Human gaze control during real-world scene perception. *Trends in Cognitive Sciences*, *7*(11), 498–504.

Henderson, J. M., Brockmole, J. R., Castelhano, M. S., & Mack, M. L. (2007). Visual saliency does not account for eye movements during search in real-world scenes. In R. P. G. van Gompel, M. H. Fischer, W. S. Murray, & R. L. Hill (Eds.), *Eye movements: A window on mind and brain* (pp. 537–562). Oxford, UK: Elsevier.

Itti, L., & Koch, C. (2000). A saliency-based search mechanism for overt and covert shifts of visual attention. *Vision Research*, *40*(10–12), 1489–1506.

Itti, L., Koch, C., & Niebur, E. (1998). A model of saliency-based visual attention for rapid scene analysis. *IEEE Transactions on Pattern Analysis and Machine Intelligence*, 20, 1254–1259.

Jovancevic-Misic, J., & Hayhoe, M. (2009). Adaptive Gaze Control in Natural Environments. *Journal of Neuroscience*, *29*(19), 6234–6238. doi:10.1523/JNEUROSCI.5570-08.2009.

Judd, C. H. (1905). The Müller-Lyer illusion. *Psychological Monographs*, *7*(1), 55–81.

Judd, C. H., & Courten, H. C. (1905). The Zöllner illusion. *Psychological Monographs*, *7*(1), 112–139.

Kanan, C., Tong, M. H., Zhang, L., & Cottrell, G. W. (2009). SUN: Top-down saliency using natural statistics. *Visual Cognition*, *17*(6–7), 979–1003.

Koch, C., & Ullman, S. (1985). Shifts in selective visual-attention—towards the underlying neural circuitry. *Human Neurobiology*, *4*(4), 219–227.

Land, M. F., & Furneaux, S. (1997). The knowledge base of the oculomotor system. *Philosophical Transactions of the Royal Society of London Series B-Biological Sciences*, *352*(1358), 1231–1239.

Land, M. F., & Lee, D. N. (1994). Where we look when we steer. *Nature*, *369*(6483), 742–744. doi:10.1038/369742a0.

Land, M. F., & McLeod, P. (2000). From eye movements to actions: How batsmen hit the ball. *Nature Neuroscience*, *3*(12), 1340–1345. doi:10.1038/81887.

Land, M. F., & Tatler, B. W. (2001). Steering with the head: The visual strategy of a racing driver. *Current Biology, 11*(15), 1215–1220.
Land, M. F., & Tatler, B. W. (2009). *Looking and acting: Vision and eye movements in natural behaviour*. Oxford: OUP.
Land, M. F., Mennie, N., & Rusted, J. (1999). The roles of vision and eye movements in the control of activities of daily living. *Perception, 28*(11), 1311–1328.
McAllister, C. N. (1905). The fixation of points in the visual field. *Psychological Monographs, 7*(1), 17–53.
Mehta, B. & Schaal, S. (2002). Forward models in visuomotor control. *Journal of Neurophysiology* 88, 942–953.
Miall, R.C. & Wolpert, D.M. (1996). Forward models for physiological motor control. *Neural Networks*, 9, 1265–1279.
Montague, P., & Hyman, S. (2004). Computational roles for dopamine in behavioural control. *Nature, 431*, 760–767.
Navalpakkam, V., & Itti, L. (2005). Modeling the influence of task on attention. *Vision Research,* 45, 205–231.
Navalpakkam, V., & Itti, L. (2010). A goal oriented attention guidance model. *Biologically Motivated Computer Vision*, 81–118.
Nuthmann, A., & Henderson, J. M. (2010). Object-based attentional selection in scene viewing. *Journal of Vision*, 10(8) 20, 1–19. doi:10.1167/10.8.20.
Nyström, M., & Holmqvist, K. (2008). Semantic override of low-level features in image viewing-both initially and overall. *Journal of Eye Movement Research*, 2, 1–11.
Parkhurst, D., Law, K., & Niebur, E. (2002). Modeling the role of salience in the allocation of overt visual attention. *Vision Research, 42*(1), 107–123.
Patla, A. E., & Vickers, J. N. (1997). Where and when do we look as we approach and step over an obstacle in the travel path? *Neuroreport, 8*(17), 3661–3665.
Patla, A. E., & Vickers, J. N. (2003). How far ahead do we look when required to step on specific locations in the travel path during locomotion? *Experimental brain research Experimentelle Hirnforschung Expérimentation cérébrale, 148*(1), 133–138. doi:10.1007/s00221-002-1246-y.
Pelz, J. B., & Canosa, R. (2001). Oculomotor behavior and perceptual strategies in complex tasks. *Vision Research, 41*(25–26), 3587–3596.
Rayner, K. (1998). Eye Movements in Reading and Information Processing: 20 Years of Research. *Psychological bulletin, 124*(3), 372–422.
Reinagel, P., & Zador, A. M. (1999). Natural scene statistics at the centre of gaze. *Network (Bristol, England)*, 10(4), 341–350.
Rothkopf, C. A., Ballard, D. H., & Hayhoe, M. M. (2007). Task and context determine where you look. *Journal of Vision, 7*(14), 16.1–20. doi:10.1167/7.14.16.
Sailer, U., Flanagan, J. R., & Johansson, R. S. (2005). Eye-hand coordination during learning of a novel visuomotor task. The Journal of neuroscience: The official journal of the Society for. *Neuroscience, 25*(39), 8833–8842. doi:10.1523/JNEUROSCI.2658-05.2005.
Schultz, W. (2000). Multiple reward signals in the brain. *Nature Reviews Neuroscience, 1*(3), 199–207.
Seydell, A., McCann, B. C., Trommershäuser, J., & Knill, D. C. (2008). Learning stochastic reward distributions in a speeded pointing task. *Journal of Neuroscience, 28*, 4356–4367.
Sprague, N., Ballard, D. H., & Robinson, A. (2007). Modeling embodied visual behaviors. *ACM Transactions on Applied Perception*, 4, 11.
Steinman, R. M. (2003). Gaze control under natural conditions. In: Chalupa, L.M., & Werner, J.S. (Eds). *The Visual Neurosciences*. pp. 1339–1356. Cambridge: MIT Press.
Stritzke, M., Trommershäuser, J., & Gegenfurtner, K. R. (2009). Effects of salience and reward information during saccadic decisions under risk. *Journal of the Optical Society of America A, 26*, B1–B13.
Stratton, G. M. (1906). Symmetry, linear illusions, and the movements of the eye. *Psychological Review, 13*, 82–96.

Tassinari, H., Hudson, T. E., & Landy, M. S. (2006). Combining priors and noisy visual cues in a rapid pointing task. *Journal of Neuroscience*, *26*, 10154–10163.

Tatler, B. W. (2007). The central fixation bias in scene viewing: Selecting an optimal viewing position independently of motor biases and image feature distributions. *Journal of Vision*, 7(14):4, 1–17, http:// www.journalofvision.org/content/7/14/4, doi:10.1167/ 7.14.4.

Tatler, B. W., & Kuhn, G. (2007). Don't look now: the misdirection of magic. In: van Gompel, R., Fischer, M., Murray, W., & Hill, R. (Eds). *Eye Movement Research: Insights into Mind and Brain*. pp. 697–714. Amsterdam: Elsevier.

Tatler, B. W., Baddeley, R. J., & Gilchrist, I. (2005). Visual correlates of fixation selection: Effects of scale and time. *Vision Research*, *45*(5), 643–659.

Tatler, B. W., Hayhoe, M. M., Land, M. F., & Ballard, D. H. (2011). Eye guidance in natural vision: Reinterpreting salience. *Journal of Vision, 11(5), 5* 1–23. doi:10.1167/11.5.5.

Torralba, A., Oliva, A., Castelhano, M. S., & Henderson, J. M. (2006). Contextual guidance of eye movements and attention in real-world scenes: The role of global features in object search. *Psychological Review*, *113*(4), 766–786.

Treisman, A. M., & Gelade, G. (1980). A feature-integration theory of attention. *Cognitive Psychology*, *12*, 97–136.

Trommershäuser, J., Maloney, L. T., & Landy, M. S. (2003). Statistical decision theory and the selection of rapid, goal-directed movements. *Journal of the Optical Society of America A*, *20*, 1419–1433.

Trommershäuser, J., Maloney, L. T., & Landy, M. S. (2008). Decision making, movement planning, and statistical decision theory. *Trends in Cognitive Sciences*, *12*, 291–297.

Underwood, G., Foulsham, T., van Loon, E., Humphreys, L., & Bloyce, J. (2006). Eye movements during scene inspection: A test of the saliency map hypothesis. European Journal of. *Cognitive Psychology*, *18*(3), 321–342. doi:10.1080/09541440500236661.

Wischnewski, M., Belardinelli, A., & Schneider, W. (2010). Where to look next? Combining static and dynamic proto-objects in a TVA-based model of visual attention. *Cognitive Computation*, *2*(4), 326–343.

Wolfe, J. (2007). Guided Search 4.0: Current Progress with a model of visual search. In W. Gray (Ed.), *Integrated models of cognitive systems* (pp. 99–119). New York: OUP.

Wolfe, J. M. (1998). What can 1 million trials tell us about visual search? *Psychological Science*, *9*(1), 33–39.

Wolpert. D.M., Ghahramani, Z., & Jordan, M.I. (1995). An internal model for sensorimotor integration. *Science* 269, 1880–1882.

Yarbus, A. L. (1967). *Eye movements and vision*. New York: Plenum Press.

Zago M, McIntyre J, Patrice Senot P, & Lacquaniti F (2009) Visuo-motor coordination and internal models for object interception. *Experimental Brain Research* 192(4):571–604.

Zelinsky, G. J. (2008). A theory of eye movements during target acquisition. *Psychological Review*, *115*(4), 787–835. doi:10.1037/a0013118.

Guidance of Attention by Feature Relationships: The End of the Road for Feature Map Theories?

Stefanie I. Becker

1 Introduction

It is well known that conscious perception is severely capacity limited: At any moment in time, only a few objects can be consciously perceived. Attention is needed to select items from cluttered visual scenes for further in-depth processing. In the past, much effort has been devoted to identify the factors that guide attention and determine which item will be selected first.

The currently dominant view is that attention can be guided by two independent attentional systems: First, a *stimulus-driven system* guides attention to the most salient locations in the visual field, such as suddenly appearing items ('onsets'; e.g. Yantis 2000), or items with a high-feature contrast (e.g. Theeuwes 1994, 2010; Wolfe 1994). Importantly, attention is allocated to these items in a purely stimulus-driven fashion, that is, without or even against the goals and intentions of the observers to perform a certain task (e.g. Yantis 1993). A second attentional system is *goal-dependent* and guides attention to items that match the observer's goals and intentions to find a sought-after item (e.g. Folk et al. 1992; Wolfe 1994). For instance, when we search for a red item, such as a woman with a red skirt, attention can be involuntarily captured by other red items. Importantly, capture by target-similar items is usually much stronger than capture by salient irrelevant items that do not match our top-down settings (e.g. Folk and Remington 1998). This *similarity effect* (defined as stronger capture by target-similar than target-dissimilar distractors) has been found in numerous studies with the visual search paradigm and has usually been taken to show that top-down tuning to specific feature values can override effects of bottom-up saliency (e.g. Ansorge and Heumann 2003; Becker et al. 2009; Eimer et al. 2009; Folk and Remington 1998; Ludwig and Gilchrist 2002).

Most models of attentional guidance assume that attention is tuned to particular *feature values*. For example, in the colour dimension, attention can be biased to

S. I. Becker (✉)
The University of Queensland, Brisbane, Australia
e-mail: s.becker@psy.uq.edu.au

M. Horsley et al. (eds.), *Current Trends in Eye Tracking Research*,
DOI 10.1007/978-3-319-02868-2_3, © Springer International Publishing Switzerland 2014

select either red, green, yellow or blue items; in the orientation dimension, attention can be biased for horizontal, vertical or differently tilted orientations and in the size dimension, attention can be biased to select items of different sizes (e.g. Treisman and Gelade 1980; Treisman and Sato 1990; Wolfe 1998; Wolfe and Horowitz 2004).

Multiple different mechanisms have been proposed to describe feature-based tuning of attention. For instance, *feature similarity views* assume that attention is tuned to the target feature value (e.g. Martinez-Trujillo and Treue 2004). According to the *attentional engagement theory,* attention can be additionally biased against selecting the feature value of the nontarget(s) ('nontarget inhibition'; see; e.g. Duncan and Humphreys 1989). Although accounts of top-down selection differ with respect to the feature value that will be prioritized in a given instance, they uniformly assume that top-down selection is achieved by activating or inhibiting specific feature maps (e.g. red, green; Duncan and Humphreys 1989; Folk and Remington 1998; Koch and Ullman 1985; Maunsell and Treue 2006; Navalpakkam and Itti 2007; Treisman and Gelade 1980; Treisman and Sato 1990; Wolfe 1994). Feature maps are populations of sensory neurons that are topographically organized and respond to specific feature values ('feature detectors'). Most theories assume that visual selection is achieved by modulating the response gain of feature-specific neurons, increasing the response gain of feature detectors responding to the target feature, and/or decreasing the response gain of feature detectors responding to the nontarget features (e.g. Koch and Ullman 1985; Maunsell and Treue 2006; Navalpakkam and Itti 2007; Spitzer et al. 1988).

Contrary to these feature-based theories, it has recently been proposed that attention is guided by target–nontarget relationships, that is, information that specifies how the target differs from irrelevant items (Becker 2010). According to this new *relational view,* the target and irrelevant context are not evaluated separately—with activation applied to the target feature and inhibition to the nontarget feature. Instead, the feature relationship between the target and context is evaluated and attention is guided towards items sharing the target–context relation (e.g. redder). For example, when searching for the orange shirt of a goalkeeper, it would depend on the context how attention is top-down tuned: When the goalkeeper is embedded in a team wearing all yellow shirts, as in the left panel of Fig. 1, attention would be tuned towards all *redder* items. By contrast, if the goalkeeper is surrounded by a team wearing all red shirts, attention would be tuned to all *yellower* items (see right panel of Fig. 1).

1.1 Is There a Similarity Effect?

The relational theory is overall quite similar to the classical feature-based views, but the two theories make different predictions with regard to stimuli that can capture our attention. For example, if observers have to search for an orange target, and attention is top-down tuned to the target feature value, then only orange items should capture attention. Items of different colours should capture to the degree that they are similar to orange (feature similarity view; e.g. Folk and Remington 1998). By contrast, the relational account predicts that attention should be tuned to target–

Fig. 1 Example of searching for orange in different contexts: According to the relational account, colours are encoded relative to the context, and visual selection is biased to the relative attribute that distinguishes the target from the context. Hence, when searching for the orange-shirted goalkeeper, visual selection would be biased to *redder* when the goalkeeper is among a team clad in yellow (left), whereas attention would be biased to *yellower*, or the *yellowest* item when the goalkeeper is among red-shirted players (right).

nontarget relationships. Thus, if the orange target is redder than the nontargets (e.g. yellow), attention should be tuned to all redder items. A consequence of tuning to redder is that items that are redder than the target itself should be able to attract attention. This holds because a relational top-down setting specifies only the direction in which an item differs from the context, and does not contain information about the exact feature values of the target or the context. Hence, the item with the most extreme feature in the specified direction (e.g. reddest item) should always capture attention most strongly (e.g. Becker 2008, 2010).

Studies testing the relational account against the feature similarity account confirmed the predictions of the relational account (e.g. Becker 2010; Becker et al. 2010): When observers had to search for an orange target among irrelevant yellow nontargets, a red distractor captured attention more strongly than an orange distractor—despite the fact that the red distractor was more dissimilar from the target. When the orange target was embedded among red irrelevant nontargets, a yellow distractor captured more strongly than a target-similar orange distractor, consistent with a top-down setting for yellower (e.g. Becker et al. 2010). Across all conditions, the colour contrasts of all stimuli were controlled, so that stronger capture by the red or yellow distractors could not be explained by bottom-up factors such as feature contrast or visual saliency (Becker et al. 2010; Becker and Horstmann 2011).

These results were diametrically opposite to the often-reported similarity effect, that items can only attract attention when they are similar to the sought-after target feature (e.g. Folk and Remington 1998), and demonstrated, for the first time, that a target-dissimilar distractor can capture attention more strongly than a target-similar distractor. This is an important finding, as the similarity effect can be regarded as the most direct evidence for feature-based theories of top-down guidance. Although these first results ruled out similarity theories of attentional guidance, they are still consistent with a less well-known group of feature-based theories, viz., optimal tuning accounts.

2 The Relational Account Versus Optimal Tuning Accounts

More capture by target-dissimilar distractors may still be consistent with feature-based accounts, when we assume that attention can be top-tuned tuned towards a different feature value than the target feature value (e.g. tuning to red in search for orange). *Optimal tuning accounts* assume that attention is always tuned towards the feature value that optimally distinguishes the target from the nontarget features, and would allow for such a shift in top-down tuning. Especially when the feature value distributions for the target and nontarget features overlap to a large extent, tuning attention to a more extreme feature that is shifted away from the nontarget features can enhance the signal-to-noise ratio and provide a more optimal setting than tuning to the target feature itself (e.g. Lee et al. 1999; Navalpakkam and Itti 2007; Scolari and Serences 2010). With this, optimal tuning accounts could provide an alternative explanation for the finding that a red distractor captured more than a target-similar orange distractor, simply by assuming that attention was tuned more towards red than orange.

The optimal tuning account and the relational account both predict that a distractor with an 'exaggerated' target colour (i.e. a colour shifted away from the nontarget colour) should capture more strongly than a target-similar distractor. However, optimal tuning accounts assume that a distractor with the *nontarget colour* could *not* capture attention, because attention has to be biased away from the nontarget feature (to provide a better signal-to-noise ratio). By contrast, the relational account would *allow* capture by a nontarget colour (provided that the set-up allows this colour to differ from all other colours and to be the most extreme in the target-defining direction). This holds because visual selection of the target is thought to depend only on its relationship to other items, not a specific feature value, so any item with the same relationship(s) as the target should capture.

In the visual search paradigm, the target and distractors are always presented in the same display and hence it is impossible to create targets and distractors that have different feature values and yet both have the most extreme feature value (e.g. reddest item in the display). To critically test whether a nontarget-coloured distractor can capture attention, it is necessary to present the distractors ('cues') prior to the target in a separate cueing display, so that the distractor (cue) features can be manipulated independently of the target and nontarget features. In a recent study (Becker et al. 2013), observers were asked to search for an orange target among three yellow-orange nontargets (target redder condition), and to ignore differently coloured cues that were briefly flashed (100 ms) prior to the target display. The cueing displays consisted of four cues, three of which constituted the context for the differently coloured singleton cue. In one condition, the singleton cue had the same colour as the nontargets (yellow-orange), and the other three cues were yellow, rendering the singleton cue redder than the cue context (same relative colour as the target). The results showed that the yellow-orange cue with the nontarget colour captured attention (Becker et al. 2013), as reflected in faster response times when it was presented at the target location (valid cue) than when it was presented at a nontarget location (invalid cue; e.g. Posner 1980). These results demonstrate that, contrary to the optimal tuning account, a distractor with the nontarget colour can still capture attention, provided that the cue–context relations match the relationship between the target and the nontargets.

The study of Becker et al. (2013) included multiple control conditions to ensure that capture by a nontarget-coloured cue is not due to bottom-up factors (e.g. the specific colours used in one condition). In one experimental block, the target and nontarget colours were also reversed, so that observers now had to search for a yellow-orange target among orange nontargets (target yellower condition). In this condition, the yellow-orange cue (among yellow-other cues) failed to capture attention—despite the fact that it had the same colour as the target. This outcome was predicted by the relational theory: As the task required tuning to a yellower target, a target-similar cue should not capture if it is itself redder than the cue context. These findings strongly support the relational account that capture is largely independent of the absolute feature values of target and nontargets, and instead depends on the relationships of target and distractor.

Taken together, the current evidence invalidates the prevalent doctrine that capture necessarily depends on similarity to the target feature (i.e. 'similarity effect'; e.g. Folk and Remington 1998; Becker et al. 2008), or similarity to the exaggerated target feature (i.e. 'optimal feature'; e.g., Lee et al., 1999, Scolari & Serences, 2010) and suggests that capture is instead determined by feature relationships. Previous studies did not vary the similarity of distractors independently of their relative features: Hence, it is possible that the often-reported similarity effect in previous studies was due to the distractor matching the relative, not absolute, colour of the target. In fact, one of the strengths of the relational theory is that it seems consistent with all results that were previously interpreted in support of a feature-specific top-down setting (cf. Becker 2010).

3 Can Guidance by Relationships Be Explained by Feature-Based Theories of Attention?

Is guidance of attention by feature relationships really inconsistent with common feature-based accounts of attention, or can feature based-accounts somehow account for top-down tuning to relative features? This question is unfortunately not easy to answer, as the top-down tuning component as well as the feature maps, 'channels' or 'filters' are not clearly specified in the mainstream models of visual attention (e.g. Mozer and Baldwin 2008). The view most commonly found in the literature seems to be that attention is guided by separate, disconnected feature maps that basically act like feature detectors and signal the location of items with particular feature values (e.g. Itti and Koch 2000; Koch and Ullman 1985; Treisman and Gelade 1980; Treisman and Sato 1990; Wolfe 1994). With this, feature map theories do not seem to be able to account for guidance by feature relationships, as it is impossible to specify feature relationships within feature maps. For example, an orange feature detector cannot signal whether the item in its receptive field is redder or yellower than other items in the context. To obtain information about feature relationships, we would need additional feature detectors that signal, for example, the presence of red and yellow items. Then, feature relations could be computed by another layer of neurons (i.e. 'comparator maps') that receive input from the orange, yellow and red feature detectors. However, to account for guidance by feature relationships, attention would have to be guided by comparator maps—not feature maps or feature detectors, as is proposed by current feature-based models of visual attention.

Adding another layer of neurons to feature map models may also not seem the most parsimonious way to account for guidance by feature relationships. In fact, one rather problematic aspect of feature-based theories is that they have to propose a feature map or population of feature-specific neurons for each feature that can be top-down selected. As noted by Maunsell and Treue (2006), the number of neurons required by feature map theories may very well exceed the number of neurons in the brain that are actually involved in the guidance of attention. Adding comparator maps to the already proposed feature maps would aggravate this problem and hence does not seem advisable.

An alternative solution would be to propose a small number of categorical channels that respond to an entire range of feature values (e.g., all colour values from yellow to red), with a red detector responding maximally to red (e.g., Wolfe, 1994). Such broadly tuned feature detectors could indeed explain many relational effects (e.g., Becker, 2010). However, such an account could not explain our ability to select intermediate colours (e.g., orange among red and yellow). If the account is equipped with an extra channel to explain fine-grained selectivity (e.g., an extra orange channel), it loses the ability to explain relational effects such as capture by nontarget-coloured items.

In sum, the only practical and parsimonious approach seems to be to abandon the idea of separate feature detectors, and to ponder the idea that sensory neurons can directly encode the direction in which features differ from one another, and that this information can be used to direct attention towards items that have the same feature relationship(s) (e.g. Becker 2010; Becker et al. 2013). In the following, it will be argued that the current neurophysiological evidence also supports the possibility of 'relational feature detectors'.

4 Are Sensory Neurons Feature Detectors or Relational?

The idea that attention is guided by independently working feature maps or feature detectors has dominated the theoretical landscape for a long time, and the reason that the concept has been so very successful is probably because feature detectors seem to have the most basic response characteristics, and we can easily imagine how they work (e.g. Nakayama and Martini 2011). However, the idea that attention is guided by feature detectors is not strongly supported by neurobiological evidence. For feature detectors or 'channels' to be operating *independently* of one another, it appears that different colours, for example, must be encoded by entirely different populations of neurons. Contrary to this claim, neurophysiological studies have found many neurons in the visual cortex of the monkey that respond to two colours, in the fashion of an opponent-colour mechanism (e.g. Hubel and Wiesel 1967; De Valois et al. 2000; Gouras 1974). For example, some opponent-colour cells increase their firing rate in response to input from L-cones (e.g. red) but significantly decrease their firing rate in response to input from M-cones (e.g. green). These neurons cannot be regarded as feature detectors, because they do not respond to a specific colour. Opponent cells also often show different responses depending on whether they receive input from the same cone in the centre or in the surround of their receptive fields (RFs). L–M opponent-colour cells, for example, fire in response to an L-cone increment in the centre but also to an M-cone decrement in the surround of the RF (e.g. Conway 2001; De Valois et al. 2000). These cells and their counterparts (e.g. M–L cells) are predominantly found in the lateral geniculate nucleus (LGN) and could be regarded as 'purely relational cells', because they signal that the centre of the receptive field is occupied by something 'redder' than the surround—without specifying whether the centre contains red or whether the surround contains green. Also at later cortical stages, there are still 'relational neurons' that relate inputs from different cone types to one another, in line with the hypothesis that neurons can signal feature relationships (e.g. Conway 2001; De Valois et al. 2000).

Admittedly, it is at present unclear what proportions of colour-sensitive neurons should be classified as relational cells vs. feature detectors, and whether and to what extent these different classes of neurons are involved in the guidance of attention. Of note, studies investigating the neurophysiological underpinnings of attention also cannot distinguish between a relational and feature-specific account of guidance: Several studies reported that the response gain of some feature-specific neurons increased in expectation of the target (e.g. Martinez Trujillo and Treue 2004; Motter 1994). However, none of the studies systematically varied the target–background relations and/or the distractor-background relations (e.g. Atiani et al. 2009; David et al. 2008; Kastner et al. 1999; Luck et al. 1997; Motter 1994; Scolari and Serences 2010; Spitzer et al. 1988). Hence, the available neurophysiological evidence which has always been interpreted in support of a feature-based account also seems consistent with a relational account.

In sum, the current state of evidence does not seem to provide good reasons to claim that relational neurons could not exist or that they could not guide attention. Naturally, the existence of neurons signalling feature relationships also does not preclude that feature detectors exist or that they can guide attention as well (e.g. Conway and Tsao 2009). However, at least in the monkey, early processing of colour (e.g. in the LGN) seems to initially rely mostly on L–M type of cells that are context dependent (e.g. De Valois et al. 2000). All information that can be used at later, cortical stages has therefore to be extracted from relational, context-dependent information.

It seems possible that colour processing initially proceeds relational and becomes feature specific only at later stages of visual processing. This is also in line with the observation that colour perception shows both relational and absolute characteristics. Relational or context-dependent characteristics are reflected in the fact that the visual system remains susceptible to different surrounding colours, so that, for example, a grey patch can look slightly green when surrounded by a large green area ('simultaneous colour contrast'; e.g. Conway et al. 2002). Absolute characteristics are, for instance, reflected in our ability to categorize colours, and to recognize a specific colour in many different lighting conditions ('colour constancy'). Given that all colour processing has to be based on the initially relational information, it is an interesting question to what extent colour perception may still be relational at very late stages of visual processing. In fact, Foster (2003) argued that there is insufficient evidence for colour constancy, as colour constancy could itself be relational.

5 Can the Relational Theory Explain Feature-Specific Tuning?

Above, it was argued that the current neurophysiological evidence does not appear to support a feature detector account more strongly than a relational account of attention. In addition, it could be asked whether there is any more direct evidence that a relational theory can account for tuning to specific feature values. A first problem with this question is that it is not entirely clear what would qualify as an instance for 'clearly feature-specific tuning'.

As outlined above, the perhaps most compelling evidence for feature-specific tuning was the similarity effect, that is, the finding that target-similar distractors capture attention most strongly. This finding has been called into question by findings demonstrating that target-dissimilar distractors can capture as well, provided that relative attributes as the target (e.g. Becker et al. 2013, in press). Another finding that can be interpreted as evidence for feature-specific tuning is our ability to select items that have intermediate feature values, that is, features that are directly sandwiched between more extreme features. For example, visual search studies have shown that attention can be tuned to a medium-sized target when half of the nontargets are larger and half are smaller than the target (e.g. Hodsoll and Humphreys 2005). Similarly, we can also select an orange item when half of the nontargets are red and the other half is yellow (e.g. Bauer et al. 1995; Navalpakkam and Itti 2006). In these examples, the target feature does not differ in a single direction from the nontarget features; rather, as half of the nontargets is smaller (or redder) than the target and the other half is larger (or non-redder), the target differs in two opposing directions from the nontargets ('non-linearly separable';

e.g. Bauer et al. 1995). Previous studies show that search efficiency is decreased for such intermediate features; however, search is not random, indicating that attention can still be top-down guided to intermediate features (Bauer et al. 1995; Hodsoll and Humphreys 2005; Navalpakkam and Itti 2006).

To explain these results from a relational perspective, it has to be assumed that attention can be biased to two opposing relationships simultaneously, to select the item that is 'neither the reddest nor the yellowest item' in the display. Such a twofold relational setting would result in an attentional bias for items intermediate between two more extreme feature values (i.e. reddest and yellowest) and, thus, would mimic a feature-specific setting. However, deviating from a feature-specific account, selection of the intermediate item should still strongly depend on the context; that is, in search for an orange target among red and yellow nontargets, an orange distractor should only capture attention when it is similarly embedded in a context containing both redder and yellower items. Orange distractors should, however, fail to capture when they are either the reddest or the yellowest items in their context(s).

This hypothesis was tested in a spatial cueing study, where observers had to search for an orange target among two red and two yellow nontargets (Becker et al. 2013). Prior to the target frame, a cueing frame was presented that contained five cues, four of which constituted the cue context. Capture was tested by a cue that could have either the same colour as the target (orange) or a different colour (yellow-orange). Critically, the context cues were coloured such that the uniquely coloured singleton cue sometimes had an intermediate colour relative to the cue context and sometimes an extreme colour (e.g. orange the reddest colour in the cueing display). The results showed that an orange cue captured attention only when it was embedded among redder and yellower cues, but not when the cue context rendered it the reddest cue. In addition, a target-dissimilar yellow-orange cue captured attention to the same extent as the target-similar (orange) cue when it was embedded among redder and yellower cues, but failed to capture when it was the yellowest cue in the cueing display (Becker et al. 2013).

These results indicate that attention was not top-down tuned to the exact feature value of the target (orange). Rather, capture depended on whether the context rendered the singleton cue colour intermediate between more extreme colours, indicating that attention had been tuned simultaneously towards the target feature and the features of the context. This finding has two important implications: First, it indicates that the target and nontarget colours (red, yellow and orange) were not processed independently and separately of each other—contrary to the major claims of current feature-based theories, that visual selection is achieved by a number of independent feature detectors (e.g. Treisman and Sato 1990; Wolfe 1994). Second, and even more importantly, the results demonstrate that information from different target–nontarget relations (redder/yellower) can be used to bias attention to intermediate features. With this, a relational top-down setting can mimic the effects of a feature-specific top-down setting, as it allows selection of a range of intermediate feature values that are bounded by more extreme features. In fact, it is possible that evidence previously interpreted in favour of feature-specific tuning was really due to relational tuning. Of note, the target and nontarget features were always held constant in previous studies (e.g. Bauer et al. 1995; Hodsoll and Humphreys 2005; Navalpakkam and Itti 2006). Hence, it is possible that observers did not apply a feature-specific setting, but used information provided by the nontarget context to tune attention to the target in a relational manner.

That said, it should also be noted that the most recent studies showed that we are also able to tune attention to a specific target size or target colour, when the nontargets varied randomly and tuning attention to relative features could not help localising the target (Becker et al., in press; Harris et al., 2013). Interestingly, performance was poorer in this feature-specific search mode than when observers engaged in relational search (Becker et al., in press). Taken together, there is evidence that attention can be tuned to the target attributes in a context-dependent manner, as well as to specific features in a context-independent manner. The challenge for future research will be to find out how the two search modes can be explained in a unified theory of attention, and what their respective neural substrates are.

6 Conclusion

The present chapter introduced a new relational account of attention and argued that it can provide an alternative to current feature-based theories of attention. Among the findings in favour of the relational account were results showing that feature relationships account for (1) capture by irrelevant distractors and (2) selection of a target with an intermediate feature, among heterogeneous nontargets. Some of the findings in favour of the relational account were clearly inconsistent with feature-specific views of guidance, amongst them the finding that a nontarget-coloured distractor can capture attention. Moreover, it has been argued that the current neurophysiological evidence also does not unequivocally support the feature detector concept. From a theoretical stance, top-down tuning to feature relationships seems to be able to account for the majority of findings that were previously interpreted as evidence for feature-specific tuning of attention, including neurophysiological evidence (but see Becker et al., in press; Harris et al., 2013). Whether a relational account can eventually explain all the instances of feature-specific guidance of attention, and may eventually replace feature-based accounts of attention remains to be determined by future research. What is clear from the present review is that the features in the context play an important role in top-down tuning of attention. Thus, studies focussing on attention shifts to single stimuli in isolation are unlikely to provide the insights necessary to advance current theories of visual search.

Acknowledgements This research was supported by an ARC postdoctoral fellowship awarded to Stefanie I. Becker.

References

Ansorge, U., & Heumann, M. (2003). Top-down contingencies in peripheral cueing: The roles of color and location. *Journal of Experimental Psychology: Human Perception and Performance, 29*, 937–948.

Atiani, S., Elhilali, M., David, S. V., Fritz, J. B., & Shamma, S. A. (2009). Task difficulty and performance induce diverse adaptive patterns in gain and shape of primary auditory cortical receptive fields. *Neuron, 61*, 467–480.

Bauer, B., Jolicoeur, P., & Cowan, W. B. (1995). Visual search for colour targets that are or are not linearly separable from distractors. *Vision Research, 36*, 1439–1465.

Becker, S. I. (2010). The role of target-distractor relationships in guiding attention and the eyes in visual search. *Journal of Experimental Psychology: General, 139*, 247–265.

Becker, S. I. (2008). Can intertrial effects of features and dimensions be explained by a single theory? *Journal of Experimental Psychology: Human Perception and Performance, 34*, 1417–1440.

Becker, S. I., & Horstmann, G. (2011). Novelty and saliency in attentional capture by unannounced motion singletons. *Acta Psychologica, 136*, 290–299.

Becker, S. I., Ansorge, U., & Horstmann, G. (2009). Can intertrial priming account for the similarity effect in visual search? *Vision Research, 49*, 1738–1756.

Becker, S. I., Folk, C. L., & Remington, R. W. (2010). The role of relational information in contingent capture. *Journal of Experimental Psychology: Human Perception and Performance, 36*, 1460–1476.

Becker, S. I., Folk, C. L., & Remington, R. W. (2013). Attentional capture does not depend 2 feature similarity, but on target-nontarget relations. *Psychological Science*. 24, 634–647.

Becker, S.I., Harris, A.M., Venini, D., & Retell, J.D. (in press). Visual search for colour and shape: When is the gaze guided by feature relationships, when by feature values? Journal of Experimental Psychology: Human Perception and Performance.

Conway, B. R. (2001). Spatial structure of cone inputs to color cells in alert macaque primary visual cortex (V1). *The Journal of Neuroscience, 21*, 2768–2783.

Conway, B. R., & Tsao, D. Y. (2009). Color-tuned neurons are spatially clustered according to color preference within alert macaque posterior inferior temporal cortex. *PNAS, 106*, 18034–18039.

Conway, B. R., Hubel, D. H., & Livingstone, M. S. (2002). Color contrast in macaque V1. *Cerebral Cortex, 12*, 915–925.

David, S. V., Hayden, B. Y., Mazer, J. A., & Gallant, J. L. (2008). Attention to stimulus features shifts spectral tuning of V4 neurons during natural vision. *Neuron, 59*, 509–521.

De Valois, R. L., Cottaris, N. P., Elfar, S., Mahon, L. E., & Wilson, J. A. (2000). Some transformations of color information from lateral geniculate nucleus to striate cortex, *PNAS, 97*, 4997–5002.

Duncan, J., & Humphreys, G. W. (1989). Visual search and stimulus similarity. *Psychological Review, 96*, 433–458.

D'Zmura, M. (1991). Color in visual search. *Vision Research, 31*, 951–966.

Eimer, M., Kiss, M., Press, C., & Sauter, D. (2009). The roles of feature-specific task set and bottom-up salience in attentional capture: An ERP study. *Journal of Experimental Psychology: Human Perception and Performance, 35*, 1316–1328.

Folk, C. L., & Remington, R. (1998). Selectivity in distraction by irrelevant featural singletons: Evidence for two forms of attentional capture. *Journal of Experimental Psychology: Human Perception and Performance, 24*, 847–858.

Folk, C. L., Remington, R. W., & Johnston, J. C. (1992). Involuntary covert orienting is contingent on attentional control settings. *Journal of Experimental Psychology: Human Perception and Performance, 18*, 1030–1044.

Foster, D. H. (2003). Does color constancy exist? *TICS, 7*, 439–443.

Gouras, P. (1974). Opponent-color cells in different layers of foveal striate cortex. *Journal of Physiology, 238*, 583–602.

Harris, A.M., Remington, R.W., & Becker, S.I. (2013). Feature specificity in attentional capture by size and colour. Journal of Vision, 13:12, 1–15.
Hodsoll, J. P., & Humphreys, G. W. (2005). The effect of target foreknowledge on visual search for categorically separable orientation targets. *Vision Research*, *45*, 2346–2351.
Hubel, D. H., & Wiesel, T. N. (1969). Receptive fields and functional architecture of monkey striate cortex. *Journal of Physiology*, *195*, 215–243.
Itti, L., & Koch, C. (2000). A saliency-based search mechanism for overt and covert shifts of visual attention. *Vision Research*, *40*, 1489–1506.
Kastner, S., Pinsk, M. A., De Weerd, P., Desimone, R., Ungerleider, L. G. (1999). Increased activity in human visual cortex during directed attention in the absence of visual stimulation. *Neuron*, *22*, 751–761.
Koch, C., & Ullman, S. (1985). Shifts in visual selective attention: Towards the underlying neural circuitry. *Human Neurobiology*, *4*, 219–227.
Lee, D. K., Itti, L., Koch, C., & Braun, J. (1999). Attention activates winner-take-all competition among visual filters. *Nature Neuroscience*, *2*, 375–381.
Luck, S. J., Chelazzi, L., Hillyard, S. A., Desimone, R. (1997) Neural mechanisms of spatial selective attention in areas V1, V2, and V4 of macaque visual cortex. *Journal of Neurophysiology*, *77*, 24–42.
Ludwig, C. J. H., & Gilchrist, I. D. (2002). Stimulus-driven and goal-driven control over visual selection. *Journal of Experimental Psychology: Human Perception and Performance*, *28*, 902–912.
Martinez-Trujillo, J. C., & Treue, S. (2004). Feature-based attention increases the selectivity of population responses in primate visual cortex. *Current Biology*, *14*, 744–751.
Maunsell, J. H. R., & Treue, S. (2006). Feature-based attention in visual cortex. *TICS*, *29*(6):317–322.
Motter, B. C. (1994). Neural correlates of attentive selection for color and luminance in extrastriate area V4. *The Journal of Neuroscience*, *14*, 2178–2189.
Mozer, M. C., & Baldwin, D. S. (2008). Experience-guided search: A theory of attentional control. In J. Platt, D. Koller, & Y. Singer (Eds.), *Advances in Neural Information Processing Systems 20* (pp. 1033–1040). Cambridge: MIT Press.
Navalpakkam, V., & Itti, L. (2006). An integrated model of top-down and bottom-up attention for optimizing detection speed. Proceedings IEEE Conference on Computer Vision and Pattern Recognition (CVPR), pp. 2049–2056.
Navalpakkam, V., & Itti, L. (2007). Search goals tunes visual features optimally. *Neuron*, *53*, 605–617.
Nakayama, K., & Martini, P. (2011). Situating visual search. *Vision Research*, *51*, 1526–1537.
Posner, M. I. (1980) Orienting of attention. *Quarterly Journal of Experimental Psychology*, *32*, 3–25.
Scolari, M., & Serences, J. T. (2010). Basing perceptual decisions on the most informative sensory neurons. *Journal of Neurophysiology*, *104*, 2266–2273.
Spitzer, H., Desimone, R., & Moran, J. (1988). Increased attention enhances both behavioral and neuronal performance. *Science*, *240*, 338–340.
Theeuwes, J. (1994). Stimulus-driven capture and attentional set: Selective search for color and visual abrupt onsets. *Journal of Experimental Psychology: Human Perception and Performance*, *20*, 799–806.
Theeuwes, J. (2010). Top-down and bottom-up control of visual selection. *Acta Psychologica*, *135*, 97–135.
Treisman, A., & Gelade, G. (1980). A feature integration theory of attention. *Cognitive Psychology*, *12*, 97–136.
Treisman, A., & Sato, S. (1990). Conjunction search revisited. *Journal of Experimental Psychology: Human Perception and Performance*, *16*, 459–478.
Wolfe, J. M. (1994). Guided Search 2.0: A revised model of visual search. *Psychonomic Bulletin & Review*, *1*, 202–238.

Wolfe, J. M. (1998). Visual search. In: H. Pashler (Ed.), *Attention* (pp. 30–73). London: University College London Press.
Wolfe, J. M., & Horowitz, T. S. (2004). What attributes guide the deployment of visual attention and how do they do it? *Nature Reviews Neurosience*, *5*, 495–501.
Yantis, S. (1993). Stimulus-driven attentional capture and attentional control settings. *Journal of Experimental Psychology: Human Perception and Performance*, *19*, 676–681.
Yantis, S. (2000). Goal-directed and stimulus-driven determinants of attentional control. In S. Monsell & J. Driver (Eds.), *Attention & performance, XVIII* (pp. 73–103). Cambridge: MIT Press.

Gaze and Speech: Pointing Device and Text Entry Modality

T. R. Beelders and P. J. Blignaut

1 Introduction

Communication between humans and computers is considered to be a two-way communication between two powerful processors over a narrow bandwidth (Jacob and Karn 2003). Most interfaces today utilize more bandwidth with computer-to-user communication than vice versa, leading to a decidedly one-sided use of the available bandwidth (Jacob and Karn 2003). An additional communication mode will invariably provide for an improved interface (Jacob 1993) and new input devices which capture data from the user both conveniently and at a high speed are well suited to provide more balance in the bandwidth disparity (Jacob and Karn 2003). In order to better utilize the bandwidth between human and computer, more natural communication which concentrates on parallel rather than sequential communication is required (Jacob 1993). The eye tracker is one possibility which meets the criteria for such an input device. Eye trackers have steadily become more robust, reliable and cheaper and, therefore, present themselves as a suitable tool for this use (Jacob and Karn 2003). However, much research is still needed to determine the most convenient and suitable means of interaction before the eye tracker can be fully incorporated as a meaningful input device (Jacob and Karn 2003).

Furthermore, the user interface is the conduit between the user and the computer and as such plays a vital role in the success or failure of an application. Modern-day interfaces are entirely graphical and require users to visually acquire and manually manipulate objects on screen (Hatfield and Jenkins 1997) and the current trend of Windows, Icons, Menu and Pointer (WIMP) interfaces have been around since the 1970s (Van Dam 2001). These graphical user interfaces may pose difficulties to users with disabilities and it has become essential that viable alternatives to mouse and keyboard input should be found (Hatfield and Jenkins 1997). Specially designed applications which take users with disabilities into consideration are available but these do not necessarily compare with the more popular applications. Disabled users

T. R. Beelders (✉) · P. J. Blignaut
University of the Free State, Bloemfontein, South Africa
e-mail: beelderstr@ufs.ac.za

M. Horsley et al. (eds.), *Current Trends in Eye Tracking Research*,
DOI 10.1007/978-3-319-02868-2_4, © Springer International Publishing Switzerland 2014

should be accommodated in the same software applications as any other computer user, which will naturally necessitate new input devices (Istance et al. 1996) or the redevelopment of the user interface. Eye movement is well suited to these needs as the majority of motor-impaired individuals still retain oculomotor abilities (Istance et al. 1996). However, in order to disambiguate user intention and interaction, eye movement may have to be combined with another means of interaction such as speech. This study aims to investigate various ways to provide alternative means of input which could facilitate use of a mainstream product by disabled users. These alternative means should also enhance the user experience for novice, intermediate, and expert users. The technologies chosen to improve the usability of the word processor are speech recognition and eye tracking. The goal of this study is, therefore, to determine whether the combination of eye gaze and speech can effectively be used as an interaction technique to replace the use of the traditional mouse and keyboard within the context of a mainstream word processor. This will entail the development of a multimodal interface which will allow pointing-and-clicking, text entry, and document formatting capabilities.

The many definitions for multimodal interfaces (for example, Coutaz and Caelen 1991; Oviatt 1999; Jaimes and Sebe 2005; Pireddu 2007) were succinctly summarized for the purposes of this study as:

> A **multimodal interface** uses several human modalities which are combined in an effort to make human–computer interaction easier to use and learn by using characteristics of human–human communication.

Multimodal interfaces themselves date back to 1980, when Richard Bolt, in his seminal work entitled *Put That Here* (Bolt 1981), combined speech and gestures to select and manipulate objects. A distinct advantage of multimodal interfaces is that they offer the possibility of making interaction more natural (Bernhaupt et al. 2007). Furthermore, a multimodal interface has the potential to span across a diverse user group, including varying skill levels, different age groups as well as increasing accessibility for disabled users whilst still providing a natural, intuitive and pleasant experience for able-bodied users (Oviatt and Cohen 2000). For the purposes of this study, a multimodal interface was developed for a popular word processor application and tested as both a pointing device as well as for use as a text entry modality.

Both eye gaze (for example, Hansen et al. 2001; Wobbrock et al. 2008) and speech recognition (for example, Klarlund 2003) have been used in the past for the purpose of text entry. The current study will include eye gaze as an input technique but will require the use of an additional trigger mechanism, namely speech, in order to determine whether the accuracy and speed of the text entry method can be increased in this manner. The multimodal interface should also allow targets to be selected; thus, the viability of a number of pointing options was first investigated. Document formatting capabilities were provided through speech commands but the analysis thereof is beyond the scope of this chapter.

2 Background

Using a physical input device in order to communicate or perform a task in human–computer dialogue is called an interaction technique (Foley et al. 1990 as cited in Jacob 1995). However, for the purposes of this study, the definition will be modified and used in the following context:

> An **interaction technique** is the use of any means of communication in a human–computer dialogue to issue instructions or infer meaning.

Using eye gaze as an interaction device, specifically in the form of a pointing device, could seem natural as users tend to look at objects they are interacting with. However, the use of eye gaze does present some problems such as the Midas touch problem (Jacob 1991). Some of the associated problems of using gaze as a pointing device can be overcome through the use of an additional modality, such as speech.

Psycholinguistic studies have shown that there is a temporal relationship between eye gaze and speech (for example, Just and Carpenter 1976; Tanenhaus et al. 1995), often referred to as the eye–voice span. The eyes move to an object before the object is mentioned (Griffin and Bock 2000) with an approximate interval of 500 milliseconds between the eye movement and speech (Velichkovsky et al. 1997 as cited in Kammerer et al. 2008). However, recently it has been shown that these fixations on objects of interest could occur anywhere from the start of a verbal reference to 1500 milliseconds prior to the reference (Prasov et al. 2007). While the relationship between eye gaze and speech could be confirmed in a separate study, a large variance in the temporal difference between a fixation and a spoken reference to an object was also found (Liu et al. 2007) which could explain the various temporal differences reported on in different texts. This could lead to misinterpretation when attempting to react to verbal and visual cues in synchrony based on gaze position at the time a verbal command is uttered. However, eye gaze has been successful in resolving ambiguities when using speech input (Tanaka 1999) as it has been found that for the majority of verbal requests, users were looking at the object of interest when the command was issued. In order to maximize the disambiguation of both eye gaze and speech in this study, the user will be expected to maintain eye gaze on the desired object whilst issuing the verbal command to interact with that object.

The combination of eye gaze and speech has been used in the past for data entry purposes. For example, in a study conducted in the UK, eye gaze and speech could be used to complete a television license application (Tan et al. 2003a). In this instance, eye gaze was used to establish focus on a particular entry field and then dictation was used to complete the field which currently had focus. Users of the system much preferred using the eye gaze and speech to complete the application form even though it was neither the fastest nor the most accurate means of form completion tested.

The RESER and SPELLER (Tan et al. 2003b) systems used single-character entry mechanisms as opposed to dictation of complete words. The former application required users to gaze at the required key on a cluster keyboard and then to utter the letter that they wished to type. Suggestions were given to complete the word currently being typed which the user could then accept or reject. The SPELLER application requires the entire word to be typed out character by character. For text entry, users preferred the mouse and the keyboard while speech and eye gaze were the preferred means of data recovery.

Dasher is a text entry interface which uses continuous pointing gestures to facilitate text entry (Ward et al. 2000). Speech Dasher extends the capabilities of Dasher even further by including speech recognition as well (Vertanen and MacKay 2010). Speech Dasher uses the same selection technique as the original Dasher but allows the user to zoom through entire words as opposed to single characters. The word set is obtained through speech recognition where the user speaks the text they would like to enter. With an error recognition rate of 22 %, users were able to achieve typing speeds of 40 WPM (Vertanen and MacKay 2010) which is similar to keyboard text entry. Speech Dasher is an example of a multimodal interface where gaze is used to enhance the capabilities of speech recognition.

The current study built on the idea that eye gaze can be used to establish which keyboard button is required by the user. However, instead of relying on the inaccurate or time-consuming methods of eye gaze only, an additional modality is suggested. The use of the look-and-shoot method with a physical trigger assumes that the user may have some mobility although it may be possible to use a triggering mechanism such as blowing in a pipe. Instead, this study will remove the reliance on physical dexterity and will build on the idea proposed by Tan et al. (2003b) that speech could be used to activate the focused key. However, it also assumes that some users may have limited vocabularies and may not be able to vocalize all alphabetic letters. Therefore, a single command, which can be customized to meet the abilities of the user, will be used to activate the key which currently has focus. Through this means, it will be possible to provide text entry capabilities using eye gaze and speech.

3 Eye Gaze and Speech as a Pointing Device

In order to use the proposed modality for text entry, it must first be established how eye gaze and speech can best be used as a pointing device, since in this context pointing forms the basis of text entry. Furthermore, if the interface is to be used within a word processor, the user must be able to select targets such as buttons.

The most commonly used metrics to evaluate pointing devices are speed and accuracy (MacKenzie et al. 2001) which give a good indication as to whether there is a difference between the performance of pointing devices (Hwang et al. 2004). ISO ratified a standard, ISO 9241-9, for determining the speed and accuracy of pointing devices for comparison and testing purposes. The ISO standard uses a throughput metric which encapsulates both speed and accuracy (ISO 2000) in order to compare pointing devices and is measured using any one of six tasks including three point-and-click tasks which conform to Fitts' law (Carroll 2003).

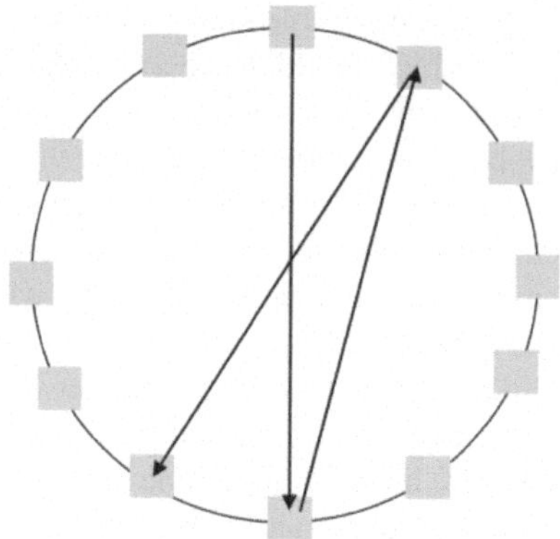

Fig. 1 Multidirectional tapping task

The one-directional tapping test requires the participant to move from a home area to a target and back. In contrast, the multidirectional tapping test (Fig. 1) consists of 24 boxes placed around the circumference of a circle. The participant is then required to move from the centre of the circle to a target box. From there the participant must move to and click in the box directly opposite that box and then proceed in a clockwise direction around the circle until all the targets have been clicked and the user is back at the first selected target box.

The ISO standard has been used to test eye tracking as an input device (Zhang and MacKenzie 2007). This test used the multidirectional tapping test across four conditions, namely (a) a dwell time of 750 ms, (b) dwell time of 500 ms, (c) look-and-shoot method which required participants to press the space bar to activate the target they were looking at and (d) the mouse (Zhang and MacKenzie 2007). A head-fixed eye-tracking system with an infrared camera and a sampling rate of 30 Hz was used for the study. The look-and-shoot method was the best of the three eye-tracking techniques with a throughput of 3.78 bps compared to the mouse with 4.68 bps.

The fact that the look-and-shoot method is the most efficient activation mechanism is not surprising since the selection time of a target is not dependent on a long dwell time and theoretically target acquisition times for all interaction techniques should be similar. Target acquisition in this chapter refers to when the target receives focus to such an extent that visual feedback is given. This does not imply that the target has been selected yet. Therefore, when a fixation is detected on a target or when the mouse enters the bounds of a target, the target is said to be acquired. The time required to press the space bar, particularly if users can keep their hand on it, should be shorter than the dwell time, which was confirmed by the results of the aforementioned study (Zhang and MacKenzie 2007). Recommendations stemming from the study included that a dwell time of 500 ms seemed the most appropriate so as to avoid the Midas touch problem whilst simultaneously ensuring that participants did not get impatient waiting for system reaction (Zhang and MacKenzie 2007). Increasing the width of the target reduced the number of errors made but had no effect on the throughput.

In a comparable study, the ISO standard was used to compare four pointing devices which could serve as a substitute mouse for disabled users (Man and Wong 2007). The four devices tested were the (1) CameraMouse, which was activated by body movements captured via a Universal Serial Bus (USB) webcam, (2) a Head-Array Mouse Emulator, an Adaptive Switch Laboratories, Inc. (ASL) mouse emulator that can provide solutions for power mobility, computer interfacing and environmental control for people with severe disabilities, (3) a CrossScanner, which has a mouse-like pointer activated by a single click and an infrared switch and (4) a Quick Glance Eye Gaze Tracker which allows cursor placement through the use of eye movement (Man and Wong 2007). Targets had a diameter of 20 pixels and the distance between the home and the target was 40 pixels. Two disabled participants, both with dyskinetic athetosis and quadriplegia, were tested over a period of eight sessions with two sessions per week. Each participant was analyzed separately and it was found that the CrossScanner was suitable for both participants although the ASL Head-Array was also suitable for use by one of the participants.

While ISO9241-9, similar to Fitts' law, is undoubtedly a step in the right direction, allowing researchers to establish whether there are differences in speed and accuracy between various pointing devices, it does, however, fail to determine why these differences exist (Keates and Trewin 2005). MacKenzie et al. (2001) propose seven additional measures which will provide more information as to why differences are detected between performance measures of pointing devices. These measures are designed to complement the measures of speed, accuracy and throughput and to provide more insight into why differences exist between pointing devices. The seven measures as proposed by MacKenzie et al. (2001) are as follows:

1. **Target re-entry**
   a. If the pointer enters the area of the target, leaves it and then re-enters it, a target re-entry has occurred.
2. **Task axis crossing**
   a. A task axis crossing is recorded if the pointer crosses the task axis on the way to the target. The task axis is normally measured as a straight line from the centre of the home square to the centre of the target (Zhang and MacKenzie 2007).
3. **Movement direction change**
   a. Each change of direction relative to the task axis is counted as a movement direction change.
4. **Orthogonal direction change**
   a. Each change of direction along the axis orthogonal to the task axis is counted as an orthogonal direction change.
5. **Movement variability**
   a. This "represents the extent to which the sample points lie in a straight line along an axis parallel to the task axis".
6. **Movement error**
   a. This is measured as the average deviation of the sample points from the task axis, regardless of whether these sample points are above or below the task axis.

7. **Movement offset**
   a. This is calculated as the mean deviation of sample points from the task axis.

The ISO9241-9 multidirectional tapping task was used to verify these metrics with 16 circular targets, each 30 pixels in diameter and placed around a 400-pixel-diameter outer circle (MacKenzie et al. 2001). These seven metrics, as well as throughput, movement time and missed clicks were used in a study to determine the difference in cursor movement for motor-impaired users (Keates et al. 2002).

A further six metrics, which could assist in determining why a difference exists, were specifically designed for use with disabled users and were proposed by Keates et al. (2002). These measures were not used during this study as they were not considered relevant. An additional metric measuring the number of clicks outside the target is also suggested in order to measure the performance of pointing devices (Keates et al. 2002).

4 Methodology

4.1 Experimental Design

The ISO test requires that the size of the targets and the distance between targets be varied in order to measure the throughput. In this study, however, variable size targets were used, but in order to reduce the time required to complete a test the distance between targets was not adjusted during this testing.

Standard Windows icons are 24 × 24 (visual angle ≈ 0.62°) pixels in size. This was, therefore, used as the base from which to start testing target selection with speech recognition and eye gaze. Miniotas et al. (2006) determined that the optimal size for targets when using speech recognition and eye gaze as a pointing device was 30 pixels. This was determined using a 17″ monitor with a resolution of 1,024 × 768. Participants were seated at a viewing distance of 70 cm. This translated into a viewing angle of 0.85°. The eye tracker used in the current study was a Tobii T120 with a 17″ monitor where the resolution was set to 1,280 × 1,024. In order to replicate the viewing angle of 0.85° obtained by Miniotas et al. (2006), a 30-pixel target could be used but at a viewing distance of 60 cm from the screen. Therefore, the next size target to be tested in the trials was determined to be a 30 × 30 pixel button. It was decided to also test a larger target than that established by Miniotas et al. (2006). Following the example set by Miniotas et al. (2006) of testing target sizes in increments of 10 pixels, the final target size to be used was 40 pixels (visual angle ≈ 1.03°).

The multidirectional tapping task used in this study had 16 targets situated on a circle with a diameter of 800 pixels. The square targets were positioned on the edges of the circle—thereby creating an inner circle with a diameter of 800 pixels.

Target acquisition was either via eye tracking and speech recognition (denoted by ETS for the purpose of this chapter) or the mouse (M). The mouse was used to establish a baseline for selection speed. When using a verbal command to select a target, the subjects had to say "go" out loud in order to select the target that they were looking at. This method of pointing can, therefore, be considered analogous to look and shoot.

Magnification (ETSM) and the gravitational well (ETSG) were used to combat various shortcomings of using eye gaze for target selection, namely the instability of the eye gaze and the difficulties experienced in selecting small targets. The default zoom factor for the magnification enlarged the area to double its actual size within a 400×300 window while the gravitational well was activated within a 50-pixel radius around each button. The target button which had to be clicked was denoted by an "X".

This resulted in a total of 14 trials per session, the number of which served as motivation for not adding more trials for the mouse as this would simply prolong the session time and might cause participants to become irritable and fatigued during the session.

A balanced Latin square for all trial conditions was obtained by following the instructions provided by Edwards (1951). Participants were randomly assigned to a Latin square condition for each session.

Together with the throughput measure of the ISO standard, additional measurements were analyzed in an effort to explain the difference in performance if such a difference exists between the interaction techniques. To this end, the total task completion time was measured as well as the task completion time from when the target was highlighted to when it was clicked, the number of target re-entries, the number of incorrect targets which were acquired during task completion and the number of incorrect clicks. This will allow efficiency and effectiveness of each interaction technique to be tested.

4.2 Participants

Participants, who were senior students at the university at which the study was conducted, volunteered to participate in the study. For each session completed, the participant received a small cash amount.

Each participant completed three sessions and each session consisted of all 14 trials. In total there were 15 participants who completed all three sessions.

Eleven of the participants were male and four were female. The average age of the participants was 22.3 (standard deviation = 1.9). The only selection criteria was that the participant have normal or corrected-to-normal vision, that they were proficient with the mouse and that they had no prior experience with either eye tracking or speech recognition.

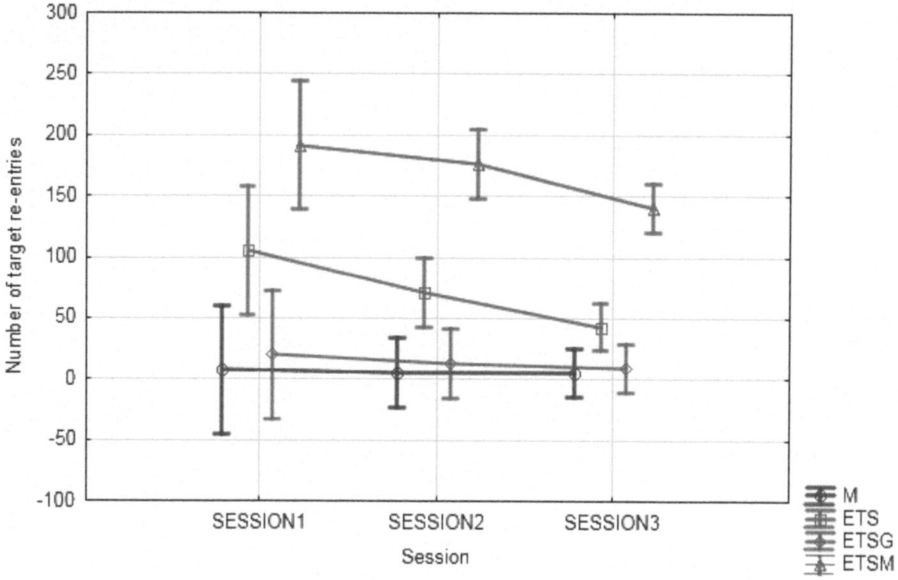

Fig. 2 Target re-entries for all interaction techniques

5 Results

5.1 Throughput and Time to Complete a Trial

As stipulated in the ISO test, the throughput was measured and analyzed for each of the interaction techniques. The results of this analysis, as well as the time to complete a trial, are discussed in detail in Beelders and Blignaut (2012). In summary, the mouse had a much higher throughput than the other interaction techniques. In terms of the time to complete a trial, the mouse and the use of the gravitational well had comparable selection times. Interestingly, when using a gravitational well, the time to select a target was much faster than when using any other interaction technique, including the mouse.

5.2 Target Re-Entries

Target re-entries were defined as the number of times the designated target was gazed upon before the user was able to click on it.

The graph in Fig. 2 plots the number of target re-entries for all interaction techniques over the three sessions.

At an α-level of 0.05, there was a significant difference between the number of target re-entries for the different interaction techniques ($F(3, 56) = 32.071$). Post-hoc

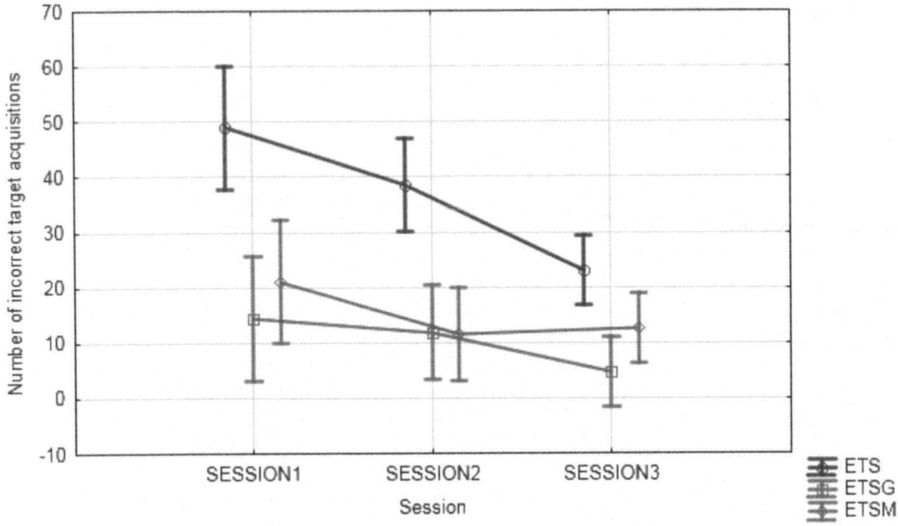

Fig. 3 Incorrect target acquisitions for all interaction techniques

tests indicated that ETSM had a significantly higher incidence of target re-entries than the other interaction techniques. This would imply that it was much more difficult to achieve a prolonged stable gaze on a button such that the required verbal command can be issued when the magnification tool was activated than for any other interaction technique. ETS also differed significantly from the mouse and ETSG. ETSG did not differ significantly from the mouse, which means that ETSG is able to perform comparably with the mouse in terms of target re-entries.

There was also a significant difference between the sessions ($F(2, 112)=4.249$).

5.3 Incorrect Target Acquisitions

Incorrect target acquisitions were defined as the number of times a target, which was not the designated target, was acquired. This means that in the event of the eye tracker and speech being used, each time a button received enough focus to give visual feedback, the incorrect target acquisitions were incremented, provided that the focused button was not the designated target. The number of incorrect target acquisitions was counted as those targets which were acquired *after* the designated target had been acquired. Therefore, the incorrect targets that were acquired could not be attributed to normal searching for the designated target. For the purposes of this measurement, only the eye gaze and speech interaction techniques will be included in the analysis as the number of incorrect target acquisitions for the mouse interaction techniques was always zero.

The graph in Fig. 3 plots the number of incorrect target acquisitions for all included interaction techniques over all sessions.

For this measure, ETSG had the best performance, although all interaction techniques exhibited some degree of improvement, most notably that of ETS.

At an α-level of 0.05, there was a significant difference between the interaction techniques ($F(2, 42) = 19.327$) as well as the sessions ($F(2, 84) = 12.046$, $p < 0.05$).

All the sessions differed significantly from one another. Since only ETSM actually increased slightly in session 3, it can be surmised that the incorrect target acquisitions lessened at a significant rate over time. ETS, in particular, had a sharp decrease and it may be beneficial to increase the number of sessions so that it can be properly analyzed whether it can ever reach the low values of ETSG or ETSM. In terms of the interaction techniques, ETS differs significantly from both ETSG and ETSM. ETSG and ETSM do not differ significantly from each other.

Observations made of the participants while they were completing the tasks could provide an explanation for this. Many participants soon realised that when struggling to focus on a button it was sometimes easier to focus on another button at a suitable distance from the designated one. It was not necessary to focus on this other button for a protracted time. Participants would then look back at the designated button and the extended movement seemed to provide more accuracy in focusing on the desired target rather than trying to "fine-tune" the selection within a small area around the designated button. The smoothing algorithm could have contributed to this as small movements within a certain radius are interpreted as a single fixation. Since the gravitational well effectively pulls the selection onto the nearest target once the "pointer" is within a certain distance, it becomes easier to focus on a target and no fine-tuning is required. This could explain the reason why ETSG has such a low number of acquisitions compared to ETS.

ETSM also has a lower rate and this could possibly be attributed to participants rather trying to fine-tune the selection when using the magnification. Since the buttons appear larger, participants may have perceived the fine-tuning process to be easier since a larger target could create the impression that it can be easily acquired. The high incidence of target re-entries coupled with the low number of incorrect target acquisitions may serve to substantiate the suspicion that fine-tuning was the preferred method for ETSM.

The similar pattern for ETS, regarding target re-entries and target acquisitions, also corroborates the claim that the participants preferred to employ the use of a shifting of their eye gaze to focus on another button and then returning to the designated button. Closer inspection of the averages for ETS shows that incorrect target acquisitions constituted approximately half the number of re-entries for each session. This could indicate that participants would attempt to re-acquire the designated target and, when they were unable to achieve a stable selection, they resorted to focusing on another target before attempting to select the designated target—in contrast to the strategy employed with ETSM.

The reason for this could be that the magnification disturbs the users while they adjust their gaze and they are unwilling to move their gaze substantially because they perceive this to require more effort when magnification is activated. Another reason for the different strategies could be attributed to the fact that the magnification tool that was used has in-built visual feedback which allows the user to get an

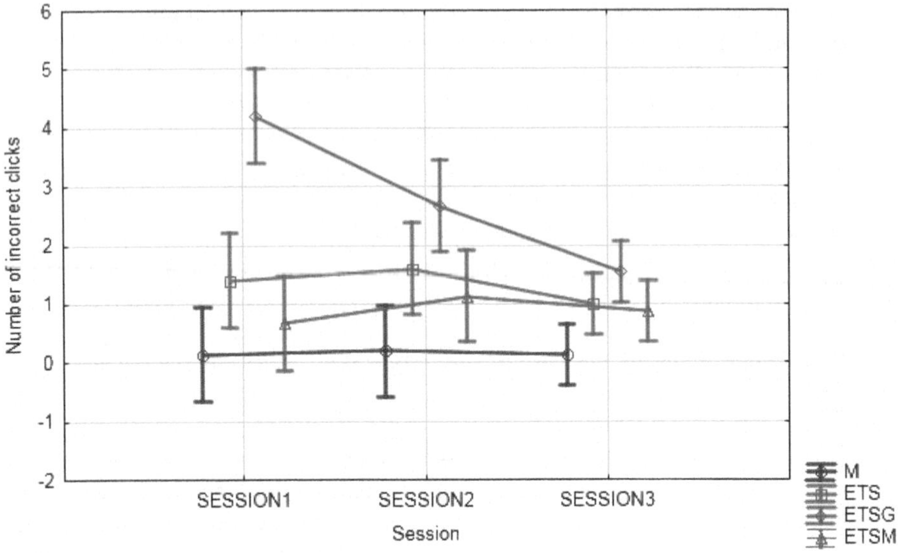

Fig. 4 Incorrect clicks for all interaction techniques

approximation of their eye gaze position, which is centred in the magnified area. Since this feedback is present, the user may feel that fine-tuning is a better option since they can determine how close they are to the target, which is not the case with ETS. With ETS, they will know they have lost the target but not how close they are to re-acquiring it; hence, they feel more secure glancing at another target, establishing position and then looking at the required target again until they can maintain a stable eye gaze. Therefore, to slave a cursor to the eye gaze may be disruptive but in this instance it could tentatively be said that it may have provided useful information to the participants. However, the evidence suggests that it in no way increased the efficiency or effectiveness of target selection and, therefore, it is not recommended for use.

The average number of target re-entries for ETSG was roughly the same as the average incorrect target acquisitions for ETSG. This could provide evidence that when using ETSG, the target was easier to acquire and it was easier to keep the focus long enough to issue the required command. Since the buttons were effectively larger, it would make sense that they were easier to focus on for a prolonged period of time.

5.4 Incorrect Clicks

Incorrect clicks were determined as the number of times a target that was not the designated target was clicked during a trial.

The graph in Fig. 4 plots the number of incorrect clicks for all interaction techniques.

Owing to the fact that there was significant interaction between the factors, each session was analyzed separately to determine whether there was a difference between the interaction techniques. It was found that in the first session, ETSG differed significantly from all other techniques; in the second session ETSG differed significantly from the mouse and ETSM and in the third session only from the mouse.

These results clearly show that ETSG resulted in the highest number of incorrect clicks. Although continued practice allowed ETSG to have a comparable number of incorrect clicks to ETS and ETSM, its performance could not match that of the mouse over the three sessions. This indicates that some learning did take place over the three sessions.

Natural eye movement may provide an explanation for the observed difference. Participants could acquire the target and then issue a verbal command while already starting to look at the next target (for all eye gaze and speech interaction techniques). Since the use of the gravitational well increases the speed with which a target can be acquired, this often meant that by the time the speech engine recognized the command, the next target had already been acquired. This could account for the high number of incorrect targets for ETSG. These findings also confirm previous findings that the fixation immediately prior to the action or command being issued usually occurs on the object of interest (Land and Tatler 2009; Maglio et al. 2000).

Since ETSG had significantly lower incorrect target acquisitions, coupling it with this finding of more incorrect clicks creates the following dilemma. The use of the gravitational well increases the possibility of correctly acquiring a target and maintaining a stable gaze on the target. This is evidenced by the fact that other eye gaze and speech interaction techniques caused participants to first glance away, acquire another target and then glance back. However, the fact that a gravitational well is present together with human tendency to start glancing at the next object of interest whilst still issuing a command to the current target, means that the next target is acquired far quicker than when no gravitational well is present. This causes the next target to be incorrectly clicked on with higher frequency for ETSG. Since participants started moving their eye gaze away from the buttons before the speech command had been executed for all eye gaze interaction techniques, it would be assumed that for ETS and ETSM, which pose greater difficulty in target acquisition, the participant would inadvertently have caused a click somewhere on the application form which was not a clickable area. Unfortunately, this measurement was not captured during these tests. Further research must be done in order to determine if this proposition is true.

6 Discussion

Incorrect clicks were experienced with all eye gaze interaction techniques although more so with ETSG. Nevertheless, this finding corresponds with the finding of Kaur et al. (2003) that the target which was acquired a certain amount of time prior to command execution is the target that must be selected. Although the interval was found to be 630 ms (Kaur et al. 2003), this interval will have to be confirmed for use with eye gaze and speech. While natural eye gaze movement appears to dictate that the target prior to command utterance must be selected, it must still be determined whether this will appear natural to the user or whether they would prefer to adapt to the use of ETSG as it was tested in this study. Clearly, practice allows them to adjust their natural behaviour to a degree to compensate for the interaction technique as is evidenced by the improvement over the sessions. However, requiring users to change their natural behaviour is not the aim of a multimodal interface. Therefore, it becomes necessary to establish the interval required for target selection and test the usability of that compared to the standard gravitational well employed in this study.

Previous studies such as the touch-sensitive mouse (Drewes and Schmidt 2009) and MAGIC pointing (Zhai et al. 1999) warped the mouse pointer to the position of the eye gaze and then users were required to use the mouse to click on the desired target. Although this exploits the high speed of eye gaze and also reduces incidences of incorrect clicks since users are not likely to click on the incorrect target when having to manually manipulate a mouse pointer, some physical dexterity is required. The solution may lie in a combination of this technique and speech. Eye gaze could be used to establish intent, a single voice command could be issued to warp the pointer to the selectable target closest to the current eye gaze, and once the user has verified that the correct target is acquired, a second command can be issued to click on the target. For fine-tuning purposes of the mouse cursor, direction- or target-based navigation can also be provided.

7 Multimodal Word Processor

The next step was to test the modality when used for text entry. For these purposes, it was decided to use the familiar environment of Microsoft Word® and to simply develop a multimodal interface for Word. Visual Studio Tools for Office (VSTO), which allows developers to create extensions to the Office Suite with customized functionality (Anderson 2009), was used to add multimodal functionality (Fig. 5).

An on-screen keyboard was available which was overlaid on the bottom of the current document. Users could then type in the Word document by focusing their gaze on the desired character on the keyboard and issuing a verbal command to trigger the keyboard key. Auditory feedback, in the form of a beep, was given to alert the user that the character had been typed. This should allow them to continue typing without having to glance back at the document for confirmation.

As can be seen in Fig. 5, a magnification tool was available which allowed the area directly under the gaze of the user to be enlarged. Typing tasks using the magnification tool were not required during this study and will, therefore, not be discussed in this chapter. Figure 5 also shows a number of other customizations which were available in the multimodal interface which was developed, most of which are beyond the scope of this chapter but it is interesting to note their inclusion.

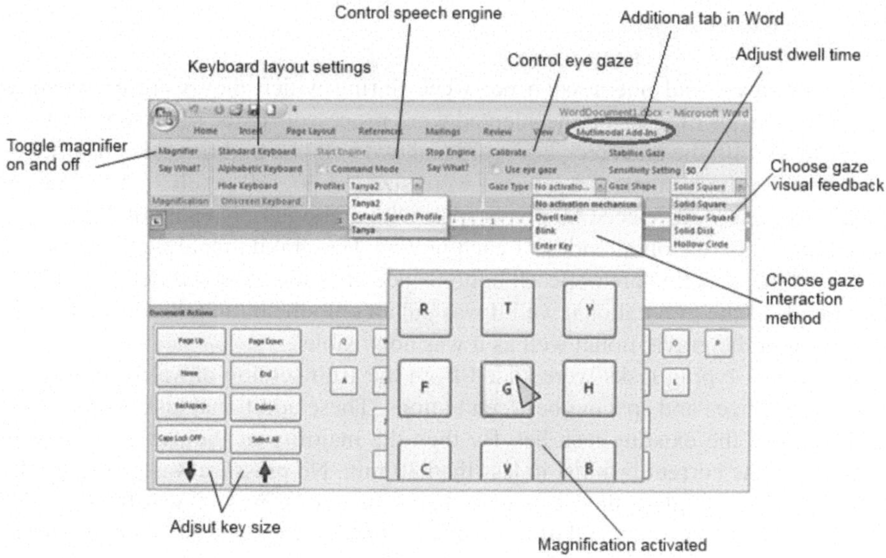

Fig. 5 Multimodal add-in for Microsoft Word

8 Analysis

8.1 Participants

A total of 25 participants participated in the 10-week long study. A prerequisite for participation in the study was sufficient computer literacy as well as word processor expertise. Forty percent of the sample was drawn from second-year computer science students who were registered for a community service module. The other 60 % of the sample was drawn from the student assistants for the computer literacy course of the university, with the proviso that they were not studying for a computer science or related degree. These students all had to complete the literacy course prior to becoming an assistant and they had to achieve at least 70 % for a competency test of Microsoft Office applications. Their proficiency with Word was verified through the completion of a questionnaire before the commencement of the study. The questionnaire evaluated the duration and frequency of use in order to determine an expertise measurement.

The first week was simply an introductory session and the data collected there were not included in the analysis. Furthermore, the data of three participants had to be discarded from the sample due to various reasons. Of the remaining 22 participants, only 8 completed all sessions on the on-screen keyboard and 14 with the traditional keyboard. These were the participants who were included in the analysis.

8.2 Tasks

Each participant had one session per week during which they were expected to complete a series of tasks on the adapted word processor. Three of these tasks were typing tasks with the on-screen keyboard and two with the traditional keyboard. For each session, the buttons of keyboard were sized at 60×60 pixels ($\approx 1.55°$ visual angle). Buttons were spaced 60 pixels apart with a gravitational well of 20 ($\approx 0.52°$ visual angle) pixels on all sides of each button. The gravitational well effectively increased the selection area of each button since once the gaze was detected within the bounds of the gravitational well it was pulled onto the button. Participants were not aware of the gravitational well as it was not visible.

Additional typing tasks were added from the fifth session onwards in order to test varying sizes and spacing between buttons. These additional tasks were added to the end of the existing task list. By then the majority of the participants were completing the current task list in less than 30 min. No pressure was placed on the participants to complete all tasks within their scheduled time so it was felt that adding additional tasks to the end of the test would not unduly cause any more anxiety or place more strain on the participants. Within these additional typing tasks, the first one had to be completed using the originally sized and spaced buttons. The next two had to be completed with buttons that were 50×50 ($\approx 1.29°$ visual angle) pixels in size and spaced 70 ($\approx 1.80°$ visual angle) pixels apart. Following this, there were another two tasks which had to be completed using buttons that were 50×50 pixels in size but were spaced 60 pixels apart. The original configuration of button will henceforth be referred to as speech-SC (small, closely spaced), the larger button configuration as speech-L and the smaller more widely spaced configuration will be referred to as speech-SW.

The typing tasks required the participant to type a phrase that was randomly selected from a set of 35 phrases. The phrase set used was a subset of the 500 as determined by MacKenzie and Soukoreff (2002) to be everyday phrases which are commonly used. The results of the typing tests from session 5 onwards will be the focus of this chapter.

8.3 Measures

The Levenshtein distance (Levenshtein 1965) between two strings measures how many insertions, deletions and substitutions have taken place between presented text and transcribed text. The sum of these errors can then divided by the number of characters to give a character error rate (CER) (Read 2005). Since there are multiple ways in which the presented text can be transformed into the transcribed text, the possible transformations or optimal alignments were identified, their mean length was calculated and then the Levenshtein distance was divided by this mean length to give an error rate (MacKenzie and Soukoreff 2003). This was how the CER was calculated for text entry in this study.

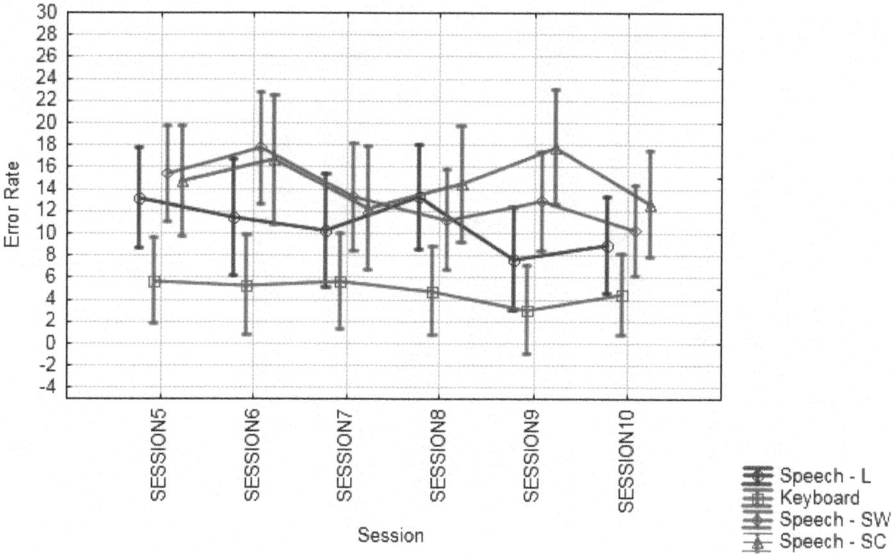

Fig. 6 Character error rate for secondary study

The second measurement analyzed during the study was characters per second (CPS). This measures the number of characters that were typed and then divides it by the time taken to type the characters, measured in seconds. In order to ensure that the time it takes to read a phrase does not unduly influence the results, the time taken to type the phrase was measured from when the first character was typed to when the last character was typed.

9 Results of Study

9.1 Character Error Rate

The graph in Fig. 6 plots the mean error rate for all sessions and for all interaction techniques.

From the graph, it can be surmised that the keyboard had the lowest error rate of all interaction techniques for all sessions. Thereafter, speech-L had the next lowest error rate while the smaller buttons, both speech-SW and speech-SC, caused the highest error rates for all sessions. The latter two seem to cause approximately the same error rates while typing; however, the widely spaced buttons have an improved error rate during the later sessions while the error rates for the closely spaced buttons increased over the same period.

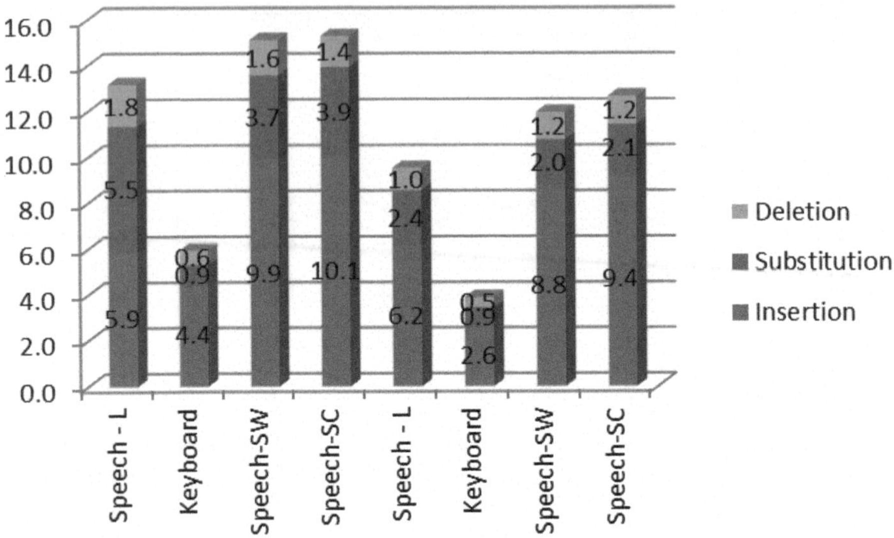

Fig. 7 Breakdown of **character error rate** (CER) for secondary study

At an α-level of 0.05, it was found that there was a significant difference between the error rate of the different interaction techniques ($F(3, 43)=7.303$). Post-hoc tests indicate that the keyboard differed significantly from both speech-SW and speech-SC. In this instance, it is encouraging to determine that speech-L does not differ significantly from the keyboard in these later sessions. This would seem to indicate that after some practice with the larger buttons, the number of errors made decreases. The same cannot be said of the smaller buttons. There was also a significant difference between the CER for the sessions ($F(5, 215)=2.530$), where session 6 differed significantly from session 10.

The errors were then categorized as insertions, deletions or substitutions and further analyzed as such. The bar graph in Fig. 7 shows the breakdown for the first session (first four stacks) using all interaction techniques and the last session (last four stacks). Deletions are on top of each stack, substitutions are in the middle and insertions constitute the lower part of each stack.

The percentage of insertion errors was the highest for all interaction techniques for both of these sessions. The interaction techniques of speech-SW and speech-SC have very similar distributions over the number of insertions, substitutions and deletions.

At an α-level of 0.05, there was a significant difference between the interaction techniques in terms of the number of insertions ($F(3, 44)=4.100$), but not for deletions ($F(3, 39)=1.638$). Owing to significant interaction between the factors, separate analyses had to be performed for the substitution errors where it was found that there was a significant difference between the interaction techniques for all sessions.

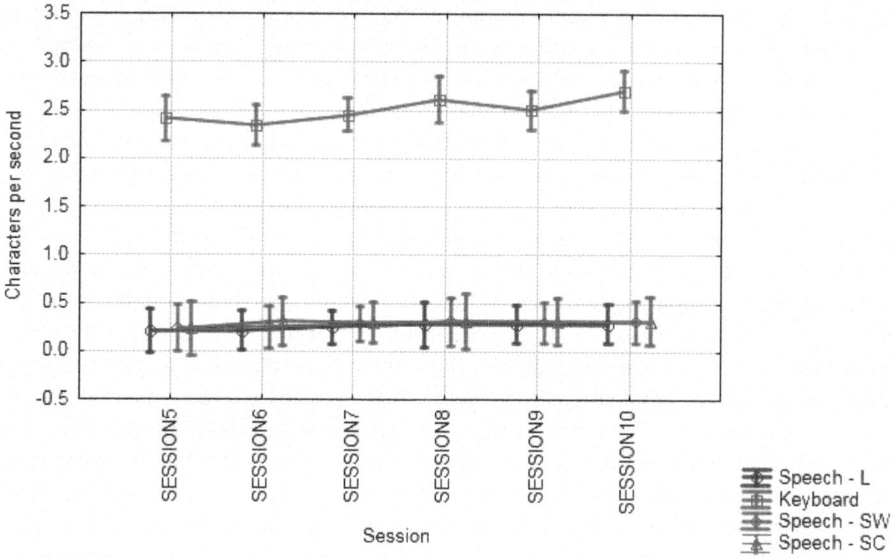

Fig. 8 Characters per second for secondary study

Post-hoc tests indicated that the use of speech-SW resulted in significantly more insertions than the keyboard. For the substitution errors, the keyboard generally had significantly less errors than a variety of the speech interaction techniques depending on the session.

9.2 Characters per Second

The CPS were measured for each interaction technique and for each session. The graph in Fig. 8 plots this measure for each session and each interaction technique.

When using the keyboard, participants were clearly able to type at a much faster rate than when using eye gaze and speech with the on-screen keyboard, which remained fairly consistent regardless of the keyboard settings. As can be expected, there was a significant difference between the interaction techniques at an α-level of 0.05 ($F(3, 44) = 148.369$). There was a significant difference between sessions ($F(5, 15) = 3.002$); in particular, session 10 differed significantly from sessions 5 and 6.

10 Discussion

It was found that the eye gaze and speech interaction technique had a significantly higher error rate than that of the keyboard, undoubtedly as a result of a higher number of insertions and substitutions. This may serve as confirmation that even when using eye gaze and speech as a text input mechanism, the user is inclined to glance away before completing the issuing of the verbal command. The average insertions are generally higher than the substitutions which would seem to indicate that users are aware that they have activated the incorrect character and attempt to correct it by inserting the correct character. This is encouraging as it indicates that users become familiarized with the system such that they can interpret the selection (indicated by audio feedback) and are able to make corrections to text entry. Further research could confirm these suppositions by capturing the correction of the text input as well so that it can be analyzed to determine whether incorrect inputs are reversed/erased before text input is continued. Whether the buttons are large, small and widely spaced or small and closely spaced seems to be of little consequence. There was no difference between the error rates of these three interaction techniques and they all differed from the keyboard at some stage. However, the interaction technique of speech-L did seem to offer the most improved error rate as it did not differ from the keyboard when analyzed for the later sessions only. In some instances, there was improvement over the sessions, which indicates some measure of learning when using the interaction technique. If the learning effect can be maintained then more practice with the eye gaze and speech could eventually lead to an effectiveness measurement which is comparable to that of the keyboard.

In terms of efficiency, the keyboard also outperformed all the eye gaze and speech interaction techniques with significantly higher numbers of CPS which could be typed. The typing speed of the eye gaze and speech also did not improve as exposure increased. This could indicate that either more practice is needed to achieve increased speeds or the typing speed quickly reaches the fastest achievable rate. Neither the size of the buttons nor the spacing between buttons affected the efficiency of the eye gaze and speech.

Therefore, in terms of effectiveness and efficiency, the three eye gaze and speech interaction techniques seem fairly interchangeable as they perform on comparable levels to each other. The keyboard is far more effective and efficient than any of the eye gaze and speech interaction techniques when used for text input.

No similar studies were found with which these results could be compared. However, the fact that speech outperforms keyboard input for young children (Read et al. 2001) indicates that the learning curve for keyboard entry is fairly steep. This could be the same for text entry with eye gaze and speech. Although there was no significant improvement in the speed of the text entry, participants clearly became more comfortable with the use of the interaction technique. Therefore, extended practice may be required to improve speeds.

The mean entry rate of eye gaze and speech fell within the range between 0.2 and 0.3 CPS. Considering that the entry rate was relatively low for context switching at 12 WPM (Morimoto and Amir 2010) and 9 WPM for symbol creator (Miniotas et al. 2003), the range in this study was much lower than these previous studies. A previous study (Majaranta 2009) showed that the use of both visual and auditory feedback increased the entry speed to 7.55 WPM which is still faster than the speeds achieved in this study. Speech Dasher achieved much higher speeds (40 WPM), while using Dasher with eye gaze also resulted in higher speeds (17 WPM).

Therefore, when comparing the text entry method to studies using only eye gaze without text predictors, speech and eye gaze performs slightly better. However, the speeds are still lower than using text prediction methods and when using speech as an activator. While these comparisons are promising since they indicate that speech and eye gaze could facilitate faster entry speeds than using eye gaze only, they are discussed with caution since the text entered in the current study required only a few short phrases to be entered and more prolonged use could have an impact on the entry speed.

11 Conclusion

As evidenced by the incorporation of the technologies used in the multimodal interface of this study, the time has perhaps dawned when they should be exploited as replacement interaction techniques. Speech recognition has become a standard feature in personal computers and is often available for dictation purposes. Similarly, there are packages available for purchase which can react to spoken commands (cf. Dragon n.d.). Furthermore, the first fully integrated eye-controlled laptop has recently been showcased at exhibitions (Tobii 2011) and bodes well for the adoption of eye tracking as a standard feature in personal computers. Cheaper, accurate eye trackers (cf. Haro et al. 2000) are also available which could function just as well as a standard interaction technique.

Therefore, the fact that a popular mainstream application can be adapted to include a highly customizable, multimodal interface could be a step in the right direction for the next generation of interfaces. The multimodal user interface displays great potential and test results indicate that the interaction techniques can be used for pointing and selecting tasks and common word processing tasks. Moreover, it has been proven that speech recognition can indeed be used for editing commands in a word processor which was contrary to theoretical beliefs (Klarlund 2003). This could mean that in the future a more diverse group of users can be accommodated and disabled users may no longer have to be relegated to using specialized applications.

The findings, therefore, suggest that the word processor is well placed to include such an interface in future developments as the technology is rapidly becoming available. As it is foreseen that access to the technologies by mainstream users is imminent, future word processors could be developed with multimodal interfaces incorporated.

The combination of eye gaze and speech could successfully be used to fulfil the needs of a pointing device, particularly when employed with a gravitational well. While text entry was slower than using a keyboard, indications are that there was an overall positive response to the interface and that it may well herald a suitable multimodal interface. The ease with which participants became accustomed to the interface is further proof of the naturalness and intuitiveness provided by speech and eye gaze. With constant progress being made in the development of the hardware required by such an interface, the proposed multimodal interface may well lay the foundation for a word processor to continue its exploitation of emerging technologies and remain a forerunner in the establishment of trends. While there is undoubtedly room for improvement and expansion, the use of eye gaze and speech has proven to be very promising.

References

Anderson, T. (2009). *Pro office 2007 development with VSTO*. United States of America: APress

Beelders, T. R., & Blignaut, P. J. (2012). Using eye gaze and speech to simulate a pointing device. In *Proceedings of the symposium on eye-tracking research and application (ETRA)*, Santa Barbara, California

Bernhaupt, R., Palanque, P., Winkler, M., & Navarre, D. (2007). Usability study of multi-modal interfaces using eye-tracking. In *Proceedings of INTERACT 2007*, 412–424

Bolt, R. (1981). Gaze-orchestrated dynamic windows. *Computer Graphics*, *15*(3), 109–119.

Carroll, J. M. (2003). *HCI models, theories, and frameworks: Towards a multidisciplinary science*. San Francisco: Morgan Kaufmann.

Coutaz, J., & Caelen, J. (1991). A taxonomy for multimedia and multimodal user interfaces. In *Proceedings of the Second East-West HCI conference*, St Petersburg, Russia, 229–240

Dragon Naturally Speaking. (nd). History of speech and voice recognition and transcription software. Retrieved 13 Feb 2009 from http://www.nuance.com

Drewes, H., & Schmidt, A. (2009). The MAGIC touch: Combining MAGIC-pointing with a touch-sensitive mouse. In *Human-Computer Interaction—INTERACT 2009. 12th IFIP TC 13 International Conference, Part II*, Uppsala, Sweden, 415–428

Edwards, A. L. (1951). Balanced Latin-square designs in psychological research. *The American Journal of Psychology*, *64*(4), 598–603.

Foley, J. D., Van Dam, A., Feiner, S. K., & Hughes, J. F. (1990). *Computer graphics: Principles and practice*. Reading, Massachusetts: Addison-Wesley.

Griffin, Z. M., & Bock, K. (2000). What the eyes say about speaking. *Psychological Science*, *11*, 274–279.

Hansen, J. P., Hansen, D. W., & Johansen, A. S. (2001). Bringing gaze-based interaction back to basics. In C. Stephanidis (Ed.), *Universal Access in HCI (UAHCI): Towards an Information Society for All—Proceedings of the 9th International Conference on Human-Computer Interaction (HCII'01)*, 325–328. Mahwah: Lawrence Erlbaum Associates

Haro, A., Essa, I., & Flickner, M. (2000). A non-invasive computer vision system for reliable eye tracking. In *Proceedings of CHI '00*, The Hague, Netherlands, 167–168

Hatfield, F., & Jenkins, E. A. (1997). An interface integrating eye gaze and voice recognition for hands-free computer access. In *Proceedings of the CSUN 1997 Conference*, 1–7

Hwang, F., Keates, S., Langdon, P., & Clarkson, J. (2004). Mouse movements of motion-impaired users: A submovement analysis. In *Proceedings of ASSETS '04*, Atlanta, Georgia, United States of America, 102–109

Istance, H. O., Spinner, C., & Howarth, P. A. (1996). Providing motor impaired users with access to standard Graphical User Interface (GUI) software via eye-based interaction. In *Proceedings of 1st European Conference on Disability, Virtual Reality and Associated Technology*, Maidenhead, United Kingdom, 109–116

ISO. (2000). *ISO 9241–9: Ergonomic requirements for office work with visual display terminals (VDTs)—Part 9: Requirements for non-keyboard input devices*. International Organization for Standardization

Jacob, R. J. K. (1991). The use of eye movements in human-computer interaction techniques: What you look at is what you get. *ACM Transactions on Information Systems*, *9*(2), 152–169.

Jacob, R. J. K. (1993). Eye movement-based human-computer interaction techniques: Toward non-command interfaces. In H. R. Hartson & D. Hix (Eds), *Advances in human-computer interaction, 4, 151–190. Norwood.* New Jersey: Ablex Publishing.

Jacob, R. J. K. (1995). Eye tracking in advanced interface design. In W. Barfield & T. A. Furness (Eds.), *Virtual environments and advanced interface design* (pp. 258–288). New York: Oxford University Press.

Jacob, R. J. K., & Karn, K. S. (2003). Eye tracking in human-computer interaction and usability research: Ready to deliver the promises (Section Commentary). In J. Hyona, R. Radach & H. Deubel (Eds), *The mind's eye: Cognitive and applied aspects of eye movement research* (pp. 573–605). Amsterdam: Elsevier Science.

Jaimes, A., & Sebe, N. (2005). Multimodal human computer interaction: A survey. *IEEE workshop on human computer interaction*, Las Vegas, Nevada, United States of America, 15–21

Just, M. A., & Carpenter, P. A. (1976). Eye fixations and cognitive processes. *Cognitive Psychology*, *8*, 441–480.

Kammerer, Y., Scheiter, K., & Beinhauer, W. (2008). Looking my way through the menu: The impact of menu design and multimodal input on gaze-based menu selection. In *Proceedings of the Symposium on Eye Tracking Research and Applications (ETRA)*, Savannah, Georgia, United States of America, 213–220

Kaur, M., Tremaine, M., Huang, N., Wilder, J., Gacovski, Z., Flippo, F., & Mantravadi, S. (2003). Where is "it"? Event synchronization in gaze-speech input systems. In *Proceedings of ICIM '03*, Vancouver, Canada, 151–158

Keates, S., Hwang, F., Langdon, P., Clarkson, P. J., & Robinson, P. (2002). Cursor movements for motion-impaired computer users. In *Proceedings of ASSETS '02*, Edinburgh, Scotland, 135–142

Keates, S., & Trewin, S. (2005). Effect of age and Parkinson's Disease on cursor positioning using a mouse. In *Proceedings of ASSETS '05*, Baltimore, Maryland, United States of America, 68–75

Klarlund, N. (2003). Editing by voice and the role of sequential symbol systems for improved human-to-computer information rates. In *Proceedings of ICASSP*, Hong Kong, 553–556

Land, M. F., & Tatler, B. W. (2009). *Looking and acting: Vision and eye movements in natural behaviour*. United States of America: Oxford University Press.

Levenshtein, V. I. (1965). Binary codes capable of correcting deletions, insertions, and reversals. *Doklady Akademii Nauk*, *163*, 845–848.

Liu, Y., Chai, J, Y., & Jin, R. (2007). Automated vocabulary acquisition and interpretation in multimodal conversational systems. In *Proceedings of the 45th Annual Meeting of the Association for Computational Linguistics*

MacKenzie, I. S., Kauppinen, T., & Silfverberg, M. (2001). Accuracy measures for evaluating computer pointing devices. In *Proceedings of SIGCHI '01*, Seattle, Washington, United States of America, 9–16

MacKenzie, I. S., & Soukoreff, R. W. (2002). A character-level error analysis technique for evaluating text entry methods. In *Proceedings of NordiCHI 2002*, Aarhus, Denmark, 243–246

MacKenzie, I. S., & Soukoreff, R. W. (2003). Phrase sets for evaluating text entry techniques. In *Extended Abstracts of the ACM Conference on Human Factors in Computing Systems—CHI 2003*, Fort Lauderdale, Florida, United States of America, 754–755

Maglio, P. P., Matlock, T., Campbell, C. S., Zhai, S., & Smith, B. A. (2000). Gaze and speech in attentive user interfaces. In *Proceedings of the Third International Conference on Advances in Multimodal Interfaces*, Vancouver, Canada, 1–7

Majaranta, P. (2009). Text entry by eye gaze. Dissertations in Interactive Technology, number 11, University of Tampere

Man, D. W. K., & Wong, M.-S., L (2007). Evaluation of computer-access solutions for students with quadriplegic athetoid cerebral palsy. *American Journal of Occupational Therapy*, *61*, 355–364.

Miniotas, D., Špakov, O., & Evreinov, G. (2003). Symbol Creator: An alternative eye-based text entry technique with low demand for screen space. In *Proceedings of Human Computer Interaction—INTERACT '03*, Zurich, Switzerland, 137–143

Miniotas, D., Špakov, O., Tugoy, I., & MacKenzie, I. S. (2006). Speech-augmented eye gaze interaction with small closely spaced targets. In *Proceedings of the 2006 Symposium on Eye Tracking Research and Applications (ETRA)*, 67–72

Morimoto, C. H., & Amir, A. (2010). Context switching for fast key selection in text entry applications. In *Proceedings of the 2010 Symposium on Eye Tracking Research and Applications (ETRA)*, 271–274

Oviatt, S. (1999). Mutual disambiguation of recognition errors in a multimodal architecture. In *Proceedings of the ACM SIGCHI 99*, Pittsburgh, Pennsylvania, United States of America, 576–583

Oviatt, S., & Cohen, P. (2000). Multimodal interfaces that process what comes naturally. *Communications of the ACM, 43*(2), 45–53.

Pireddu, A. (2007). Multimodal Interaction: An integrated speech and gaze approach. Thesis, Politecnico di Torino

Prasov, Z., Chai, J. Y., & Jeong, H. (2007). Eye gaze for attention prediction in multimodal human-machine conversation. In *Proceedings of AAAI Spring Symposium on Interaction Challenges for Intelligent Assistants*

Read, J. (2005). On the application of text input metrics to handwritten text input. Text Input Workshop, Dagstuhl., Germany

Read, J., MacFarlane, S., & Casey, C. (2001). Measuring the usability of text input methods for children. In *Proceedings of Human-Computer Interaction (HCI) 2001*, New Orleans, United States of America, 559–572

Tan, Y. K., Sherkat, N., & Allen, T. (2003a). Eye gaze and speech for data entry: A comparison of different data entry methods. In *Proceedings of the International Conference on Multimedia and Expo*, Baltimore, Maryland, United States of America, 41–44

Tan, Y. K., Sherkat, N., & Allen, T. (2003b). Error recovery in a blended style eye gaze and speech interface. In *Proceedings of ICMI '03*, Vancouver, Canada, 196–202

Tanaka, K. (1999). A robust selection system using realtime multi-modal user-agent interactions. In *Proceedings of IUI'99*, 105–108

Tanenhaus, M. K., Spivey-Knowlton, M., Eberhard, K., & Sedivy, J. (1995). Integration of visual and linguistic information during spoken language comprehension. *Science, 268*, 1632–1634.

Tobii. (2011). Tobii unveils the world's first eye-controlled laptop. Retrieved 14 March 2011 from http://www.tobii.com/en/ eye-tracking-integration/global/news-and-events/press-releases/tobii-unveils-the-worlds-first-eye-controlled-laptop/

Van Dam, A. (2001). Post-Wimp user interfaces: The human connection. In R. Earnshaw, R. Guedj, A. van Dam & J. Vince (Eds), *Frontiers of human-centred computing, online communities and virtual environments* (pp. 163–178). London: Springer-Verlag.

Velichkovsky, B. M., Sprenger, A., & Pomplun, M. (1997). Auf dem Weg zur Blickmaus: Die Beeinflussung der Fixationsdauer durch kognitive und kommunikative Aufgaben. In R. Liskowsky, B. M. Velichkowsky & W. Wünschmann (Eds), *Software-Ergonomie* (pp. 317–327)

Vertanen, K., & MacKay, D. J. C. (2010). Speech Dasher: Fast writing using speech and gaze. In *Proceedings of CHI 2010*, Atlanta, Georgia, United States of America, 595–598

Ward, D. J., Blackwell, A. F., & MacKay, D. J. C. (2000). Dasher—a data entry interface using continuous gestures and language models. In *Proceedings of UIST 2000: The 13th Annual ACM Symposium on User Interface Software and Technology*, San Diego, California, United States of America, 129–137

Wobbrock, J. O., Rubinstein, J., Sawyer, M. W., & Duchowski, A. T. (2008). Longitudinal evaluation of discrete consecutive gaze gestures for text entry. In *Proceedings of the Symposium on Eye Tracking Research and Applications (ETRA)*, Savannah, Georgia, United States of America, 11–18

Zhai, S., Morimoto, C., & Ihde, S. (1999). Manual And gaze input cascaded (MAGIC) pointing. In *Proceedings of CHI '99: ACM Conference on Human Factors in Computing Systems*, Pittsburgh, Pennsylvania, United States of America, 246–253

Zhang, X., & MacKenzie, I. S. (2007). Evaluating eye tracking with ISO 9241– Part 9. In J. Jacko (Ed.), *Human Computer Interaction*, 779–788

Improving the Accuracy of Video-Based Eye Tracking in Real Time through Post-Calibration Regression

Pieter Blignaut, Kenneth Holmqvist, Marcus Nyström and Richard Dewhurst

The fact that offsets between observed and actual gaze position show trends across the display, both with regard to direction and magnitude, and the fact that the offsets appear to be consistent for the same participant over time suggest the use of a regression approach to improve the accuracy of eye tracking based on previous visits to the same region on the display. We propose that an extra step is added to the calibration procedure in which participants are expected to click on a number of tiny blue dots spread out over the display. We show that this approach can improve the accuracy of data that are recorded by video-based systems from two leading manufacturers.

1 Introduction

Eye tracking can be used to obtain information about how people acquire and process information while they read (Rayner et al. 2007), browse a website (Goldberg et al. 2002), shop (Vikström et al. 2009), drive a motor car (Crundall et al. 2003), interpret medical images (Donovan et al. 2008) and other areas where the ability to decipher the visual world around us is critical. Eye tracking can also be used as input modality during computer interaction such as eye typing (Abe et al. 2007).

No matter what the application area, it is important that the point of gaze is recorded accurately or otherwise the information obtained or action executed might be different from what was intended. For example, the need for greater accuracy and a simpler way to achieve it are of paramount importance in reading research and other branches of cognitive science. In this regard, Rayner et al. (2007, p. 522) states that "…there can be a discrepancy between the word that is attended to even

P. Blignaut (✉)
University of the Free State, Bloemfontein, South Africa
e-mail: pieterb@ufs.ac.za

K. Holmqvist · M. Nyström · R. Dewhurst
Lund University, Lund, Sweden

M. Horsley et al. (eds.), *Current Trends in Eye Tracking Research*,
DOI 10.1007/978-3-319-02868-2_5, © Springer International Publishing Switzerland 2014

at the beginning of a fixation and the word that is recorded as the fixated word. Such discrepancies can occur for two reasons: (a) inaccuracy in the eye tracker and (b) inaccuracy in the eye-movement system."

Lack of accuracy, also known as systematic error, may not be a problem in usability studies where the areas of interest are large (e.g. 8° of visual angle) and are separated by large distances (Zhang and Hornof 2011), but in studies where the stimuli are closely spaced as in reading, or irregularly distributed as on a web page, uncertainty of as little as 0.5–1° can be critical in the correct analysis of eye-tracking data. Accuracy is also of great importance for gaze-input systems (Abe et al. 2007).

Systematic errors may result from bad calibrations, head movements, astigmatism, eyelid closure or other sources strongly dependent on the particular characteristics of the individual participant (Hornof and Halverson 2002). Systematic errors can be several degrees of visual angle which may have a serious impact on results that refer to the number of fixations or the amount of time spent on a specific area of interest, but is of much less importance for fixation and saccade detection.

While it is common practice to do recalibration between trials, for example, in reading research, this is not always possible or feasible (for example, in studies where task completion time is measured or where an interruption might impact on contextual information or the participant's thought processes). Furthermore, it has also been reported that eye trackers often maintain a systematic error even directly after careful calibration (Hornof and Halverson 2002).

A common approach to improve accuracy is to grab a fixation or group of fixations and pull it to a new position as can be done with the EyeLink system (SR-Research 2007). Hyrskykari (2006) developed a method for reading in which inaccurate data are moved in real time to the most probable line. Buscher et al. (2009) did a manual nine-point calibration after a recording to determine an offset that was then applied to all recorded data. In general, any correction procedure which applies the same correction across the entire display would be effective only if the error is of the same magnitude and direction across the display.

In general, any correction procedure with a single central calibration point would be effective only if the error is of the same magnitude and direction across the entire display. Recording accuracy data for each participant over a set of points before and after recording commences would give the researcher a measure of inaccuracy that could later be used to shift the data back accordingly.

In this chapter, we argue that since most eye trackers today claim to be drift-free or compensate for drift automatically (SensoMotoric Instruments n.d. (a); SensoMotoric Instruments n.d. (b); SR Research 2006; Tobii 2011), the initial calibration could be adjusted to achieve high accuracy before recording starts. Since the aim of this study is to evaluate a procedure for improving and reporting accuracy in video-based eye tracking, specific reference will be made to the accuracy of these types of eye trackers as reported in the literature. The current study will be placed in context with other studies with similar goals whereafter the methodology of the current study will be outlined. Analysis of the data that was captured during participant testing will be followed by a discussion about the feasibility of a regression approach to improve the accuracy of video-based eye tracking.

2 Accuracy of Eye Tracking

Although there is a good deal of consensus about what accuracy of eye tracking (or lack thereof) is, the way to measure or calculate it is not yet settled. In principle, the accuracy of eye tracking refers to the difference between the measured gaze direction and the actual gaze direction for a person positioned at the centre of the head box (Borah 1998; Tobii 2010). This can differ across the screen and may vary depending on external conditions such as lighting, quality of the calibrations, and individual eye characteristics (Tobii 2010). Accuracy can be quantified in terms of the systematic error as explained above (Hornof and Halverson 2002). It is measured as the distance in degrees between the position of a known target point and the average of a set of collected raw data samples from a participant looking at the point (Hornof and Halverson 2002; Holmqvist et al. 2011). This error may be averaged over a set of target points that are distributed across the display.

3 Reported Accuracy of Video-Based Eye Trackers

Video-based eye tracking is the most widely practised eye-tracking technique (Hua et al. 2006). This technique is based on the principle that when near-infrared (NIR) light is shone onto the eyes, it is reflected off the different structures in the eye to create four Purkinje reflections (Crane and Steele 1985). With this technique, the vector difference between the pupil centre and the first Purkinje image (PI), i.e. the reflection caused by the anterior cornea (also known as the glint or corneal reflection, CR), is tracked. CR/pupil devices are largely unobtrusive and easy to operate. The combination of pupil and CR tolerates head movements to a small extent and is the most practical approach for many experimental environments (Hua et al. 2006).

In order to compare and replicate research results, it is essential that researchers report the accuracy of eye tracking. Stating the manufacturers' specifications only could be misleading since it is known that accuracy can vary considerably across participants and experimental conditions (Hornof and Halverson 2002; Blignaut and Beelders 2009). Some researchers, for instance, Tatler (2007) and Foulsham and Underwood (2008), make a point of recalibrating until the measured accuracy is below 0.5°, and only then start recording.

While an accuracy of 0.3° has been reported for tower-mounted high-end systems operated by skilled operators (Jarodzka et al. 2010; Holmqvist et al. 2011), remote systems are mostly usually less accurate. Komogortsev and Khan (2008), for example, used an eye tracker with an accuracy specification of 0.5° but found that after removing all invalid recordings, the mean accuracy over participants was 1°. They regarded systematic errors of less than 1.7° as being acceptable. Johnson et al. (2007), using an alternative calibration procedure for another commercially available eye tracker with stated accuracy of 0.5°, found an azimuth error of 0.93° and an elevation error of 1.65°. Zhang and Hornof (2011) also found accuracy of

1.1° which is well over the 0.5° as stated by the manufacturer. Using a simulator eyeball, Imai et al. (2005) found that a video-based eye tracker has an x-error of 0.52° and a y-error of 0.62°. Van der Geest and Frens (2002) reported an x-error of 0.63° and a y-error of 0.72°. They remarked that the results from a video-based system "should be treated with care when the accuracy of fixation position is required to be smaller than 1°" (Van der Geest and Frens 2002, p. 193). Chen et al. (2008) propose a "robust" video-based eye-tracking system that has an accuracy of 0.77° in the horizontal direction and 0.95° in the vertical direction which they deem acceptable for many HCI applications that allow natural head movements. Brolly and Mulligan (2004) also proposed a video-based eye tracker with accuracy in the order of 0.8°. Hansen and Ji (2010) provided an overview of remote eye trackers and reported the accuracy of most model-based gaze estimation systems to be in the order of 1–2°. According to them, accuracy of less than 1° can only be achieved with multiple cameras and light sources.

Hornof and Halverson (2002) thoroughly studied the nature of systematic errors in a set of eye-tracking data collected from a visual search experiment. They found that the systematic error tends to be constant within a region of the display for each participant. Specifically, the magnitude of the disparities between the target visual stimuli and the corresponding fixations were "somewhat evenly distributed around 40 pixels (about 1° of visual angle) and most were between 15 and 65 pixels" (Hornof and Halverson 2002, p. 599). Horizontal and vertical disparities remained constant to a certain degree for each participant. Thus, accuracy was not randomly distributed across all directions or sizes but was, as the name implies, systematic.

This study proposes an intervention prior to recording to improve the accuracy of a variety of video-based systems from different manufacturers in real time while recording a representative sample of participants under typical experimental conditions.

4 Calibration

The output from eye-tracking devices varies with individual differences in the shape or size of the eyes, such as the corneal bulge and the relationship between the eye features (pupil and CRs) and the foveal region on the retina. Ethnicity, viewing angle, head pose, colour, texture, light conditions, position of the iris within the eye socket and the state of the eye (open or closed) all influence the appearance of the eye (Hansen and Ji 2010) and, therefore, the quality of eye-tracking data (Holmqvist et al. 2011). In particular, the individual shapes of participant eyeballs and the varying positions of cameras and illumination require all eye trackers to be calibrated.

Calibration refers to a procedure to gather data so that the coordinates of the pupil and one or more CRs in the coordinate system of the eye video can be converted to x- and y-coordinates that represent the participant's point of regard in the stimulus space. The procedure usually consists of asking the participant to look at a number of predefined points at known angular positions while storing samples of the measured quantity (Abe et al. 2007; Kliegl and Olson 1981; Tobii 2010). There is no consensus on exactly when to collect these samples, but Nyström et al. (2011) showed that participants know better than the operator or the system when they are looking at a target.

The transformation from eye position to point of regard can be either model based (geometric) or interpolation based (Hansen and Ji 2010). With model-based gaze estimation, a model of the eye is built from the observable eye features (pupil, CR, etc.) to compute the gaze direction. In this case, calibration is not used to determine the actual gaze position but rather to record the eye features from different angles. See Hansen and Ji (2010) for a comprehensive overview of possible transformations.

Interpolation might involve, for example, a linear regression between the known data set and the corresponding raw data, using a least squares estimation to minimize the distances between the observed points and the actual points (Hoorman 2008). Other examples of two-dimensional interpolation schemes can be found in McConkie (1981) as well as Kliegl and Olson (1981) while a cascaded polynomial curve fit method is described in Sheena and Borah (1981).

Theoretically, the transformation should remove any systematic error, but the limited number of calibration points that are normally used limits the accuracy that can be achieved. Typical calibration schemes require 5 or 9 predefined points, and rarely use more than 20 points (Borah 1998).

While it is true that the calibration procedure should not distract from the main study and should preferably not be time consuming (Brolly and Mulligan 2004), it is also important that a short calibration period should not be conducted at the cost of accuracy. This study proposes that an extra step that takes about 80–120 s to complete is added to the calibration procedure to report and improve the accuracy of eye tracking. This period is probably less than the total time taken to perform a drift correction between trials as is often done during cognitive and oculomotor research. Furthermore, our method follows a methodological approach and is not as crude as manual corrections that rely on the facilitator's subjective judgement. There is also no disruption to the participants between trials.

5 Improving the Accuracy of Eye-Tracking Data

Post-calibration corrections have been done before (Buscher et al. 2009) although the details of the procedure and the magnitude of the errors have not always been reported. Also, it is not always clear how the corrections were done and if variability in the error from one region on the display to the other was taken into account.

Hornof and Halverson (2002) reasoned that because the systematic error is relatively constant within a region on the display, it is possible that the effect of systematic error on each recorded fixation can be offset by adjusting the fixation location based on the weighted average of the error vectors that are closest to that recorded fixation, with heavier weights assigned to closer error vectors. They introduced the idea of required fixation locations (RFLs)—screen locations that the participant must look at within a certain time frame in order to successfully complete a task, but without explicit instructions to fixate those locations. They showed how the disparity between the fixations recorded by the eye tracker and RFLs can be used to monitor the accuracy of the eye tracker and to automatically invoke a recalibration procedure when necessary. They also demonstrated how the disparity varies across screen regions and participants and how each participant's unique error signature can be used to reduce the systematic error in the eye movement data that are collected for that participant.

6 Motivation for This Study

The study by Hornof and Halverson (2002) triggered us to do something similar in order to confirm their results for a wider range of eye trackers running at higher frequencies, while adapting the methodology somewhat in order to avert some of the shortcomings of their approach. Therefore, there are some agreements between the above-mentioned studies and the procedure that is reported in this chapter, but there are also some important differences. In this study, we also expect a participant to look at a specific target in order to successfully complete a task and we also employ the principle that systematic error can be reduced by adjusting each recorded point based on the closest error vectors.

In their research, Hornof and Halverson (2002) used only the 60 Hz Eyegaze system from LC Technologies while this study implements the regression procedure on several eye trackers from two other manufacturers. Hornof and Halverson themselves acknowledge that "the systematic error observed here is in the range typically seen with our brand and model of eye tracker" (Hornof and Halverson 2002, p. 600). It is important to confirm that the approach is feasible on different eye trackers since manufacturers may apply some filtering on the raw data prior to delivering them to the user which might impact on the accuracy of the eye tracker.

In this study, we present the participant with a set of points on a grid (or explicit RFLs) before recording starts. This eliminates the need for identification of implicit RFLs by the researchers and ensures that the procedure can be followed even when no obvious RFLs can be identified. Zhang and Hornof (2011) also recognized this shortcoming and proposed that the error vectors are based on the most probable fixation locations with the consequent possibility of wrongly assigning fixations to targets. The solution of Zhang and Hornof (2011) is also quite time consuming as the entire stimulus must be investigated to identify the most probable fixation locations.

The above-mentioned studies applied their corrective measures to fixations as identified by some or other algorithm while this study applies the corrections to the raw data prior to aggregating them into fixations. While a fixation algorithm hides the variability in the data and serves to identify clusters of raw data points as fixations, it is known that they are not always 100 % correct with regard to position and duration (Blignaut 2009). Any corrective approach should be based on the original data and not on derived data that are dependent on the algorithm that was used. A fixation algorithm should rather be applied after the raw data have been adjusted.

Furthermore, this study focuses on real-time adjustments of raw eye-tracking data whereas the above-mentioned studies focused on post hoc corrections of fixations. Although the Hornof and Halverson approach could possibly be based on raw data and implemented in real time, no results are available to prove its feasibility. Since modern remote eye trackers are capable of achieving 500 Hz or more, it needs to be confirmed that the computer will be able to keep up with recalculating the regression coefficients for every data point as they are captured. If the approach cannot be implemented in real time, it means that it is of no use in gaze-contingent or gaze-controlled systems.

In this study, we expect a participant to fixate on small targets (diameter about 0.2°) and then click on them, whereas the RFLs of Hornof and Halverson (2002) had dimensions of about 1° horizontal × 0.5° vertical which implies an uncertainty about the exact fixation position that is larger than the expected improvement in accuracy. By forcing the participants to look very closely at a small target before a click will be accepted, very little uncertainty exists about the exact fixation position.

Both the studies of Hornof and Halverson (2002) and Zhang and Hornof (2011) applied a chin rest to stabilize participants' gaze and fix the distance to the display. Although the procedure could be executed without a chin rest, no results were reported for free head movement and it cannot be assumed that the correction procedure will be robust enough under these circumstances. This study allowed free head movement within the head box as specified by the manufacturers (approximately 30 cm × 20 cm × 30 cm) of the four remote eye trackers that were used. Consequently, this study was more representative of typical experimental conditions where remote eye trackers are utilized.

The studies of Hornof and Halverson (2002) and Zhang and Hornof (2011) used 16 and 12 participants, respectively, without mention of participant characteristics. In order to confirm that the correcting approach works for a wide variety of participants, we recruited many more participants without applying any pre-screening.

In the light of the above mentioned, the aim of this study was, therefore, to confirm previous findings that systematic error can be reduced by adjusting each recorded point based on the closest error vectors. In addition, we wanted to prove that a spatial-dependent regression formula can be used to improve the real-time accuracy of a variety of video-based eye trackers from different manufacturers running at different frequencies while recording a representative sample of participants under typical experimental conditions.

It is important to note that the intention was not to compare the accuracy of the eye trackers with one another. In fact, according to manufacturer specifications (SensoMotoric Instruments nd (a); SensoMotoric Instruments nd (b); Tobii 2010) and other studies (Nyström et al. 2011), the accuracy of these eye trackers could be better under ideal conditions than is reported here. It should also be noted that two of the systems were not yet officially released at the time of the study and manufacturers would probably have improved the accuracy since then.

7 Method

7.1 Equipment

Four different video-based eye trackers with five configurations were used in the study:

(i) A Senso Motoric Instruments (SMI) high-speed 1,250 Hz system which made use of a built-in chin rest to stabilize the participant's head. Data were recorded binocularly on this system at a rate of 500 Hz. A 24″ display (resolution 1,920×1,200) was used and the chin rest was set up such that the gaze distance to the middle of the screen was 68 cm.

(ii) An SMI RED 250 Hz system recording data at 250 Hz. A 22″ screen (resolution 1,680×1,050) was used and participants were requested to keep their heads as still as possible at a distance of 68 cm as advised by a manufacturer's representative who was present during the study.

(iii) An SMI RED 500 Hz beta prototype system recording at 250 Hz. The same screen was used as above.

(iv) The same SMI RED 500 Hz system as in (iii) recording at 500 Hz. The same screen was used as above.

(v) A Tobii TX300 beta prototype system recording at 300 Hz and running firmware version 0.0.53. The stimuli were displayed on a 22″ screen (resolution 1,920×1,080) and participants were requested to keep their heads as still as possible at a distance of 65 cm as advised by the manufacturer.

7.2 Participants

The study was conducted in the Humanities Laboratory at Lund University, Sweden. Participants were recruited from students and staff at the university, without any pre-screening. Participants for the SMI RED 500 (500 Hz), RED 500 (250 Hz) and RED 250 eye trackers were largely mutually exclusive. In other words, a specific participant was tested on only one of these three eye trackers. A large majority of these participants were tested on the Tobii TX300 eye tracker as well. Participants were tested in pairs—one participant on the Tobii system and the other on one of the SMI systems. After being tested on one eye tracker, the participants switched eye trackers. This had the effect that half of the participants were tested first on the Tobii system and the other half on one of the SMI systems. Participants for the SMI high-speed system were recruited and tested separately since this system required more time to be configured properly for each participant. Table 1 shows the number of participants together with some other aggregates for each one of the systems. Although participants were not grouped in any way and no comparison between groups was done, about 20 % of the participants proved to be difficult to track in some way or other.

Table 1 Number of participants per eye tracker

	Participants	Glasses	Lenses	Age	
				Average	Standard deviation
SMI RED 250 (250 Hz)	18	9	0	31.3	12.2
SMI RED 500 (250 Hz)	24	2	4	25.0	4.9
SMI RED 500 (500 Hz)	13	3	1	27.5	11.1
SMI high-speed (500 Hz)	18	3	2	28.6	6.5
Tobii TX300 (300 Hz)	53	12	5	27.4	9.0

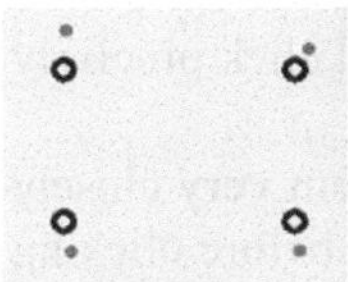

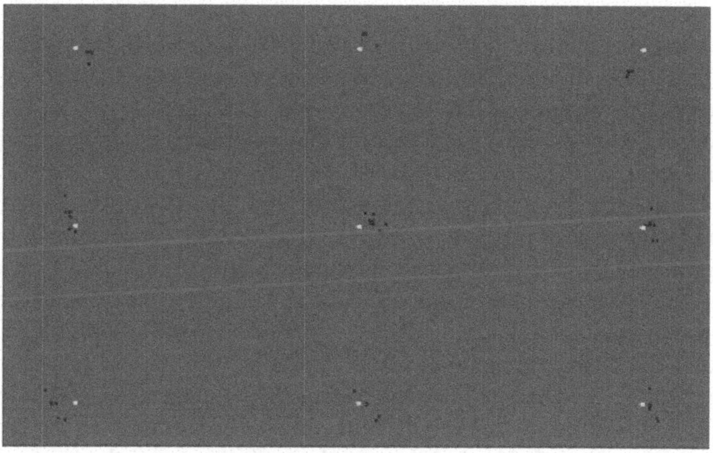

Fig. 1 Calibration validation for SMI (*left*) and Tobii (*right*). The screen prints are shown in the ratio that they appeared on the screen. For SMI, the targets are displayed as black circles while the recorded fixations are shown as red dots. For Tobii, the targets are shown as yellow dots with the blue and red dots indicating raw data samples for the left and right eyes, respectively.

7.3 *Initial Calibration*

Every participant was calibrated using the manufacturer's calibration routine with recommended settings for the best results. The calibration procedure was repeated until an acceptable calibration was obtained as determined by means of the manufacturer's validation procedure. For SMI, this procedure expected participants to look at four small circles near the middle of the screen. The calibration was considered to be acceptable if a fixation was shown for each circle and no fixation appeared in an obvious outlier position (example in Fig. 1). For Tobii, a calibration plot was displayed that showed the original calibration points together with the raw data samples that were collected during calibration. The calibration was accepted if data were recorded for both eyes at all points (example in Fig. 1).

Although both manufacturers provide a numerical value for the average systematic error, these could not be used to compare with a set upper limit since they are not comparable. Specifically, SMI averages the error over the four points in the middle of the screen while Tobii averages over all calibration points with most of them displayed at the edges of the screen where the errors are normally larger. This was not considered to be a problem since the aim of the study was not to determine the initial accuracy of the eye trackers but to determine if the regression approach would succeed to improve the accuracy to an acceptable level from any initial value.

7.4 Data Capturing

A software application was developed in C# with which the stimuli were generated. This application utilized the software development kits (SDK) as provided by the respective manufacturers to record data. The applications for the SMI and Tobii eye trackers were the same in all respects but for the implementation details of the manufacturer-specific SDK.

The stimuli on all eye trackers and for all participants consisted of two sets of tiny (diameter 8 pixels or 0.2°) blue dots that appeared in random order in a 8×5 grid such that the grid spans a horizontal angle of 38.4° and a vertical angle of 24° at the optimum gaze distance for the specific eye tracker regardless of the screen size or resolution. Participants were instructed to take their time and click precisely on a dot before the next dot would appear.

The dots were small enough to require participants to look at them very closely while making sure that the tip of the mouse was exactly on the dot before clicking (cf. Khan et al. 2011). In this way, the position of actual gaze could be established at the time of clicking. A similar procedure was followed by Koh et al. (2009) and Komogortsev and Khan (2008) although they had 17 points and participants were not requested to click on the dots but to fixate on them only.

It is argued that expecting participants to click on a dot instead of just looking at it, would improve the probability that they looked at the target at the time of data sampling. As mentioned above, Nyström et al. (2011) confirmed that participants know better than the system or the operator when they are looking at a target. Hornof and Halverson (2002, p. 594) also argued that *"the target of a mouse movement, if small enough, would make a very good implicit RFL (Required Fixation Locator)"*. Furthermore, eye movement studies confirm that, even when the hand starts moving before the eyes, the eyes arrive at the target before the hand and stay on the target until completion of the action (Abrams et al. 1990; Helsen et al. 1998).

Recorded gaze data for the left and right eyes were averaged to obtain a gaze point for every sample. The observed gaze position was calculated as the average of the valid (i.e. excluding blinks and periods where no data were recorded) raw gaze points for the 250-ms period immediately prior to the mouse click (button down) at each position.

Figure 2 shows a screen print after the last dot was clicked by a participant. Upon every click, a cluster of raw data points from the 250 ms prior to the click appeared

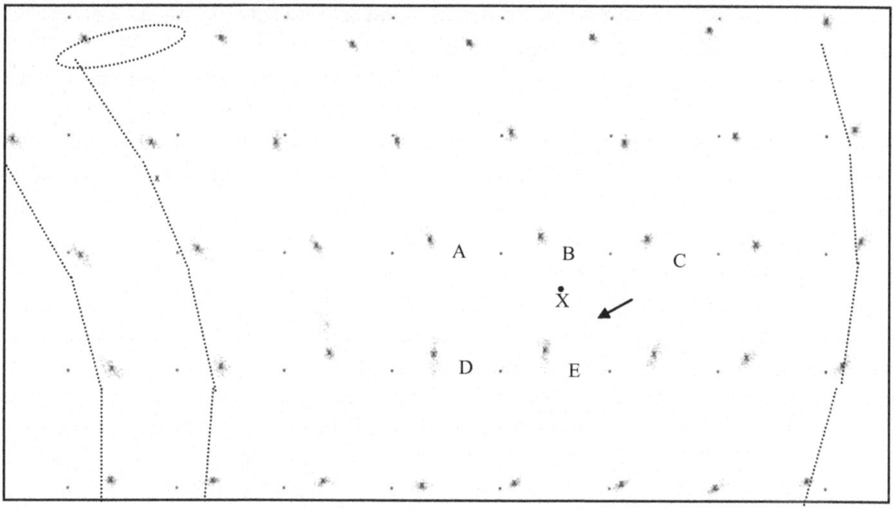

Fig. 2 Screen print showing small blue dots on which participants had to click

on the screen. The average position of the raw data points in every cluster is indicated with a black ×.

For reference purposes, the target points were indexed from 1 through 8 (left to right) and 1 through 5 (top to bottom). Every target point can, thus, be referred to as *(X, Y)* where $X \in [1,8]$ and $Y \in [1,5]$.

If the average of gaze samples in the 250 ms prior to a click was not within 5° from the target point, the click was not accepted. This ensured that participants did not look away during a click or pre-empt the position of the next dot during the period when data for the current point were captured. It also increased the probability that all samples in this interval belong to the same fixation. While this threshold is fairly large, it allowed participants to continue with the procedure even if the initial accuracy was not very good. Later, during the analysis of data, all samples further than 3° from the target point were removed.

Figure 2 also shows clusters of raw data as recorded by the eye tracker during the 250 ms prior to every click. Note that this is a scaled-down version of the original stimulus. The letters, dotted lines and circles are annotations that are referred to in the text.

7.5 *Correction of Observed Gaze Position through Regression*

Based on the offsets between the actual and observed gaze positions recorded with the first set of 40 dots, a regression formula was applied for the second set of 40 dots. This means that three sets of gaze coordinates were available for analysis, namely the original coordinates from the first and second sets of dots as well as the corrected coordinates for the second set of dots.

For the purpose of compensating for the deviations, it was reasoned that points nearer to the observed gaze position should dictate the amount and direction of correction and that not all available points should be taken into account. This is similar to the approach of Hornof and Halverson (2002) who used a weighted average based on the four error vectors that are closest to the recorded fixation, with heavier weights assigned to closer error vectors.

Unlike Hornof and Halverson (2002), however, a regression approach was followed to determine the direction and magnitude of adjustments. Because of the extra load placed on the central processing unit (CPU) during high-speed eye tracking, the regression approach should be simple and straightforward to execute in real time. In this study, an approach was followed where two separate linear regressions were carried out, one in the horizontal dimension and one in the vertical dimension (cf. Hoorman 2008). In Fig. 2, for example, an observed gaze point at position X should be adjusted as indicated by the arrow. The five closest points from the first set of dots will be identified in real time (A, B, C, D and E in the example) and a set of regression coefficients will be calculated based on the offsets of the identified points. For the adjusted x-coordinate of a point, the following least squares algorithm was used. The adjusted y-coordinate was calculated similarly:

$$m_x = \frac{5 * \sum (Obs\, x_i * Act\, x_i) - \sum (Obs\, x_i) * \sum (Act\, x_i)}{5 * \sum (Obs\, x_i^2) - \left(\sum (Obs\, x_i)\right)^2},$$

where Obs *xi* and Act *xi* refer to the observed and actual x-coordinates respectively at the five points from the first set that are nearest to the observed point:

$$c_x = \frac{\sum (Act\, x_i) - m_x * \sum (Obs\, x_i)}{5}$$

$$Adjusted\ x = m_x * Observed\ x + c_x,$$

where *Observed x* refers to the x-coordinate of the observed point (O in the example).

8 Analysis

The analysis was done in four stages. At first, the data were cleaned to remove possibly erroneous data points. Secondly, trends in the offsets between actual and observed gaze positions were identified. If it was discovered that the magnitude and direction of these offsets were constant over the display for a specific participant, a fixed adjustment could be applied to the recorded data and the regression approach would not be necessary. Thirdly, it was important to confirm that the offsets were consistent for a specific participant. In other words, the magnitude and direction of the offsets should be the same for subsequent fixations in the same region of the display or else a regression approach to improve data which are based on previous recordings would not be feasible. Finally, the improvement that could be achieved with the regression approach for various eye trackers and various regions of the display was quantified and compared with the initial offsets.

9 Preparation of the Data

All raw data samples within the 250 ms prior to a mouse click were regarded to be part of a single fixation. Raw data points that were more than 3 × the standard deviation away from a fixation centre were regarded as outliers and removed.

10 Trends in the Offsets

Some of the fixations in Fig. 2 are connected with dotted lines to highlight the trend of deviations between actual and observed gaze position for the specific participant. It is clear that a consistent, albeit not constant, difference exists between the blue dots (representing the actual gaze position) and the observed gaze positions. This agrees with the error signatures of Hornof and Halverson (2002). Contrary to the assumption of Zhang and Hornof (2011), however, it is clear that every fixation does not necessarily belong to the nearest target (as indicated, for example, by the dotted oval that connects the dot at *(2,1)* and its corresponding fixation).

In order to examine the influence of gaze position on the magnitude and position of the offsets (observed gaze position–actual gaze position), the average offset in both the x- and y-dimensions were calculated per eye tracker and grid point. Figure 3 shows the directional x-offset (blue line) and absolute value of the x-offset (black line) for the Tobii TX300 at every point in the grid, averaged over all participants. The 95 % confidence intervals are indicative of the variance caused by individual differences amongst participants.

It is clear that the errors in the horizontal direction range from about 0.5° to about 1.1°. The best accuracy (minimum absolute accuracy) is obtained in the middle of the screen ($X \in [4,5]$, $Y \in [1,5]$) while the largest errors occur at the edges of the screen ($X = 1$ or $X = 8$, $Y \in [1,5]$). The offsets are more or less the same for different vertical positions of the eye gaze although the spread is somewhat smaller at the top of the screen ($Y = 1$).

The fact that the average directional offset is about 0° in the centre of the screen ($X \in [4,5]$, $Y \in [1,5]$) indicates equal probability of the offset being to the

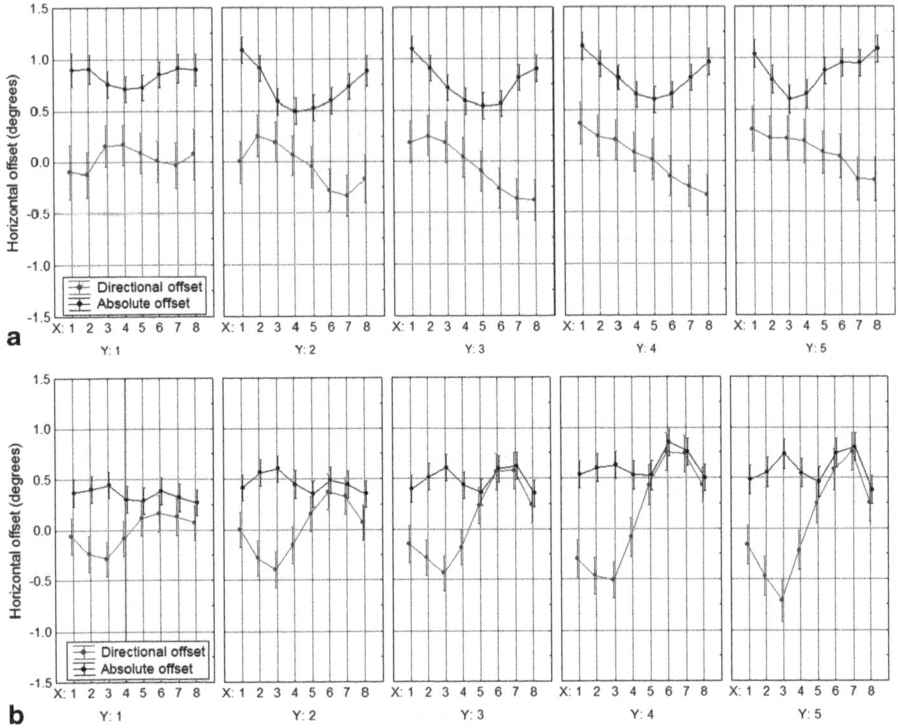

Fig. 3 Absolute and directional horizontal offsets for **a** the Tobii TX300 and **b** the SMI tower-mounted system

left or right of the target. On the bottom left-hand side of the screen (*X* ∈ *[1,3], Y* ∈ *[3,5]*), the general trend is that the observed gaze position is to the right of the target (positive directional average) while at the bottom right-hand side of the screen (*X* ∈ *[6,8], Y* ∈ *[3,5]*) the observed position is mostly to the left of the target (negative directional average).

Figure 3 shows a comparable graph for the tower-mounted SMI high-speed system. The magnitude of the horizontal offset varies between 0.3° at the top of the display (*X*=*5, Y*=*1*) to about 0.8° at the bottom of the display (*X*=*7, Y*=*5*). The S-shaped distribution of the directional offsets is once again very striking. The fact that the magnitude of the directional offsets is so close to the absolute offsets at some positions of the display (e.g. (*X* ∈ *[5,8], Y* ∈ *[3,5]* and (*X*=*3, Y*=*5*)) is indicative that the observed gaze position was at a specific side of the target for just about all participants. The trends for the SMI remote systems are very similar to the tower-mounted system.

It can, therefore, be concluded that gaze position has an influence on the offset between observed and actual gaze positions for all the eye trackers. Therefore, any attempt to correct for the offsets should take the gaze position into account and a constant adjustment over the entire display would not be appropriate.

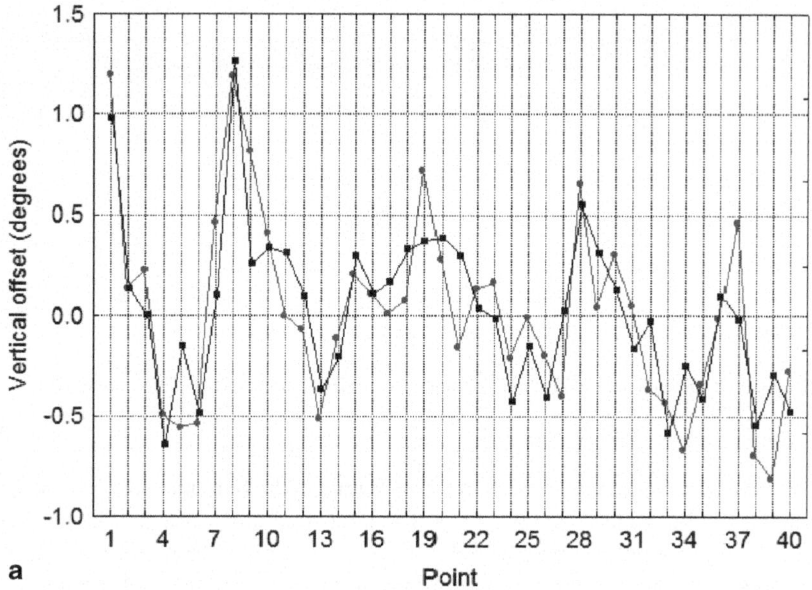

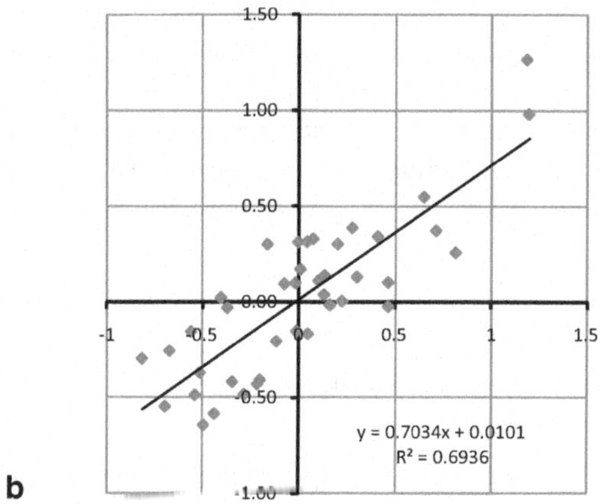

Fig. 4 **a** Vertical offset for the two sets of 40 dots for one participant on the SMI RED 500 (250 Hz) eye tracker. **b** Correlation plot of the data in (**a**)

11 Consistency Per Participant

If the recorded offsets between actual and observed gaze positions are to be used as basis for improvement of future recordings, these offsets should be replicable for a specific participant at a specific region of the display. In other words, prior to any regression, it should be proven that the offsets have the same direction and magnitude for subsequent fixations in the same region of the display.

Figure 4 shows a graph of the offset in the vertical direction (Δy) for one specific participant on the SMI RED 500 (250 Hz) for the first and second sets of 40 dots.

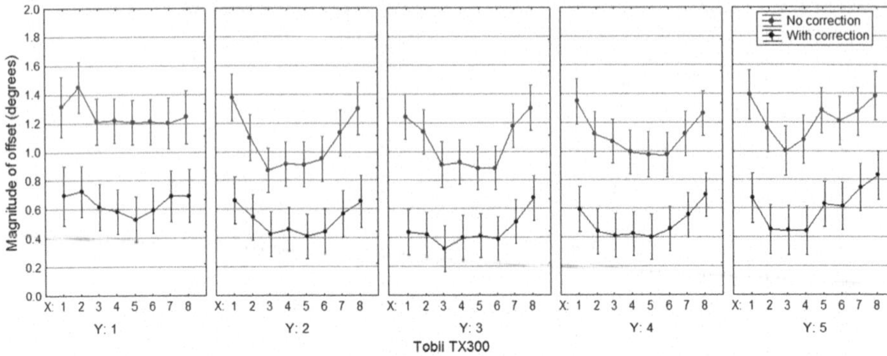

Fig. 5 Offset magnitude per grid point averaged over participants for the Tobii TX300

In order to quantify the apparent correlation between the two data sets, Fig. 4 shows the same data plotted as offsets for the first data set against offsets for the second data set. A regression line with value of R^2 is shown as well. A similar procedure was followed for all eye trackers and all participants for offsets in both the *x*- and *y*-directions. The correlation between offsets for the two sets of dots was significant ($\alpha = 0.05$) for all participants on all eye trackers in the *x*-direction. In the *y*-direction, the correlation was not significant ($\alpha = 0.05$) for one participant on each of the SMI RED 250, RED 500 (250 Hz) and high-speed eye trackers. Since the offsets are consistent with regard to direction and magnitude for the large majority of participants, an investigation into the improvement that a regression might bring about is justified (next section).

12 Improvement with the Regression Approach

The offsets that were recorded with the first set of 40 dots were used in a regression as previously explained to adjust the observed gaze position in real time. The graph in Fig. 5 shows the magnitude of the offsets $\left(\sqrt{(\Delta x)^2 + (\Delta y)^2}\right)$ (averaged over participants) at each point on the grid before and after correction for the Tobii TX300. The magnitude of the 95 % confidence intervals is indicative of the variance caused by individual differences at each grid point. It is clear that the accuracy was improved with about 0.6° over all regions of the display. It is interesting to note that, in general, the accuracy was best in the centre of the screen ($X \in [4,5]$, $Y \in [2,4]$) with some offset towards the edges of the display.

Figure 6 shows a graph for the magnitude of offsets for individual participants on the Tobii TX300, averaged over all points in the grid. The regression approach succeeded to improve the accuracy for all participants although not to the same extent for each person and not always by a significant margin.

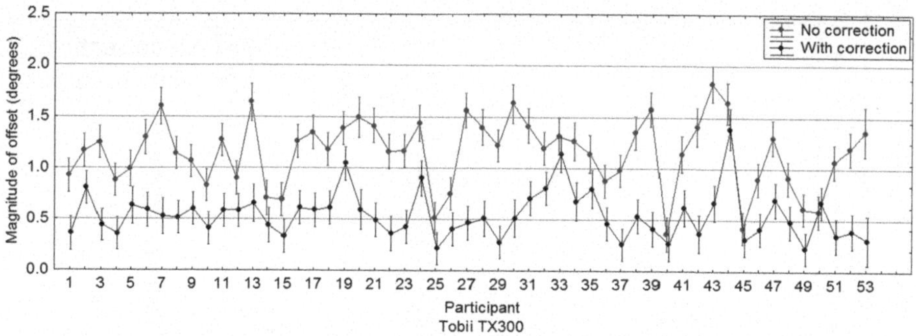

Fig. 6 Offset magnitude per participant averaged over all grid points for the Tobii TX300

Table 2 Results of a series of one-way analyses of variance for the effect of correction through regression on the magnitude of the horizontal, vertical and overall offsets per eye tracker

	N	Before correction	After correction	Improvement	F	p
Horizontal offsets						
SMI RED 250 (250 Hz)	582	0.62	0.49	0.13	15.2	<0.001
SMI RED 500 (250 Hz)	868	0.69	0.44	0.25	129.0	<0.001
SMI RED 500 (500 Hz)	415	0.73	0.52	0.21	27.5	<0.001
SMI high-speed (500 Hz)	625	0.50	0.30	0.20	101.7	<0.001
Tobii TX300 (300 Hz)	1851	0.79	0.31	0.48	754.8	<0.001
Vertical offsets						
SMI RED 250 (250 Hz)	582	0.85	0.63	0.22	32.4	<0.001
SMI RED 500 (250 Hz)	868	0.62	0.47	0.15	41.4	<0.001
SMI RED 500 (500 Hz)	415	0.75	0.58	0.17	18.8	<0.001
SMI high-speed (500 Hz)	625	0.51	0.31	0.20	88.4	<0.001
Tobii TX300 (300 Hz)	1851	0.64	0.35	0.29	409.4	<0.001
Combined offsets						
SMI RED 250 (250 Hz)	582	1.15	0.88	0.27	34.2	<0.001
SMI RED 500 (250 Hz)	868	1.03	0.72	0.31	131.5	<0.001
SMI RED 500 (500 Hz)	415	1.17	0.87	0.30	38.3	<0.001
SMI high-speed (500 Hz)	625	0.80	0.49	0.31	175.3	<0.001
Tobii TX300 (300 Hz)	1851	1.13	0.53	0.60	1021.9	<0.001

Table 2 shows the results of a series of one-way analyses of variance on correction as independent variable and magnitude of offset ($|\Delta x|$, $|\Delta y|$ and $\left(\sqrt{(\Delta x)^2 + (\Delta y)^2}\right)$ as dependent variables while controlling for the eye tracker.

Figure 7 provides a graphical visualization of these results. The confidence intervals are indicative of the variance caused by individual differences and the variance over different regions of the display (grid points). It is clear that the regression approach succeeded to improve the accuracy for all eye trackers significantly.

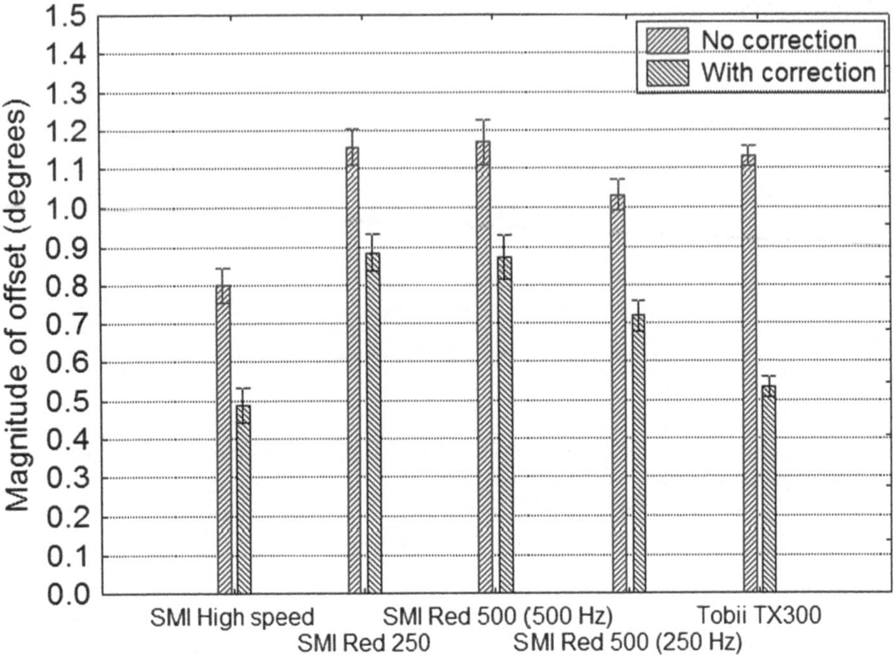

Fig. 7 Magnitude of overall offset between actual and observed gaze positions per eye tracker, averaged over participants and display region. The *vertical bars* denote 95 % confidence intervals

Table 3 Correlations between average improvement and initial systematic error

Eye tracker	N	Correlation of improvement with initial offset			Correlation of relative improvement with initial offset		
		r	R^2	p	r	R^2	p
SMI RED 250 (250 Hz)	582	0.404	0.163	<0.001	0.305	0.093	<0.001
SMI RED 500 (250 Hz)	868	0.658	0.433	<.001	0.370	0.134	<0.001
SMI RED 500 (500 Hz)	415	0.457	0.209	<0.001	0.291	0.085	<0.001
SMI high-speed (500 Hz)	625	0.694	0.482	<0.001	0.363	0.132	<0.001
Tobii TX300 (300 Hz)	1851	0.772	0.596	<0.001	0.397	0.158	<0.001

13 Correlation of Improvements with Initial Offsets

Table 3 shows the correlations between the average improvement (initial offset−adjusted offset) over all participants and grid points per eye tracker with the initial offset, i.e. the systematic error after calibration with the manufacturer's procedure only. Table 3 also shows the correlations between the relative improvements, i.e. *(initial offset−adjusted offset)/initial offset*, with the initial offset. All correlations are significant ($\alpha = 0.001$), meaning that the procedure of improvement through regression works better if the initial accuracy is poor.

14 Discussion

The calibration procedures that are provided by manufacturers aim to remove any deviation between the actual gaze position of a participant and the gaze position as recorded by the eye tracker. Since it is reasoned that this calibration routine distracts from the main task at hand, manufacturers try to shorten this procedure as much as possible, which may inadvertently negatively affect the accuracy. In this chapter, it is proposed that an extra step that takes about 80–120 s to complete is added to the calibration procedure to report and improve the accuracy of CR/pupil eye tracking. It is argued that the improved accuracy makes this investment in testing time worthwhile.

Several sources of literature reported that the accuracy of video-based eye tracking can vary across participants, experimental conditions and over time (Blignaut 2009; Van der Lans et al. 2010). Most studies report an accuracy of 1–2° to be typical while the accuracy for individual participants can be even worse (Hansen and Ji 2010). Some researchers pre-screen their participants so as to exclude those with glasses, lenses, mascara and downward eyelashes to ensure a high and even data quality, but pre-screening limits the number of available participants and prolongs the time it takes to record the data. The procedure that is suggested in this chapter is, therefore, particularly useful in projects where pre-screening is not an option but where accuracy is nevertheless of great importance.

It was shown that participant recordings are highly consistent with the same offset (magnitude and direction) being recorded repetitively at a specific display region. It was also shown that there are definite trends with regard to the direction and magnitude of the offsets across the display and it is possible that the direction of the offset could be in opposite directions in different areas of the display. While the procedure offered by EyeLink (SR Research 2007) to do a drift correction based on a single dot in the centre of the display is a step in the right direction, it does not take the gaze position into account and applies a fixed correction to all fixations across the display. The procedure discussed in this chapter, however, allows for a localized correction where the observed gaze position during future visits to a specific display region can be corrected based on a previous visit to the same area.

The procedure expects a participant to click on a grid of tiny blue dots which requires him to look very carefully at the mouse pointer and position it very steadily prior to a click. The data that are recorded during the 250 ms prior to a click are regarded as being representative of the participant's eye gaze during that period. The deviations between the recorded gaze positions and the corresponding blue dots (actual gaze positions) are used in the correction of the gaze position during subsequent recordings. This correction is done through two separate linear regressions with the regression coefficients being recalculated in real time for every recorded raw data point based on the error vectors of the five nearest calibration points at the time. In other words, a different regression formula is applied to every recorded gaze data point, depending on its x and y coordinates. This ensures that the correlation between the spatial offsets and position on the screen (Figs. 2 and 3) is taken into account.

When this procedure is to be followed as part of the calibration procedure during an eye-tracking study, a participant would be requested to click only one set of dots, but for comparison purposes two sets of blue dots were displayed in this study. The first set of dots was used to capture the offsets between actual and recorded gaze data. The second set of dots served (1) to verify that the offsets were consistent per participant and display region and (2) to evaluate the corrections that were made. For the sake of (1) above, it was necessary to display the second set of dots in the same position as the first set. While it would seem to be more convincing to have the second set of dots in different positions for the sake of (2), it does, in fact, not matter since the observed gaze positions are in any case hardly ever exactly on one of the dots and may occur anywhere in between.

While it is acknowledged that better accuracies can be achieved with the latest hardware and firmware, careful experimental set-up, pre-screening of participants and experienced operators, the order of magnitude of deviations between actual and observed gaze positions that was found in this study (see Fig. 7) agrees with the literature. It is important, however, to note that the reported accuracy values of the selected eye trackers are not important for this study—only the improvements that could be achieved with the regression procedure.

It has been shown that systematic error can be reduced significantly for a variety of video-based CR/pupil eye trackers and for virtually every single participant. Looking at Fig. 7, it is clear that there is no substantial difference in the initial accuracy that can be acquired by the four remote eye trackers as the average deviation between actual gaze position and observed gaze position lies in a band between 1.0° and 1.2° for these eye trackers. The SMI high-speed system recorded an initial accuracy of about 0.8° but after correction, the average accuracy of this eye tracker was improved to less than 0.5°. For the Tobii TX300, the average systematic error of more than 1.1° was reduced to about 0.5°. Although the improvement was smaller for the remote SMI systems, a significant improvement was still achieved.

In this study, good initial accuracies were achieved for some participants only and then only in specific regions of the display (Figs. 5 and 6). If the systematic error for a specific participant is constant with regard to direction and magnitude across the entire display, the manufacturers' calibration routines seem to work fairly well as they apply the same offset across the entire display. The procedure that was applied in this study succeeded to improve the accuracy of eye tracking also for participants with varying offsets across the display as in Fig. 2.

While the accuracy of individual recordings could be as bad as 1.5–2° (Fig. 6) before correcting, recordings with a systematic error of more than 0.8° were an exception after correction. With correction, several individual recordings, especially for the SMI high-speed and Tobii TX300, had an average accuracy across the entire display of 0.3–0.4° which are generally accepted as being exceptionally good for video-based eye trackers.

The fact that the improvement is better when the initial accuracy is bad, has important implications for difficult experimental conditions, when participant-specific physiological factors make calibration difficult or when the operator has limited experience with the equipment.

In conclusion, it was shown that existing calibration routines are not necessarily perfect and can be improved upon. The fact that offsets between observed and actual gaze position show trends across the display both with regard to direction and magnitude and the fact that the offsets appear to be consistent for the same participant over time, allows the use of a regression approach to improve the accuracy of eye tracking based on previous visits to the same region on the display. It was shown that such an approach can successfully be implemented on remote systems from two leading manufacturers as well as for a high-precision tower-mounted system.

15 Limitations and Future Possibilities

The regression model in this study made use of two separate linear least squares algorithms, one in the horizontal dimension and one in the vertical dimension. Using this regression model, the gaze position was adjusted in real time based on the five nearest calibration points to the gaze position at the time. No other models were investigated and no comparison was done to determine if a regression based on a different number of calibration points would provide better results. Also, the initial grid consisted of 40 points in 5 rows of 8 points each and no comparison was done to determine if grids with fewer points would produce equally good results. This study was done only to confirm in principle if a regression model is applicable to a variety of eye trackers and further research needs to be done to determine if and how it can be improved.

While the core of the procedure that is discussed in this chapter is the regression, it would be valuable to investigate the effect of clicking on the dots instead of just watching them would have on the accuracy that is obtained. According to the motor-control literature, some people may be better at pointing and clicking than others (Chaparro et al. 1999; Keates and Trewin 2005) and it might be important to determine if this would have an artificial effect on accuracy.

Except for the SMI high-speed system, this study allowed free head movement within the head box. The regression formula did not take any head movements into account and more research will be needed to determine if the accuracy can further be improved by incorporating dynamic three-dimensional (3D) head-position coordinates into the regression model.

Acknowledgements We wish to thank the two manufacturers who provided us with prototypes of their latest remote eye-tracking models at the time.

References

Abe, K., Ohi, S. & Ohyama, M. (2007). An eye-gaze input system using information on eye movement history. In C. Stephanides (Ed.), *Universal access in HCI, Part II. HCII2007, LNCS 4555*, 721–729. Berlin: Springer-Verlag.

Abrams, R. A., Meyer, D. E., & Kornblum, S. (1990). Eye-hand coordination: Oculomotor control in rapid aimed limb movements. *Journal of Experimental Psychology: Human Perception & Performance, 16*(2), 248–267.

Blignaut, P. (2009). Fixation identification: the optimum threshold for a dispersion algorithm. *Attention, Perception and Psychophysics, 71*(4), 881–895.

Blignaut, P., & Beelders, T. (2009). The effect of fixational eye movements on fixation identification with a dispersion-based fixation detection algorithm. *Journal of Eye Movement Research, 2*(5):4, 1–14.

Borah, J. (1998). Technology and application of gaze based control. RTO Lecture series on Alternative Control Technologies, 7–8 October 1998, Brétigny, France and 14–15 October 1998, Ohio, USA.

Brolly, X. L. C., & Mulligan, J. B. (2004). Implicit calibration of a remote gaze tracker. Proceedings of the 2004 Conference on Computer Vision and Pattern Recognition Workshop (CVPRW'04), 8, 134–146.

Buscher, G., Cuttrell, E., & Morris, M. R. (2009). *What do you see when you're surfing? Using eye tracking to predict salient regions of web pages*. CHI2009, 4–9 April 2009, Boston, Massachusetts.

Chaparro, A., Bohan, M., Fernandez, J., Choi, S. D., & Kattel, B. (1999). The impact of age on computer input device use: Psychophysical and physiological measures. *International Journal of Industrial Ergonomics, 24*, 503–513.

Chen, J., Tong, Y., Gray, W., & Ji, Q. (2008). *A robust 3D gaze tracking system using noise reduction*. Proceedings of the Eye-tracking Research Applications Conference (ETRA), Savannah, Georgia, 26–28 March 2008.

Crane, H. D., & Steele, C. M. (1985). Generation-V Dual-Purkinje-image eyetracker. *Applied Optics, 24*, 527–537.

Crundall, D., Chapman, P., Phelps, N., & Underwood, G. (2003). Eye movements and hazard perception in police pursuit and emergency response driving. *Journal of Experimental Psychology Applied, 9*(3), 163–174.

Donovan, T., Manning, D., & Crawford, T. (2008). *Performance changes in lung nodule detection following perceptual feedback of eye movements*. Medical Imaging 2008 (SPIE).

Foulsham, T., & Underwood, G. M. (2008). What can saliency models predict about eye movements? Spatial and sequential aspects of fixations during encoding and recognition. *Journal of Vision, 8*(2), 1–17.

Goldberg, J., Stimson, M., Lewenstein, M., Scott, N., & Wichansky, A. (2002). Eye tracking in web search tasks: design implications. In Proceedings of the 2002 symposium on eye tracking research & applications, 51–58.

Hansen, D. W., & Ji, Q. (2010). In the eye of the beholder: A survey of models for eyes and gaze. *IEEE Transactions on Pattern Analysis and Machine Intelligence, 32*(3), 478–500.

Helsen, W. F., Elliott, D., Starkes, J. L., & Ricker, K. L. (1998). Temporal and spatial coupling of point of gaze and hand movement in aiming. *Journal of Motor Behavior, 30*, 249–259.

Holmqvist, K., Nyström, M., Andersson, R., Dewhurst, R., Jarodzka, H., & Van de Weijer, J. (2011). *Eye tracking: A comprehensive guide to methods and measures*. London: Oxford University Press.

Hoorman, J. (2008). The effect of calibration errors on the accuracy of eye movement recordings. *Journal of Eye Movement Research, 1*(2):3, 1–7.

Hornof, A. J., & Halverson, T. (2002). Cleaning up systematic error in eye-tracking data by using required fixation locations. Behavior Research Methods. *Instruments & Computers, 34*(4), 592–604.

Hua, H., Krishnaswamy, P., & Rolland, J. P. (2006). Video-based eyetracking methods and algorithms in head-mounted displays. *Optics Express, 14*(10), 4328–4350.

Hyrskykari, A. (2006). Utilizing eye movements: Overcoming inaccuracy while tracking the focus of attention during reading. *Computers in Human Behavior, 22*(4), 657–671.

Imai, T., Sekine, K., Hattori, K., Takeda, N., Koizuka, I., Nakamae, K., Miura, K., Fujioka, H., & Kubo, T. (2005). Comparing the accuracy of video-oculography and the sclera search coil system in human eye movement analysis. *Auris, Nasus, Larynx, 32*, 3–9.

Jarodzka, H., Balslev, T., Holmqvist, K., Nyström, M., Scheiter, K., Gerjets, P., & Eika, B. (2010). Learning perceptual aspects of diagnosis in medicine via eye movement modeling examples on patient video cases. In S. Ohlsson & R. Catrambone (Eds.), *Proceedings of the 32nd Annual Conference of the Cognitive Science Society*, Portland, 1703–1708.

Johnson, J. S., Liu, L., Thomas, G., & Spencer, J. P. (2007). Calibration algorithm for eyetracking with unrestricted head movement. *Behavior Research Methods, 39*(1), 123–132.

Keates, S., & Trewin, S. (2005). Effect of Age and Parkinson's Disease on Cursor Positioning Using a Mouse. Proceedings of the 7th International ACM SIGACCESS conference on Computers and Accessibility (ASSETS'05), 9–12 October 2005, Baltimore, MD, USA.

Khan, A., Song, J., & McPeek, R. (2011). The eye dominates in guiding attention during simultaneous eye and hand movements. *Journal of Vision, 11*(1), 1–14.

Kliegl, R., & Olson, R. K. (1981). Reduction and calibration of eye monitor data. *Behavior Research Methods and Instrumentation, 13*, 107–111.

Koh, D. H., Gowda, S. M., & Komogortsev, O. V. (2009). Input evaluation of an eye-gaze-guided interface: Kalman filter vs. velocity threshold eye movement identification. EICS '09, 15–17 July 2009, Pittsburgh, Pennsylvania, USA.

Komogortsev, O. V., & Khan, J. I. (2008). Eye movement prediction by Kalman filter with integrated linear horizontal oculomotor plant mechanical model. Proceedings of the 2008 Symposium on Eye Tracking Research and Applications (ETRA), 26–28 March 2008, Savannah, Georgia, 229–236.

McConkie, G. W. (1981). Evaluating and reporting data quality in eye movement research. *Behavior Research Methods & Instrumentation, 13*(2), 97–106.

Nyström, M., Andersson, R., Holmqvist, K., Van de Weijer, J. (2011). Participants know best - the influence of calibration method and eye physiology on eye tracking data quality. *Journal of Neuroscience Methods,* 00 (2011) 1- 26.

Rayner, K., Pollatsek, A., Drieghe, D., Slattery, T., & Reichle, E. (2007). Tracking the mind during reading via eye movements: Comments on Kliegl, Nuthmann, and Engbert (2006). *Journal of Experimental Psychology: General, 136*(3), 520–529.

SensoMotoric Instruments. (n.d. (a)) RED/RED250/RED500. http://www.smivision.com/en/gaze-and-eye-tracking-systems/products/red-red250-red-500.html.

SensoMotoric Instruments. (n.d. (b)) IVIEW XTM HI-SPEED. http://www.smivision.com/en/gaze-and-eye-tracking-systems/products/iview-x-hi-speed.html.

Sheena, D., & Borah, J. (1981). Compensation for some second order effects to improve eye position measurements. In D. F. Fisher, R. A. Monty, & J. W. Senders (Eds.), *Eye Movements: Cognition and Visual Perception*. Hillsdale: Lawrence Erlbaum Associates.

SR-Research. (2006). EyeLink 1000. http://www.inition.co.uk/inition/pdf/mocap_sr_eyelink_1000.pdf.

SR-Research. (2007). EyeLink User Manual 1.3.0 [Computer software manual]. Mississauga, Ontario, Canada.

Tatler, B. W. (2007). The central fixation bias in scene viewing: Selecting an optimal viewing position independently of motor biases and image feature distributions. *Journal of Vision, 7*(14), 1–17.

Tobii Technology. (2010). Product description, Tobii T/X series eye trackers, rev. 2.1, June 2010. Tobii Technology AB.

Tobii Technology. (2011). Tobii TX300 Eye Tracker. April 2011. Tobii Technology AB.

Van der Geest, J. N., & Frens, M. A. (2002). Recording eye movements with video-oculography and scleral search coils: a direct comparison of two methods. *Journal of Neuroscience methods, 114*, 185–195.

Van der Lans, R., Wedel, M., & Pieters, R. (2010). Defining eye-fixation sequences across individuals and tasks: the Binocular-Individual Threshold (BIT) algorithm. *Behavior Research Methods*, *43*(1), 239–257.

Vikström, K., Wallin, A., Holmqvist, K. (2009). Yarbus goes shopping. Proceedings of the Scandinavian Workshop on Applied Eye-Tracking., Stavanger, Norway, 5-7 May 2009.

Zhang, Y., & Hornof, A. J. (2011). Mode of disparities error correction of eye tracking data. *Behavior Research Methods*, *43*(3), 834–842.

Gaze Shifts and Pen Velocity Minima During Line Copying with Consideration to Signature Simulation

Avni Pepe and Jodi Sita

1 Introduction

There has been only one study that has previously investigated the eye and hand behaviours during signature copying (see Pepe et al. 2012). Although eye–hand behaviours have been previously reported on during other copying tasks (see Tchalenko and Miall 2009; Tchalenko 2007, 2009), none have investigated the timing between gaze shifts (saccades) and pen velocity minima (reductions in pen velocity) during copying of simple concatenated lines.

Previous drawing studies have reported that gaze shifts are frequently made at moments that the pen slows to a velocity minimum during a stroke (see Reina and Schwartz 2003; Gowen and Miall 2006). Similar behaviours have also been reported on in aiming tasks (see Helsen et al. 1998; Helsen et al. 2000), shape drawing (see Ketcham et al. 2006) and other tasks (Neggers and Bekkering 2000). Gowen and Miall (2006) recorded the eye movements and pen kinematics of subjects during shape drawing and tracing. For these tasks, respectively, they found that 43 and 47 % of saccades were made within ± 100 msec of a velocity minima. They propose this eye–hand behaviour could indicate attention returning to eye movement planning before the hand movement has been fully programmed, whilst others suggest this behaviour could be related to the eyes receiving feedforward information about the upcoming movement (Ketcham et al. 2006; Reina and Schwartz 2003).

Pepe et al. (2012) explored the eye movements and pen kinematics of simulators (forgers) during signature simulations of two types of signatures with different complexity ratings (see Found and Rogers 1998). Although the authors found that copying these signatures evoked different eye and pen kinematic behaviours, the study did not report on temporal behaviours during copying, the nature of these behaviours and their relevance to signature simulation. Therefore, the current study was, in part, conducted to address this issue by exploring the eye–hand behaviour in a task similar to, but not synonymous with, signature copying. More specifically,

A. Pepe (✉) · J. Sita
La Trobe University, Melbourne, VIC, Australia
e-mail: A.Pepe@latrobe.edu.au

M. Horsley et al. (eds.), *Current Trends in Eye Tracking Research*,
DOI 10.1007/978-3-319-02868-2_6, © Springer International Publishing Switzerland 2014

it aimed to determine the nature of the temporal relationship between the movement of the eye and hand during a copying task and to determine the implications of this relationship with respect to signature simulation.

2 Methods

2.1 Participants

Twenty healthy subjects with normal or corrected to normal visual acuity levels participated in the study. All subjects confirmed the absence of any neurological conditions, disorders or injuries known to affect eye or hand movements, or to taking any medications known to affect eye or hand movements. Of the 20 subjects, five subjects whose data could facilitate an accurate analysis of eye–hand interactions were selected. Of the five subjects, four were female and one was male. The age range was 23–44 years with a mean of 30 years.

2.2 Materials and Apparatus

Subjects sat on a kneeling chair at a desk with a horizontal surface and an attached PTZ-1230 12 × 12 Wacom Intuos 3 digitising tablet. An inking pen was used for recording subjects' raw and digital kinematic data (FFT low-pass 12 Hz filter). A Tobii X-120 eye tracker attached under the table facing upwards at an angle of 68° captured the eye movements of subjects. Subject's viewing distance from the eye tracker ranged between 54 and 62 cm, depending on their height. A fixation was classified using the Tobii fixation filter, which was customised to a velocity threshold of 30 pixels/window and a distance threshold of 15 pixels. This filter best matched the raw sampling data. The sampling rate of the eye tracker was 60 Hz and had an accuracy of 0.5° and the digitising tablet recorded at 200 Hz with an accuracy of 0.25 mm in the x and y direction. An external Basler scA640-120gc digital camera was used to record the trials and the video footage was superimposed with eye-gaze position using Tobii Studio version 2.0.2 software. The camera was positioned directly above the centre of the display screen and had a resolution capture of 658 × 492 pixels. The camera was set to 91 frames per second and the video feed was encoded using the default Microsoft video 1 codec within Tobii studio. A headrest was used to keep subject's head position constant throughout testing to improve the spatial accuracy of eye tracking.

2.3 Data Gathering

Subjects were first calibrated to the eye tracker using a five-point calibration grid positioned on the digitising tablet surface. Once properly calibrated, subjects were

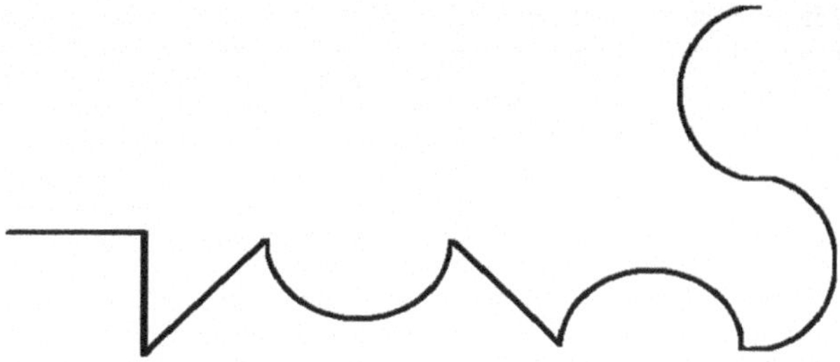

Fig. 1 Exemplar image copied by subjects

then required to have a practice attempt at copying the word 'practice' to insure desired eye-tracker accuracy and reliability, as well as subject comfort.

Subjects were required to copy a line image called the exemplar image, which was made up of eight connected lines, as shown in Fig. 1.

The image was presented and copied three times consecutively. Subjects were instructed to copy to the best of their ability. All copies were drawn at least 5° below the exemplar image to control for any potential effects arising from parafoveal vision.

3 Results

Gaze data was explored qualitatively in order to determine the extent of any temporal relationships between the eye and hand during copying. This required frame-by-frame analysis of the recordings taken using Tobii Studio version 3.0.3. Camera frames were recorded every 11 milliseconds. Of the three repeated trials, the trial that best allowed for an accurate temporal analysis between the eye and pen was analysed. This was determined by the number of pen velocity minima and their positions, as these factors affected the ability to determine the time elapsed between gaze shifts and velocity minima. Trials with higher writing velocities and therefore more automated movements tended to be selected. The coinciding of gaze shifts and velocity minima were identified by reviewing camera recordings of the drawn copy and an image of where the copies' velocity minima occurred.

It was found that the majority of gaze shifts coincided with velocity minima, regardless of whether gaze was directed at the exemplar image or the copy. Figure 2 is an image of a copy trial produced by one of the subjects.

Velocity minima that were overlapping spatially (see Fig. 2) were counted as one velocity minima. During the trial in Fig. 2, a gaze shift was made at the fifth (from the left) velocity minima. This can be viewed in Fig. 3, which shows the occurrence of a gaze shift as the pen reached that position.

Figures 2 and 3 illustrate that as the pen reached a velocity minima, the subject's gaze shifted from left to right. This gaze shift was represented by two fixation circles, the former (left) and the latter (right). Figure 3 also shows that as the pen reached the equivalent position of the eye gaze, the gaze shifted—a behaviour

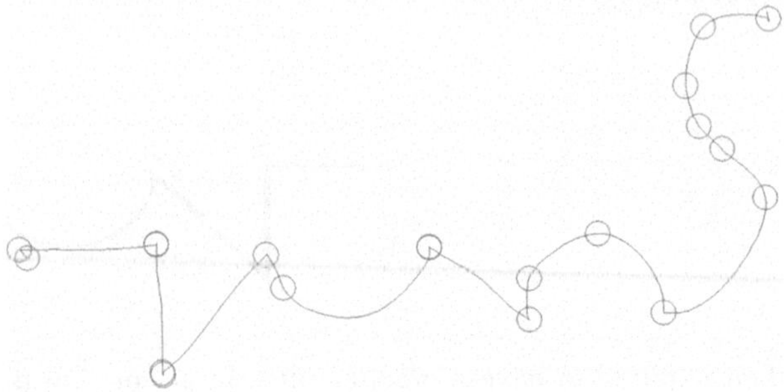

Fig. 2 Example image of a subject's copy attempt showing the locations of pen velocity minima (*circles*)

Fig. 3 An observed gaze shift (from *left* to *right*) at the moment of a pen velocity minima during copying

reported previously in drawing (see Tchalenko 2007; Gowen and Miall 2006), pointing (see Neggers and Bekkering 2000) and in face copying (see Tchalenko 2009).

Table 1 is a summary of the number of velocity minima and gaze shifts and the percentage of gaze shifts that coincided within a 33-millisecond interval.

As there tended to be more velocity minima than number of gaze shifts, not all velocity minima coincided with a gaze shift. On average, almost two thirds of gaze shifts coincided with a velocity minima, with a large proportion of these instances occurring at moments when the pen was still.

Table 1 Summary of gaze shifts, velocity minima and percentage of gaze shifts coinciding

Subject	Total gaze shifts	Total velocity minima	Total Coinciding (within 33 msec)	Percentage of gaze shifts coinciding
1	11	15	7	63.64
2	15	16	11	73.33
3	23	20	18	78.26
4	6	11	3	50.00
5	14	16	9	64.29
Mean	13.8	15.6	9.6	65.90

4 Discussion and Conclusions

The aim of the current study was to determine the nature of the temporal relationship between the movements of the eye and hand during the task of line copying. The results show that there was a close relationship between the eye and hand, with an average of two-thirds of gaze shifts occurring at moments the pen was spatially at or temporally within 33 milliseconds of a velocity minima—the former being more common than the latter. A possible explanation for this finding is that subjects were prioritizing the order of cognitive processes relevant to the task and also attempting to reduce the impact of delays in cognitive processing associated with making gaze shifts (see Rayner 1998). The advantage of making gaze shifts when the pen is still or moving slowly is that it would allow for extra time, with relation to pen distance covered, for visual processing of the next stroke from the exemplar image, or checking of spatial output of the drawn image—a process important for maintaining the spatial quality of the copy (see Smyth and Silvers 1987). This idea is consistent with the view that at these moments, there is more relative time to gather and integrate visual information (Ketcham et al. 2006). The presumed problem with making gaze shifts during moments the pen is moving quickly is that relevant cognitive processes may become suspended during this time (see Irwin et al. 1996). In addition, if further time is then required to process the visual information of the newly fixated area, large distances could be covered by the pen in the absence of relevant, or necessary cognitive processing. Given the pen can move relatively large distances over short periods of time, these delays to processing could be enough to affect the quality of output. Therefore, it is proposed that gaze shifts can come at a cost during this task. This cost can be somewhat mediated by the simulator by using certain eye–hand behaviours, and it is these behaviours that suggest tactical decision making about the execution of processes relating to copying.

Although simulators may be able to mediate costs associated with making gaze shifts during line copying, this may not be true for signature simulation. The problem associated with copying signatures is that the same strategy may not be as viable. The reason for this is that pausing during a signature simulation task can compromise line quality—a characteristic important for determining authenticity of specimen signatures. Because authentic signatures tend to be executed fluently (tend not to show abrupt pen pauses, unnecessary pen lifts or moments of dysfluency caused by slow pen movements and hesitations), the line quality is important in determining the authenticity of a signature simulation. Given line quality must also be appropriately maintained by the simulator, the strategy of simulators taking advantage of momentary pauses during line copying is unlikely to be as viable for signature simulation. This may be the reason why spatial accuracy and line fluency are thought to trade off (see Hilton 1982) during signature simulations.

The present study has provided evidence of a close temporal relationship existing between the eye and hand in the task of line copying. It is thought that this relationship may serve to help mediate the costs associated with making gaze shifts that can impact upon the ability to accurately copy an image. Future studies should determine if a similar relationship is present in the task of signature copying. In addition, future studies should aim to investigate the effects of line length and fluency on the eye–hand behaviour and quality of output during copying.

Acknowledgments I would like to thank the continuing support of my colleagues Dr. Doug Rogers and Dr. Jodi Sita, friends and family who have supported me throughout the entirety of my PhD.

References

Found, B., & Rogers, D. (1998). A Consideration of the Theoretical Basis of Forensic Handwriting Examination: The application of "Complexity Theory" to understanding the basis of handwriting identification. *International Journal of Forensic Document Examiners, 4*(2), 109–118.

Gowen, E., & Miall, R. C. (2006). Eye-hand interactions in tracing and drawing tasks. *Human Movement Science, 25,* 268–585.

Helsen, W. F., Elliott, D., Starkes, J. L., & Ricker, K. L. (1998). Temporal and spatial coupling of point of gaze and hand movements in aiming. *Journal of Motor Behavior, 30*(3), 249–259.

Helsen, W. F., Elliott, D., Starkes, J. L., & Ricker, K. L. (2000). Coupling of eye, finger, elbow, and shoulder movements during manual aiming. *Journal of Motor Behavior, 32*(3), 241–248.

Hilton, O. (1982). *Scientific examination of questioned documents (Rev. ed.).* New York: Elsevier.

Irwin, D. E., & Carlson-Radvansky, L. A. (1996). Cognitive Suppression during Saccadic Eye Movements. *Psychological Science, 7*(2), 83–88.

Ketcham, C. J., Dounskaia, N. V., & Stelmach, G. E. (2006). The role of vision in the control of continuous multijoint movements. *Journal of Motor Behavior, 38*(1), 29–44.

Neggers, S. F., & Bekkering, H. (2000). Ocular gaze is anchored to the target of an ongoing pointing movement. *Journal of Neurophysiology, 83*(2), 639–651.

Pepe, A., Sita, J., & Rogers, D. (2012). A consideration of signature complexity using simulators' gaze behaviour. *Journal of Forensic Document Examination, 22,* 5–13

Rayner, K. (1998). Eye movements in reading and information processing: 20 years of research. *Psychological Bulletin, 124*(3), 372–422.

Reina, G. A., & Schwartz, A. B. (2003). Eye-hand coupling during closed-loop drawing: Evidence of share motor planning? *Human Movement Science, 22,* 137–152.

Smyth, M. M., & Silvers, G. (1987). Functions of vision in the control of handwriting. *Acta Psychologica, 65*(1), 47–64.

Tchalenko, J. (2007). Eye movements in drawing simple lines. *Perception, 36*(8), 1152–1167.

Tchalenko, J. (2009). Segmentation and accuracy in copying and drawing: Experts and beginners. *Vision Research, 49*(8), 791–800.

Tchalenko, J., & Chris Miall, R. (2009). Eye-hand strategies in copying complex lines. *Cortex, 45*(3), 368–376.

Degree of Subject's Indecisiveness Characterized by Eye Movement Patterns in Increasingly Difficult Tasks

Yannick Lufimpu-Luviya, Djamel Merad, Bernard Fertil, Véronique Drai-Zerbib and Thierry Baccino

1 Introduction

Indecisiveness is, for some people, a character trait that manifests as a difficulty in settling between several simple or complex alternatives. It is an irritating character trait from a theoretical point of view, as it can be difficult to explain or decipher; it also becomes irritating from a practical point of view, as it seems particularly resistant to modification, sometimes even increasing while one tries to reduce it. Indecisiveness has been the topic of a large body of work. For example, relative to decisive people, Frost and Shows (1993) found that indecisive people take more time to make simple decisions; Rassin and Muris (2005) noticed that indecisive people seek more information before making decisions; and Veinott (2002) noticed that indecisive people more often postpone more difficult choices.

This study of indecisiveness is part of the ANR[†] project ORIGAMI2 ("Observation du Regard et Interprétation du Geste pour une Analyse Marketing Non Intrusive," or "observation of Gaze and Interpretation of Gesture for a Nonintrusive Marketing Analysis"). ORIGAMI2 aims at a complete analysis of the customer's decision-making process through the combination of various data acquisition tools. The analysis of a decision-making process involves three steps. First, the stimuli with which the customer is interacting must be identified. Second, the customer's behavior must be interpreted based on his hesitations, time spent staring at different objects, etc. Finally, behavior patterns that take place during the decision-making process need to be identified. Our present work is related to the second and third

† Agence Nationale de la Recherche, funded by the French Ministry of Research.Website: http://www.agence-nationale-recherche.fr/en/

Y. Lufimpu-Luviya (✉) · D. Merad · B. Fertil
LSIS-UMR CNRS 7296, Marseille, France
e-mail: yannick.lufimpu-luviya@lsis.org

V. Drai-Zerbib · T. Baccino
Lutin Userlab, Paris, France

M. Horsley et al. (eds.), *Current Trends in Eye Tracking Research*,
DOI 10.1007/978-3-319-02868-2_7, © Springer International Publishing Switzerland 2014

steps, with the goal of building a predictive model of a person's degree of indecisiveness through the use of eye-tracking methodology. First, we needed to extract relevant descriptors.

Several authors have proposed scales to quantify indecisiveness. We used Frost and Shows' Indecisiveness Scale (IS). This tool is composed of 15 statements (e.g., "I have a hard time planning my free time"); for each statement, the subject must choose from five answers, which range from "strong disagreement" to "strong agreement." The choice of the IS as a reference is a way of reaching our *gold standard:* the segmentation of the population into two groups, "decisive people" and "indecisive people." We explore later how we made such a distinction.

2 Experimental Procedure

In the first part of this chapter, the operating procedure is presented. Next, the preprocessing of the data and the construction of descriptors are described. In the third part, the results regarding the selection of the best descriptors are presented. Finally, a comparison with other results in the literature is made. The experimental procedure for each subject was standardized. First, the subject had to fill out Frost and Shows' multiple-choice questionnaire. Next, he was asked to complete three tasks in a random order. Finally, he had to fill out the multiple-choice questionnaire described by Zaichkowsky (1984), which measures the degree of motivation. We used a corneal reflection-based eye tracker†. The subject's head rested on a headrest that was located 70 cm from a monitor that was 47.7 by 29.7 cm^2. Stimuli were displayed with a resolution of 1680 by 1050 square pixels, and data were acquired at 500 Hz.

During the experiment, the subjects gave their answers orally while the experimenter took notes.

2.1 Measuring the Degree of Indecisiveness with Frost and Shows' Scale

As described in the introduction, Frost and Shows' questionnaire gives us a way of segmenting the population into two classes: decisive subjects and indecisive subjects. It is a 15-statement questionnaire, and each response is chosen within the following parameters:

- Strong disagreement (score = 1)
- Disagreement (score = 2)
- Reasonable agreement (score = 3)
- Agreement (score = 4)
- Strong agreement (score = 5)

Table 1 Distribution of the ages of male and female subjects

	18–20	21–30	31–40	41–50	51–60	61–70	Total
Women	1	9	0	2	2	1	15
Men	1	4	1	7	0	0	13
Total	2	13	1	9	2	1	28

For 6 of the 15 statements, the scoring is reversed (score for strong disagreement =5, score for disagreement =4, etc.). Frost and Shows define the subject's degree of indecisiveness as the mean value of the 15 scores.

According to Frost and Shows, subjects whose scores are less than 2.5 are labeled "decisive," while those whose scores are greater than 2.5 are labeled "indecisive." In contrast, Patalano et al. (2009) used the median score of the population as a threshold for segmentation. We, however, questioned which solution, either Frost and Shows' value of 2.5 or the median score, was more statistically relevant. The value of 2.5 is logical, but the actual statistical distribution of our population's degree of indecisiveness cannot be known; therefore, choosing the median score, as proposed by Patalano, appears to be more appropriate but may not always be an absolutely correct choice. We believe that one should also consider whether the variance should be taken into account. In section 2.6, an alternative way of segmenting the population is presented that is based on intra-class variance, as described by Otsu (1979).

2.2 *Participants*

The goal of this work was to describe and characterize each subject's degree of indecisiveness. It was necessary to recruit a large number of subjects, which can be difficult in a controlled environment. In particular, it was necessary to recruit enough subjects to obtain reliable data. Germeijs and De Boeck (2002) used 291 subjects, while Rassin et al. (2007) used 39, 56, and 62 subjects; the number of subjects depends on the requirements of the experiment. We recruited 28 French-speaking subjects. The subjects had no ophthalmological problems nor any particular difficulties in reading and understanding documents displayed on a screen. Table 1 shows the distribution of the ages of female and male subjects.

2.3 *Description of Tasks*

After filling out Frost and Shows' questionnaire, each subject was given three tasks to perform. We chose to arrange each task's alternatives in columns. The complexity of each task was determined by the number of alternatives: two alternatives in task 1, three in task 2, and four in task 3. Accordingly, task 1 was the least complex

Table 2 Stimulus on the screen for task 1

Menu 1	Menu 2
Starter	Hot dish
Hot dish	Dairy product
Dairy product or dessert	Dessert
Bread	Bread

Table 3 Stimulus on the screen for task 2

Menu 1	Menu 2	Menu 3
Salad with a side dish	Cold sandwich 30 cm	Hot sandwich 15 cm
Dairy product Bread	Dessert or dairy product	Dairy product Dessert

Table 4 Stimulus on the screen for task 3

Program 1	Program 2	Program 3	Program 4
Sports: compulsory module	Sports: compulsory module	Sports: compulsory module	Sports: noncompulsory module
Frequency: 12 h per	Frequency: 24 h per	Frequency: 48 h per	Frequency: as you wish
semester	semester	semester	
BONUS =+1 points on the final general semi-annual mark, if and only if practicing 12 h in the semester	BONUS =+2 points on the final general semi-annual mark, if and only if practicing 24 h in the semester	BONUS =+4 points on the final general semi-annual mark, if and only if practicing 48 h in the semester	BONUS=+6points on the final general semi-annual mark, if and only if practicing> 60 h in the semester

task and task 3 the most complex. In tasks 1 and 2, the subject had to choose between menus (two menus in task 1 and three menus in task 2). In task 3, the choice was between four academic programs. Tables 1, 2, and 3 show the stimuli for each task.

Because of the relatively small number of participants, we chose to increase the number of tasks. In addition to the advantage of acquiring more data, increasing the number of tasks allowed us to reduce intra-subject variability, which is an important parameter to consider when one is working with human subjects and needs to generalize features. The robustness of certain descriptors was tested relative to task complexity, and the results are presented in section 4.3. Similarly, increasing the number of tasks allowed us to filter our subjects, allowing us to avoid "noise" (section 2.5) even before building the database.

The three tasks were randomly presented to the subjects. We chose to arrange the alternatives in columns. The instructions given for tasks 1 and 2 were as follows:

> You are a student. It is lunch time. Every day, information about the day's menus is displayed on a touch screen in the campus restaurant. A menu selection can be made by touching the screen. Please indicate which menu would best suit you (menus can be vegan).

The instructions for task 3 were as follows:

You are a student. Sports are becoming an important component of academic programs. Each student may choose between 4 programs, each with a different way of practicing sports. Please indicate which program would best suit you.

2.4 *Measuring the Degree of Motivation with Zaichkowsky's Scale*

After completing the three tasks, the subject was required to complete Zaichkowsky's multiple-choice questionnaire. The purpose of Zaichkowsky's questionnaire is to measure the degree of motivation, on a scale that ranges from "low motivation" to "high motivation." The questionnaire is made up of 20 statements. The answers to each of the 20 statements are scored from 1 to 7; for 10 of the statements, the scores that correspond to each answer are reversed. Final scores can range from 20 to 140. A person whose score is under 69 is considered to have a low degree of motivation. A person whose score is between 70 and 111 is considered to have a medium degree of motivation. If his score is higher than 112, he is considered to have a high degree of motivation.

As was done by Zaichkowsky (1984), we scored the answers to each of the 20 statements from 1 to 7. Accordingly, the final scores ranged from 20 to 140.

2.5 *Filtering and Database Construction*

Before analyzing the eye-tracking data, it was necessary to first filter the population.

2.5.1 Filtering based on Zaichkowsky's scale

To avoid distorting the results, or to at least minimize any distortion, it is necessary to exclude subjects whose motivation for the experiment is too low. To do so, we started by examining the distribution of the population based on Zaichkowsky's scale.

Figure 1 shows this distribution. Among the 28 subjects, 3 had a medium degree of motivation based on their effort during the experiment; the other 25 had a low degree of motivation. If we applied this criterion, we would have had to exclude 25 subjects from our database. We could not rely solely on Zaichkowsky's scale.

2.5.2 Filtering based on both Scanpaths and Zaichkowsky's Degree of Motivation

In addition to the distribution of the degree of motivation, we also examined each subject's scanpaths during the tasks. Figure 2 shows the scanpaths of two example subjects during the three tasks. Each rectangle stands for a text field on the stimulus.

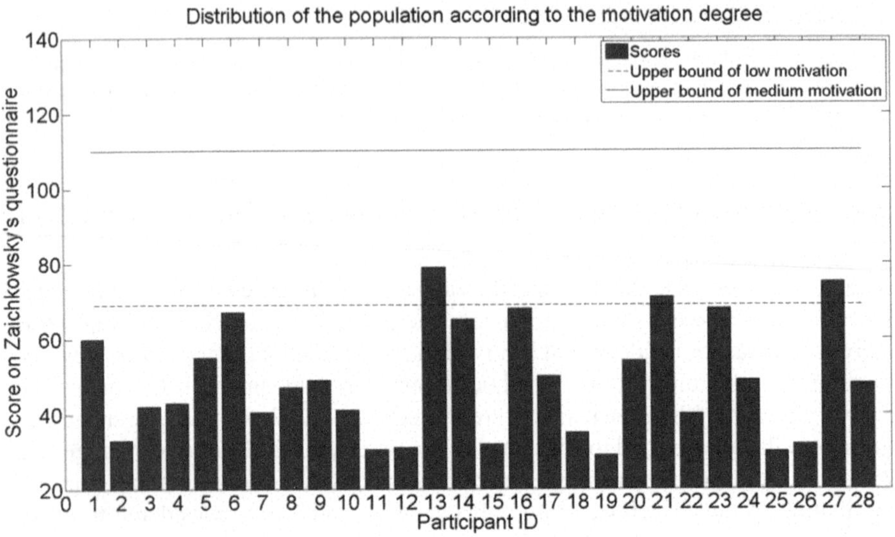

Fig. 1 Degree of motivation

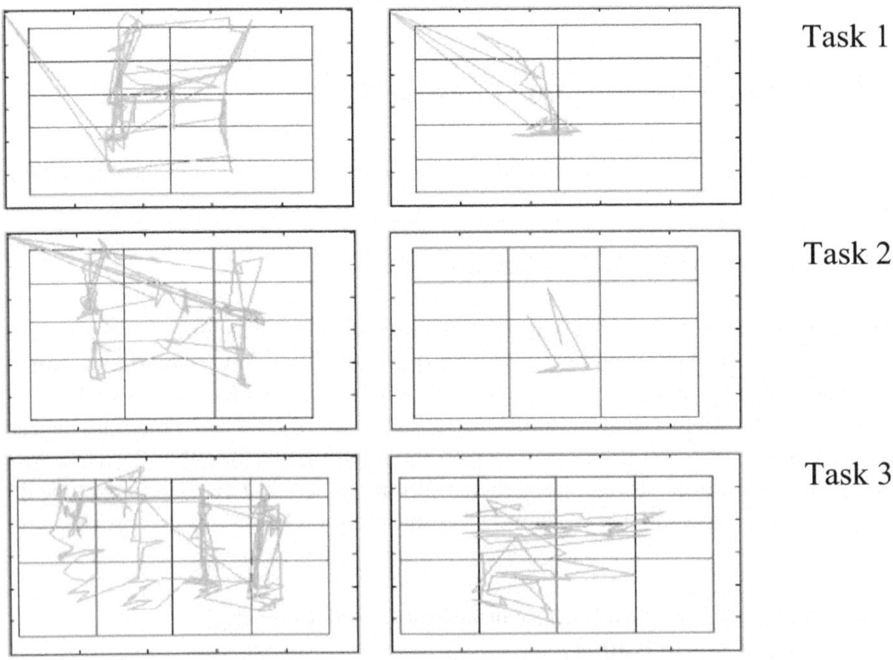

Fig. 2 Scanpaths. A good subject (*left*) and a "noise" subject (*right*)

In the first subject's scanpaths (left), his eyes look over the three columns and come back several times to several locations. Based on these scanpaths, we could infer that this subject followed the instructions and seriously participated in the experiment. In contrast, the second subject (right) did not correctly view all of the alternatives; moreover, his scanpaths are less complex than those of the first subject. This subject was labeled "noise" in the database, and his participation was not taken into account. Altogether, among the 28 participants, 6 were labeled as "noise."

The mean score of the "noise" outliers on Zaichkowsky's questionnaire was 44.91, with a standard deviation of 15.75. According to Zaichkowsky's scale, they had a low degree of motivation with the experiment. The mean score of the other 22 subjects was 49.75, with a standard deviation of 15.36. The fact that these tasks were intended for students may explain these low scores. Despite the low degrees of motivation, Zaichkowsky's scale supports the application of this filtering step.

After excluding the unreliable participants from the database, segmentation based on Frost and Shows' scale was performed.

2.6 *Segmentation based on Frost and Shows' indecisiveness scale*

Both the value of 2.5 given by Frost and Shows and Patalano's median value seemed to us to be arbitrary and only justified for a large number of subjects. We used a segmentation method based on the between-class variance, which was first proposed by Otsu (1979). This method is easy to implement and consists of two main steps:

- For each threshold s, ranging from s_{min} to s_{max} (here, 1 and 5), with a predefined step size, the intra-class variance $V_a(s)$ is given by the formula:

$$V_a(s) = p(C_1^s)*(M_1^s - M)^2 + p(C_2^s)*(M_2^s - M)^2,$$

where C_i^s is the ith class, M is the mean score of the population, M_i^s is the mean score of the ith class, and $p(C_i^s)$ is the probability of the ith class, equal to the number of subjects in the ith class divided by the total number subjects.

Minimizing the intra-class variance is equivalent to maximizing the between-class variance:

$$V_b(s) = p(C_1^s)*p(C_2^s)*(M_1^s - M_2^s)^2.$$

The optimal threshold s_0 is given by the expression:

$$V_b(s_0) = \max_{s\in[1;5]}\{V_b(s)\}.$$

Figure 3 shows the distribution of the scores of the 22 subjects. Notably, our threshold and the median value (2.51) are similar. The numbers of subjects per class were quite similar: 9 decisive subjects and 13 indecisive subjects. The first class ("decisive") had a mean value of 2.13 with a standard deviation of 0.24, and the second class had a mean of 2.78 with a standard deviation of 0.19.

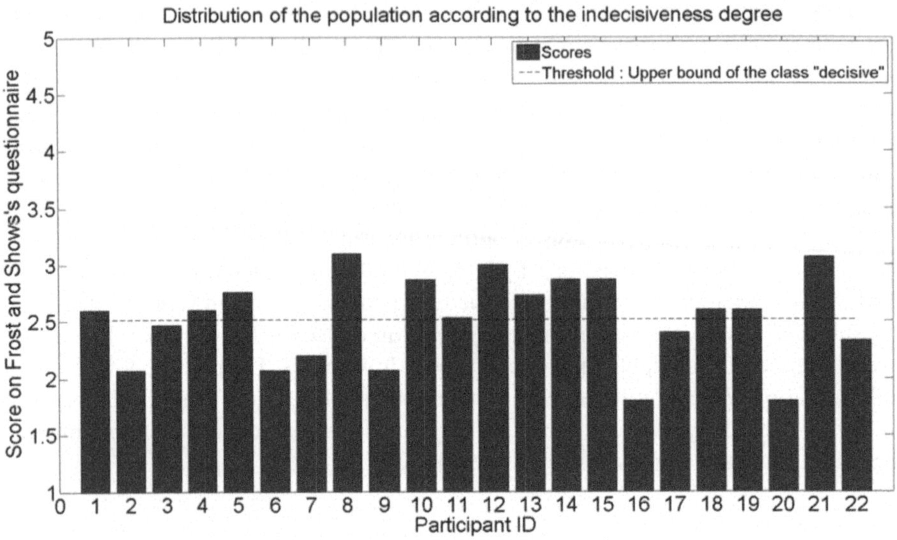

Fig. 3 Degree of indecisiveness

After segmenting the population into two classes, we next extracted the eye-tracking descriptors.

3 Eye-Tracking Descriptors

Eye-tracking descriptors were divided into two groups: descriptors based on fixation and descriptors based on saccades. We decided to focus not only on information about the alternatives that were chosen or looked at the most but also on information about the areas outside these regions. In an indecisive state, at least two alternatives must be taken into account; we wanted to study behaviors directed towards the three alternatives at which subjects spent the most time looking (or towards the two alternatives, when only two alternatives were available).

Altogether, we extracted 24 descriptors for each subject: Of these, 19 were based on fixation data (D1–D19) and 5 on saccade data (D20–D24). The percentage of time spent on noninformative areas, the chosen alternative or the i-th most observed alternative ($1 \leq i \leq 3$), is proportional to the ratio between the time spent on noninformative areas, the chosen alternative or the i-th most observed alternative, and the duration of the task. Thus, if $i = 1$, the percentage of time spent on the most observed alternative will be:

$$D9 = 100 * \frac{D6}{D1}.$$

Table 5 Descriptors

Fixation data	D1	Total time
	D2	Time awarded to the chosen alternative
	D3	Fixation's mean duration on the chosen alternative
	D4	D3/D1
	D5	Percentage of time awarded to the chosen alternative
	D6	Time awarded to the first most observed alternative
	D7	Fixation's mean duration on the first most observed alternative
	D8	D7/D1
	D9	Percentage of time awarded to the first most observed alternative
	D10	Time awarded to the second most observed alternative
	D11	Fixation's mean duration on the second most observed alternative
	D12	D11/D1
	D13	Percentage of time awarded to the second most observed alternative
	D14	Time awarded to the third most observed alternative
	D15	Fixation's mean duration on the third most observed alternative
	D16	D15/D1
	D17	Percentage of time awarded to the third most observed alternative
	D18	Time awarded to the noninformation areas
	D19	Percentage of time awarded to the noninformation areas
Saccade data	D20	Total path length
	D21	Alternative-based path length
	D22	Ratio between alternative-based path length and total path length
	D23	Ratio between the number of alternative-based saccades and the total number of saccades
	D24	Ratio between the number of saccades to the chosen alternative and the total number of saccades

It should be noted that for task 1, we did not have descriptors based on the third most observed alternative (D14, D15, D16, and D17) because task 1 consisted of only two alternatives.

The alternative-based path length (D21) was calculated as the sum of the lengths of the saccades going from one alternative to another. This descriptor should be relevant in quantifying the participant's indecisiveness.

Such a large number of eye movement patterns were gathered to build, in future works, a predictive model that can be used to determine a person's degree of indecisiveness; this list is likely to evolve.

4 Eye-Tracking Results

In this part, the selection of the most relevant descriptors is presented. Then, the average behavior of decisive and indecisive subjects is discussed in relation to task complexity. An interesting conclusion about behavior is drawn based on either the first or the second half of the task. Eventually, the results regarding the relationship between the degree of indecisiveness and the degree of personal motivation are presented.

4.1 First and Second Halves of a Task

It has been shown that every decision-making situation can be divided into two main elementary parts (Patalano et al. (2009)). First, information about the proposed alternatives must be gathered; in our case, each subject was supposed to be looking at the stimuli as a whole. Second, a decision-making process is launched. The boundary between each part depends on the subject and is not always easy to define. Therefore, to simplify our measures, and as was done in Patalano et al.'s work, we assigned the first part of the decision-making process to the first half of fixation. Further studies can be made to refine this boundary.

The 24 descriptors were calculated three times: for the entire task, for the first half, and for the second half. We then wanted to describe how a subject behaved, not only during the task as a whole but also during each half (see Part III.3).

4.2 Selecting Relevant Descriptors

To select the most relevant descriptors, a one-factor analysis of variance (ANOVA) was used. Because there were two classes ("decisive" and "indecisive"), the null hypothesis was that the patterns of each class originated from the same population; the alternative hypothesis was that they did not.

Table 6 provides the *p* values for the main hypothesis for each of the three tasks. The values in column *F* are for the first half of the task, those in column *S* are for the second half, and those in column *T* are for the entire task. A *p* value was considered significant if it was less than 0.05.

Only the descriptors whose *p* values were significant for the full duration of at least one task were included: 13 descriptors from the fixation data and 2 from the saccade data. The names of these descriptors are depicted in bold in the table. It should be noted that no descriptor was statistically significant for each part of every task (*F, S,* and *T*) and that only six were significant in two different tasks (D4, D8, D14, D17, D20, and D21). Because of this selection, we can now discuss the importance of task complexity and the relative importance of the first or second part of a task in characterizing indecisiveness.

4.3 Degree of Indecisiveness and Task Complexity

Figures 4 and 5 display the standardized total durations of the tasks (descriptor D1) and the standardized durations spent looking at the chosen alternative (D2).

Let us consider task duration. Relative to the average decisive subject, the average indecisive subject spent more time on a task (Fig. 4). Moreover, he spent more time looking at the alternative he ultimately selected (Fig. 5), with a higher mean fixation duration. We noted that the longer the duration of the task, the larger the

Table 6 The *p* values of each descriptor for the first half of the task (*F*), the second half (*S*), and the entire task (*T*)

	Task 1: complexity			Task 2: complexity			Task 3: complexity		
	F	*S*	*T*	*F*	*S*	*T*	*F*	*S*	*T*
D1	0.213	0.401	0.297	0.076	0.080	0.076	0.002*	0.003*	0.002*
D2	0.392	0.977	0.579	0.197	0.606	0.320	0.009*	0.126	0.016*
D3	0.236	0.928	0.244	0.029*	0.854	0.573	0.013*	0.020*	0.010*
D4	0.020*	0.185	0.027*	0.651	0.377	0.559	0.058	0.124	0.035*
D5	0.782	0.446	0.403	0.433	0.620	0.975	0.553	0.238	0.204
D6	0.452	0.394	0.504	0.290	0.062	0.150	0.020*	0.030*	0.005*
D7	0.371	0.342	0.245	0.700	0.394	0.940	0.735	0.901	0.744
D8	0.022*	0.025*	0.019*	0.421	0.143	0.377	0.030*	0.017*	0.031*
D9	0.422	0.726	0.207	0.087	0.914	0.563	0.026*	0.515	0.187
D10	0.307	0.464	0.221	0.170	0.172	0.074	0.001*	0.005*	0.002*
D11	0.668	0.082	0.361	0.774	0.465	0.950	0.994	0.542	0.500
D12	0.094	0.765	0.151	0.105	0.135	0.117	0.001*	0.137	0.047*
D13	0.595	0.552	0.177	0.555	0.911	0.773	0.777	0.479	0.802
D14				0.029*	0.150	0.111	0.004*	0.066	0.015*
D15				0.171	0.120	0.490	0.066	0.368	0.237
D16				0.685	0.903	0.725	0.102	0.017*	0.010*
D17				0.035*	0.781	0.655	0.047*	0.647	0.987
D18	0.401	0.696	0.774	0.045*	0.572	0.100	0.074	0.180	0.083
D19	0.693	0.775	0.834	0.140	0.823	0.219	0.098	0.507	0.174
D20	0.006*	0.003*	0.001*	0.193	0.410	0.282	0.137	0.046*	0.059
D21	0.117	0.002*	0.007*	0.181	0.168	0.148	0.431	0.023*	0.085
D22	0.410	0.154	0.187	0.273	0.297	0.142	0.373	0.271	0.889
D23	0.347	0.368	0.262	0.953	0.529	0.646	0.198	0.430	0.287
D24	0.225	0.146	0.131	0.651	0.741	0.975	0.253	0.278	0.804
	3	3	4	4	0	0	10	8	10

* p-value ≤ 0.05.

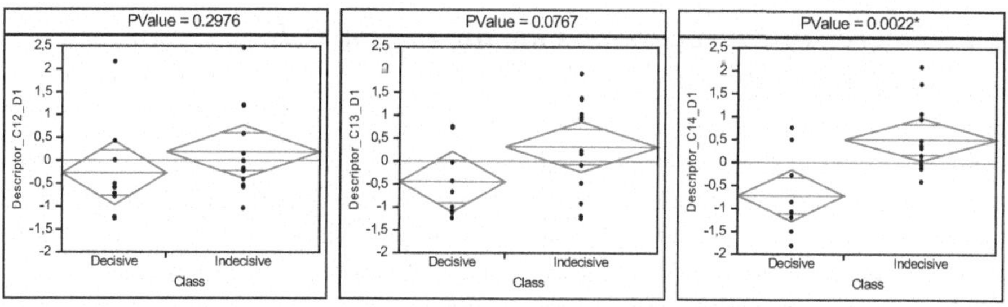

Fig. 4 Standardized total durations of the tasks (D1) for tasks 1–3 (from *left* to *right*)

difference between the indecisive class and the decisive class. However, this observation did not apply for all of the 24 descriptors.

Let us then consider all of the descriptors for the full task durations (Table 6, column *T*). We can see that for the most complex task, almost half of the descriptors

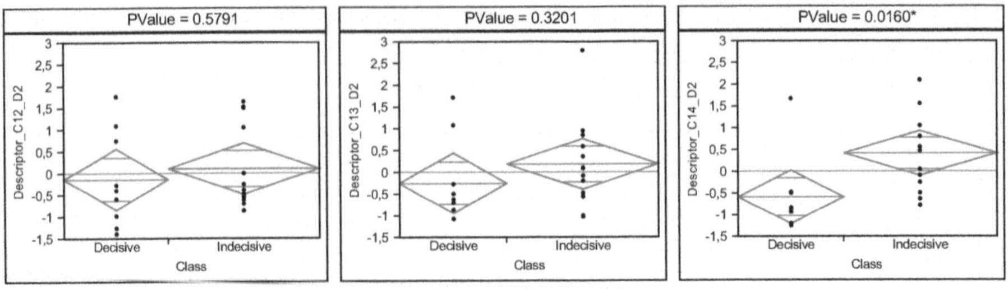

Fig. 5 Standardized durations spent looking at the chosen alternative (D2), for tasks 1–3 (from *left* to *right*)

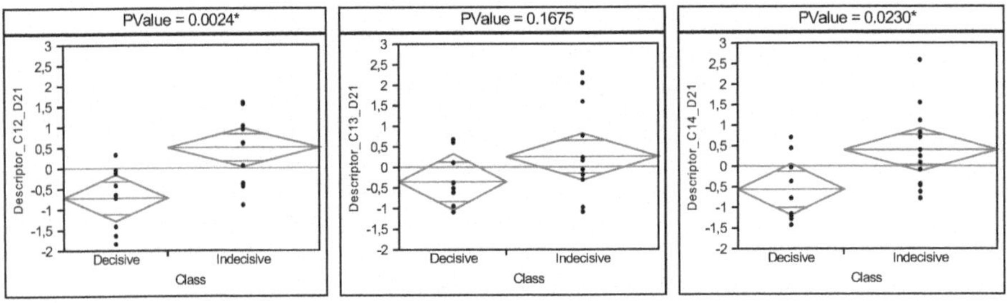

Fig. 6 Alternative-based path lengths in standardized values (D21) for parts *S* of the first task (*left*), the second task (*middle*), and the third task (*right*)

were statistically significant. For the first and second tasks, four and zero descriptors were relevant, respectively. We did not find that the relevance of the descriptors increased with task complexity. It would have been interesting to study the subjects' behaviors for a greater number of stimuli with increasing complexities.

4.4 Degree of Indecisiveness during the Parts F and S of the Tasks

If we consider the alternative-based path length (Fig. 6), we can see that during the second half of a decision-making situation, the average indecisive subject, relative to a decisive one, has a greater alternative-based path, although there is no significant difference during the first half of the task. We obtained the same results with the descriptor "total path length." These results are consistent with the idea that after reading the instructions and exploring all of the alternatives, an indecisive subject, relative to a more decisive one, browses a greater distance before orally formulating his answer.

The data presented in Fig. 7 are consistent with the hypothesis proposed by Patalano et al. (2009) that the relationship between the degree of indecisiveness and the exploratory strategy employed depends on whether the subject is in the first or second stage of the decision-making process.

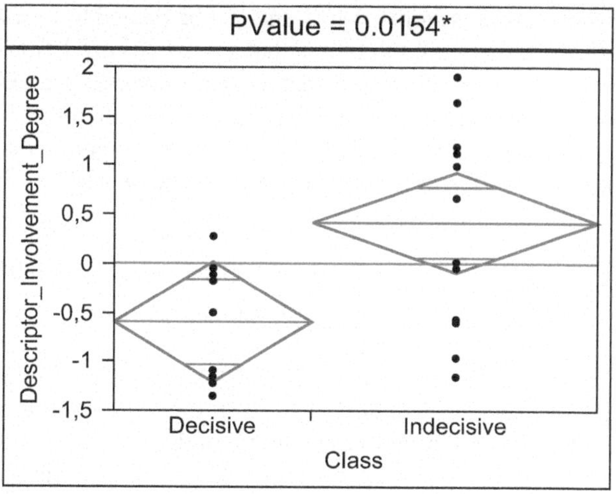

Fig. 7 A one-factor analysis of variance (ANOVA): class and degree of motivation

4.5 Degree of Indecisiveness and Degree of Personal Motivation

In this section, we present data relating to the relationship between the degree of indecisiveness and Zaichkowsky's degree of personal motivation. To analyze this relationship, we began by performing an analysis of variance and then applied Kendall's nonparametric test (τ) to more precisely identify any interactions.

The ANOVA revealed that the classes were statistically separable based on the degree of motivation: The probability that all of the 22 degrees of motivation came from the same population was 0.0154, which is less than the value of alpha (0.05). We were interested in looking more deeply at this interaction.

Kendall's rank correlation coefficient is based on Spearman's. The calculation is quite simple. First, the n degrees of indecisiveness must be sorted. The rank values of the n degrees of motivation are then sorted according to the degrees of indecisiveness. After this step, only the degrees of motivation are taken into account. For the oth observation, we counted the number of observations that were greater than the oth (weight "+ 1") or less than the oth (weight "− 1"). We then obtained a third column of $n-1$ values by summing the weights. The maximum total weight, S, is equal to $\frac{n*(n-1)}{2}$ if the order is perfect, as it would then be the sum of the n first integer numbers. If the order is the perfect opposite, S will be equal to $\frac{-n*(n-1)}{2}$. In the case of a total absence of correlation, S is equal to 0. The Kendall rank correlation is given by the expression

$$\tau = \frac{S}{\frac{n(n-1)}{2}} = \frac{2S}{n(n-1)},$$

where τ ranges from − 1 to 1 and can be related to Pearson's coefficient: The closer it is to 1, the more likely it is that a positive correlation exists, while the closer it is to − 1, the more likely it is that a negative correlation exists. Finally, if is close to 0, there is a strong probability that there is no monotonic link between the two degrees.

We calculated Kendall's τ for the entire population (n=22), for the decisive group (n=9), and for the indecisive group (n=13):

$$\tau_{Total} = 0.35, PValue_{Total} = 0.03*$$
$$\tau_{Decisive} = 0.59, PValue_{Decisive} = 0.04*$$
$$\tau_{Indecisive} = -0.16, PValue_{Indecisive} = 0.49*.$$

Thus, according to our decisive database, we can conclude that the higher the degree of indecisiveness (below 2.5), the higher the probability that the degree of motivation will be high as well. As for the indecisive subjects, we can only conclude that there is a high likelihood that there is no monotonic link between the degree of indecisiveness and the degree of personal motivation.

5 Conclusions and Discussion

We proposed an automated method for the segmentation of a population into two classes: decisive and indecisive. We emphasized a possible correlation between Frost and Shows' indecisiveness scale and Zaichkowsky's personal degrees of motivation. We successfully filtered the participants and identified 15 eye-tracking descriptors based on the ocular responses of the 22 subjects during the three different tasks.

Some of the results are consistent with the work of Frost and Shows (1993). For example, an indecisive person, relative to a more decisive one, takes more time to make simple decisions. We also found the same results as Ferrari and Dovidio (2000): An indecisive person, relative to a decisive person, seeks more information about the alternative chosen but not more information overall. It should be noted that the greatest number of relevant descriptors was found for the most complex task (task 3).

These analyses were applied to a population of 22 subjects. The lack of relevancy of certain descriptors, from a statistical point of view, may be due to the lack of power. Whether it is valid to base the first step of a decision-making process on the first half of fixation may also need to be evaluated further. Indeed, it would be interesting to identify a method of determining this threshold in each subject. For example, by applying a sliding temporal window to the saccade data, we could detect when the alternative-based saccade frequency exceeds the frequency of saccades directed to the alternatives.

As mentioned in the introduction, this indecisiveness study is part of our work on behavioral marketing. Marketing managers need to understand the behavior of customers during a purchasing act: What does first catch the customer's eye, between what products does the customer hesitate, why does he hesitate, etc.? Our work could also be translated into research fields that involve the subject's emotional state, such as motor racing competitions, fighter pilots' flights, or psychiatric disorders.

References

Ferrari, J. R., & Dovidio, J. F. (2000). Information search process by indecisives: Individual differences in decisional procrastination. *Journal of Research in Personality, 34*, 127–137.

Frost, R. O., & Shows, D. L. (1993). The nature and measurement of compulsive indecisiveness. *Behaviour Research and Therapy, 31*(7), 683–692.

Germeijs, V., & De Boeck, P. (2002). A measurement scale for indecisiveness and its relationship to career indecision and other types of indecision. *European Journal of Psychological Assessment, 18*, 113–122.

Otsu, N. (1979). A threshold selection method from gray-level histograms. *IEEE Transactions on Systems, Man and Cybernetics, 9*, 62–66.

Patalano, A. L., Juhasz, B. J., & Dicke, J. (2009). The relationship between indecisiveness and eye movement patterns in a decision making informational search task. *Journal of Behaviour Decision Making, 23*, 353–368.

Rassin, E., & Muris, P. (2005). To be or not to be... indecisive: Gender differences, correlations with obsessive-compulsive complaints, and behavioural manifestation. *Personality and Individual Differences*, 38(5), 1175-1181.

Veinott, E. S. (2002). The effect of understanding and anticipated regret on decision readiness. Unpublished doctoral dissertation, University of Michigan, Ann Arbor.

Zaichkowsky, J. L. (1984). Measuring the Involvement Construct. *Journal of Consumer Research, 12*, 341–352.

3 Methods

We performed a series of tests which includes visual acuity assessment, reading performance and eye movements recording in two occasions, after 10 weeks of wearing the prescribed spectacles and after 10 weeks of wearing the contact lens. The visual acuity assessment was performed using a LogMAR chart for far and near distances. The reading performance and eye movements were recorded at a distance of 60 cm using the Tobii T60 eye tracker with a sampling frequency of 60 Hz. The Tobii T60 is integrated with a 17-inch TFT monitor which was used to display the reading material (black text on a white background) to the patient. The reading material was carefully selected from validated Malay reading text to ensure the comprehension difficulties are suitable for his age. It consists of one paragraph displaying 50 words in N10 materials corresponding to the patient's near VA.

4 Results

The VA assessments revealed a distance VA of 6/19 (for both spectacles and contact lens). The near VA was demonstrated to be improved by one line, i.e. from N10 with spectacles to N8 after 10 weeks of wearing contact lens. Interestingly, his reading duration (1.09 min when wearing spectacles and 0.47 min when wearing contact lens) showed marked improvement. Reading rate increased to almost double from 44 words per minute (wpm, with spectacles) to 105 wpm (with contact lens) with average fixation duration increased from 0.32 to 0.44 s (Table 1). These improvements in visual functions and reading performance parameters were supported by the eye movement recording results. The eye movement pattern with spectacles showed a modified staircase reading pattern which is due to the nystagmus (Fig. 1), whereas the eye movement pattern with contact lens demonstrates clear saccades, regressions and fixations (Fig. 2), which closely resembled eye movement pattern of normals during reading. The eye movement pattern also demonstrates longer time per fixation as reflected in the mean fixation duration indicating more time for the eyes to be stable for information processing and thus improving reading rate.

5 Discussion

We are interested in exploring the possibility of evaluating and providing more quantitative reading performance results to congenital nystagmus patients wearing different types of optical corrections. It was found that with contact lens wear, patient's near visual acuity improved considerably. In addition, he demonstrated faster

Table 1 Comparisons in visual acuity and reading parameter measurements between spectacles and contact lens

Reading task	Spectacles	Contact lens	Changes (%)
Visual acuity (distance)	6/19	6/19	No changes
Visual acuity (near)	N10	N8	1 line better
Reading duration (min)	1.09	0.47	56.88
Reading rate (wpm)	43.92	105.26	139.66
Average fixation duration (ms)	0.32	0.44	−37.50

Fig. 1 Eye movement recording result for patient's reading performance with spectacles

reading rate with contact lens compared to spectacles. Previous studies reported that as reading speed is strongly associated with visual acuity, decrease in print size is reported to increase reading speed (Lovie-Kitchin et al. 1987, 2001).

The advantages of contact lens over spectacles are reducing optical distortions and providing a wider field of view. The 'contact' position with the cornea ensures a more stable vision as the nystagmoid eye is always fixating through the optical centre of the contact lens. In contrast, vision with spectacles will reduce as the eye drifts resulting in prismatic effect and distortion (Tabibi et al. 2008)

In this study, we found that mean fixation duration with contact lens was longer compared to spectacles. This suggests that the patient spent more time on average per fixation. The analysis also showed that reading duration with contact lens was reduced to half of the time taken using spectacles. This indicates that more letters/words were processed per fixation. This finding is supported by previous work reported by Tabibi et al. (2008). They found that contact lens broadened the high-foveation-quality field and improved foveation where the subject was able to

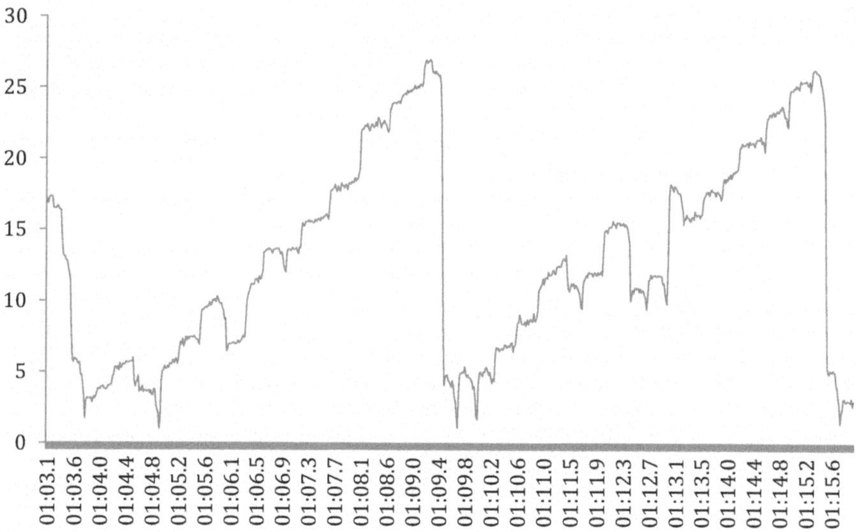

Fig. 2 Eye movement recording result for patient's reading performance with contact lens

see more area with better vision. This is supported by the fact that reading duration becomes shorter indicating that less time is needed to complete the task (Kudo et al. 1995). This could suggest that more visual input was being processed per fixation resulting in longer fixation duration. However, foveation period could not be analysed with Tobii T60 eye tracker due to low sampling rate of 60 Hz.

Wearing contact lens has been suggested to dampen nystagmus (Biousse et al. 2004). Reduction in nystagmoid eye movement results in stable vision and increased fixation time. Dampening effect from contact lens wear could affect eye movement fixation during reading. Although we only managed to analyse the average fixation duration because of the limitation of the eye tracker system, the figures presented roughly showed that refractive correction with contact lens (Fig. 2) resulted in a better fixation during reading compared to wearing spectacles (Fig. 1).

6 Conclusion

Contact lens improves reading performance in this patient. This was illustrated by increased reading duration, reading rate and eye movement recording. This case showed a transition from poor reading performance to an almost normal state. The ability to achieve normal reading rate in school children with nystagmus will facilitate their learning process. This in turn will open up more opportunities for them to pursue higher level education and increase their chance of employability.

References

Abadi, R. V. (2002). Mechanisms underlying nystagmus. *Journal of Royal Society of Medicine, 95*(5), 231–234.

Abadi, R. V., & Bjerre, A. (2002). Motor and sensory characteristics of infantile nystagmus. *British Journal of Ophthalmology, 86*(10), 1152–1160.

Abadi, R. V., & King-Smith, P. E. (1979). Congenital nystagmus modifies orientational detection. *Vision research, 19*(12), 1409–1411.

Biousse, V., Tusa, R. J., Russel, B., Azran, M. S., Das, V., Schubert, M. S., Ward, M., & Newman, N. J. (2004). The use of contact lens to treat visually symptomatic congenital nystagmus. *Journal of Neurology, Neurosurgery and Psychiatry, 75*, 314–316.

Chung, S. T. L., Mansfield, J. S., & Legge, G. E. (1998). Psychophysics of reading: XVIII The effect of print size on reading speed in normal peripheral vision. *Vision Research, 38*, 2949–2962.

Ciuffreda, K. J., Bahill, A. T., Kenyon, R. V., & Stark, L. (1976). Eye movements during reading: Case reports. *American Journal of Optometry & Physiological Optics, 53*(8), 389–395.

Dell'Osso, L. F., Traccis, S., Abel, L. A., & Erzurum, S. I. (1988). Contact lenses and congenital nystagmus. *Clinical Vision Sciences, 3*, 229–232.

Dickinson, C. M., & Abadi, R. V. (1985). The influence of nystagmoid oscillation on contrast sensitivity in normal observers. *Vision Research, 25*(8), 1089–1096.

Duchowski, A. (2003). Eye tracking methodology: Theory and practice. Springer-Verlag.

Frohman, T. C., Frohman, E. M., O'Suilleabhain, P., Salter, A., Dewey, R. B. Jr., Hogan, N., Galetta, S., Lee, A. G., Straumann, D., Noseworthy, J., Zee, D., Corbett, J., Corboy, J., Rivera, V. M., & Kramer, P. D. (2003). Accuracy of clinical detection of INO in MS: Corroboration with quantitative infrared oculography. *Neurology, 61*(6), 848–850.

Kudo, Y., Kaga, K., Ito, K., & Nakamura, M. (1995). Ocular movement during reading in patients with congenital nystagmus. *Acta Otoryngology (Stockh), 520*, 281–284.

Lovie-Kitchin, J. E., & Woo, G. C. (1987). Effect of magnification and field of view on reading speed using a CCTV. In G. C. Woo (Ed.), *Low vision: Principles and application.* New York: Springer-Verlag.

Lovie-Kitchin, J. E., Bevan, J. D., & Hein, B. (2001). Reading performance in children with low vision. *Clinical and Experimental Optometry, 84*(3), 148–154.

Reddy, S. C., & Tan, B. C. (2001). Causes of childhood blindness in Malaysia: Results from a national study of blind school students. *International Ophthalmology, 24*, 53–59.

Rubin, G. S., & Turano, K. (1994). Low vision reading with sequential word presentation. *Vision Research, 34*, 1723–1733.

Sweeney, J. A., Levy, D., & Harris, M. S. (2002). Commentary: Eye movement research with clinical populations. *Progress in Brain Research, 140*(2002), 507–522.

Tabibi, G., Wang, Z. I., & Dell'Osso, L. F. (2008). Infantile nystagmus syndrome: Broadening the high-foveation quality field with contact lenses. *Clinical Opthalmology, 2*(3), 585–589.

van der Geest, J. N., & Frens, M. A. (2002). Recording eye movements with video-oculography and scleral search coils: A direct comparison of two methods. *Journal of Neuroscience Methods, 114*(2), 185–195.

Part II
Aligning Eye Tracking and EEG Data

Triangulating the Reading Brain: Eye Movements, Computational Models, and EEG

Ronan G. Reilly

1 Introduction

Reading is a relatively recent cultural innovation, emerging in the last 5,000 years or so. It is an acquired skill involving the decoding of patterns of visual stimulation into a linguistic and, ultimately, a conceptual representation. Because it involves a considerable amount of learning and entails the interaction of several disparate brain areas, it is an ideal experimental domain to study the plasticity of the brain in a relatively pure form. As Huey (1908) observed in his seminal work *The Psychology and Pedagogy of Reading* first published just over 100 years ago, to understand reading fully would involve gaining a deep understanding of complex brain function. Huey's goal remains elusive to this day. What has changed is the availability of a powerful set of tools with which to explore the process: eye tracking, computational modelling, and electroencephalogram (EEG) recording. I will argue in this short chapter that it is only through the coordinated deployment of all three tools that we will get close to attaining Huey's goal.

2 Eye Movements in Reading

A reader's eyes move along a line of text in a sequence of fixations separated by jumps called saccades. Reading, therefore, takes the form of a series of "snapshots" during which textual information is acquired. During one such snapshot, the recognition of a single English word of average length takes about 100 milliseconds (Rayner and Pollatsek 1989). From the pioneering research of McConkie et al. (1988), it emerged that the effective target of a given eye movement is the centre of the to-be-fixated word. The goal of eye movements in reading appears to be to attain an optimal viewing position (OVP) on a word, which, if successful, facilitates rapid

R. G. Reilly (✉)
Department of Computer Science, National University of Ireland, Maynooth, Ireland
e-mail: ronan.reilly@nuim.ie

M. Horsley et al. (eds.), *Current Trends in Eye Tracking Research*,
DOI 10.1007/978-3-319-02868-2_9, © Springer International Publishing Switzerland 2014

word recognition (O'Regan 1990). In reality, fixation locations are normally distributed with means somewhere between the beginning and the centre of the word. This latter position is referred to as the preferred viewing location (PVL; Rayner 1979). The eye frequently undershoots or overshoots the optimal position or even the word boundary with the consequence that extra fixations and/or movements back to previously read words may need to be made (Nuthmann et al. 2005).

All current accounts of word targeting in reading assume that the writing system serves up unambiguously delineated word "blobs" that act as targets for the saccade programming mechanism. This solves several problems at one fell swoop. There is no need, for example, to invoke a word segmentation algorithm to extract the individual words for targeting. However, it raises the question of what happens when the words in a writing system are not so conveniently delineated. What strategies do readers adopt when reading unspaced writing systems such as Thai or Chinese? This is still very much an open research question.

3 Computational Models

Over the last 20 years, there has been a burgeoning of computational models in the field of eye movements and reading. This growth has occurred partly as a result of the wealth of data generated by modern eye-tracking technology and partly because of the need to manage theory development for what turns out to be a complex interplay of cognitive, perceptual, and motor processes. Up until the early 1990s, the main types of theory in the field were informal, verbally specified ones. Morrison's (1984) model is a good example of this genre. While providing a plausible account of the phenomena of saccade targeting and word skipping in reading, Morrison's model still omitted crucial aspects of the process. For example, it could not account for spillover effects, where the processing on one fixation affects another. It was also hard to infer reliably testable predictions from the model because of the complex parallel interaction of different processes (e.g. word recognition and saccade preparation occur in parallel in the model). The management of this complexity clearly called out for computational modelling. Moreover, a computational model was required rather than a purely mathematical one, since it effectively involves the integration of a number of distinct mathematical models into a process-based account instantiated as a computer program. As Norris (2005) put it:

In research on word recognition, models don't just resolve debates over what theories predict, they are often the only way that even the theorists themselves can be sure what their theories predict. (Norris 2005, p. 333)

While he was referring specifically to visual word recognition (VWR) models, his remarks can be applied just as emphatically to models of eye movements in reading. Tellingly, the computational instantiation of the Morrison model as E-Z Reader (Reichle et al. 1998) led to several important changes in the original model's formulation arising from inconsistencies in the relative timing of several of the component processes. These inconsistencies would have been hard or impossible to detect without the support and constraints provided by a computational implementation.

With the use of computational models now the norm in the field, a new issue arises regarding the precise relationship between model and motivating theory that it instantiates. Because of the complexity of models involved, the relationship can often be ambiguous and is frequently under-specified by the model designers. This can, in turn, lead to confusion about the testability of certain features of a model and the implications that this might have for its veridicality.

For example, in the case of the very successful interactive activation (IA) model of VWR (McClelland and Rumelhart 1981; Rumelhart and McClelland 1982), one of its features not especially deliberated over by the authors and by implication not attributed much theoretical weight was the input letter representation. The approach adopted involved having identical banks of features and letters replicated across the visual field. To most researchers, this was taken as a computational convenience and a way of skirting around what is still a hard problem in computational vision, namely position- and scale-invariant object recognition. Nonetheless, the letter representation approach adopted by the IA model became the focus of some criticism and empirical evaluation by other researchers (e.g. Mewhort and Johns 1988; Humphreys et al 1990; Davis and Bowers 2004).

The critique that certain letter-migration errors were precluded by the input's design was indeed valid and the results of Davis and Bowers' (2004) experiments were informative. Similarly, the studies of Humphreys et al. (1990) were on a much broader canvas than merely a critique of the input format for the IA model. However, one felt, to a large extent, that both critiques missed the point, since the core assumptions of the IA model were not heavily dependent on the precise nature of the letter representation used. A more central property of the whole family of IA models was that of interactivity between word, letter, and feature levels. Undermining the centrality of this property to the model's performance could be seen as much more damaging. In fact, Norris (1994) appeared to do just that with his Shortlist speech perception model. The Shortlist model, using a purely bottom-up architecture, demonstrated effects that had required top-down influences in the TRACE model of speech perception (a member of the IA family of models; McClelland and Elman 1986).

The case of the IA models is a good illustration both of how computational models can act as an important stimulus for research and of how there is little agreement about what constitutes a damaging critique. Notwithstanding Norris' (1994) findings, the IA framework of models has proved very productive in many varied cognitive modelling domains (e.g. Grainger and Jacobs 1998). The Glenmore model of reading (Reilly and Radach 2006), for example, is an IA model that can account within one mechanism for basic patterns of eye movement behaviour and accommodate a wide range of well-established empirical phenomena including parafoveal preview effects (Rayner and Pollatsek 1989).

The E-Z Reader model mentioned earlier (Reichle et al. 1998) can be regarded as a simulation of reading when higher-level linguistic processing is running smoothly (Reichle et al. 2003). The model aims to explain how lexical processing influences the progress of the eyes through the text and provides a framework for understanding word identification, visual processing, attention, and oculomotor control that determines when and where the eyes move during reading (Reichle et al. 2003).

E-Z Reader works on the hypothesis that linguistic processing affects eye movements in two different ways. First, there is a relatively low-level linguistic process that keeps the eyes moving forward. Second, higher-level processing occurs in parallel with this low-level processing and is effective when the higher-level processing is having difficulty.

The SWIFT model (Engbert et al. 2005) embodies a number of features that set it apart from E-Z Reader. Most notably, the model assumes the parallel processing of several words in a given fixation, the number of such words being constrained by the extent of the perceptual span. In contrast, E-Z Reader assumes serial processing of words. Another distinguishing feature of SWIFT is that the triggering of a saccade is autonomous from word recognition. E-Z Reader, on the other hand, assumes word recognition or at least partial recognition of a word drives the reading process forward.

There is still, however, a gulf between the modelling architectures used to account for reading data (e.g. Glenmore, SWIFT, and E-Z Reader) and those necessary to account for the neural basis of reading. Indeed, the very existence of this gulf allows the proliferation of models that are, in my view, difficult to chose among on the basis of behavioural data alone. What are required are additional constraints from the neural substrate that the models purport to abstract from.

4 Electroencephalogram

Focussing on the neural foundations of reading is a significant challenge for a number of reasons: (a) Reading is an active process, so paradigms that restrict eye movements degrade the process under investigation; (b) reading involves coordinated activity in a variety of brain regions from the retina and primary visual cortex to integration areas to language areas; and (c) both inter-subject and intra-subject variability is high across various aspects of reading. Despite these difficulties, great progress has been made using oculomotor recording, EEG, and functional magnetic resonance imaging (fMRI). The primary limitation of fMRI is that its timescale is orders of magnitude slower than the timescales of interest in the reading brain.

EEGs are recordings of minute low-frequency electrical potentials on the surface of the skull produced by neural activity within the brain (typically 10–300 μV, 0.5–40 Hz). EEGs on the surface of the scalp reflect the synchronous activity of large populations of cortical neurons—of the order of 10,000 or so—with similarly aligned current flow. Variations in the magnitude of these potentials in the same spatial location are assumed to reflect underlying cortical processing activity. The primary limitation of EEG is that it is difficult to link the noisy EEG responses to particular aspects of brain function.

All current computational models of reading tend to be behavioural rather than neuroscientific, constructed on an empirical base derived mainly from studies of eye movement patterns. Unfortunately, several successful models can account more or less equally well for the same behavioural data. The most promising source of additional constraint comes from neuroscientific data, and the best current source of such data is cognitive electro-physiological studies. Such studies have the necessary temporal resolution to provide insights into the time course of reading (Barber and Kutas 2007). Curiously, to date there have been very few attempts to use this source of constraint in the development of reading models.

The benefits of … a dynamic interplay between computational models and empirical research are clearly evident in several computational models of VWR [visual word recognition] based largely on behavioral measures (reaction time and accuracy). By contrast, on the whole, there is no similar give-and-take between computational modelers and electrophysiological researchers, perhaps because computational models have been agnostic if not silent regarding the time courses of the various neurophysiological processes or the brain areas involved in VWR. (Barber and Kutas 2007, p. 100).

While Barber and Kutas' (2007) observations refer to VWR models, their comments are even more apposite for dynamic reading. Consequently, an overarching goal of co-registration research should be to help bridge the gap between current models of reading and the complex neural basis of the process. However, building that bridge will require the significant reworking of current models and the development of new EEG paradigms compatible with the model development enterprise.

Reading and Event-Related Potentials (ERPs) ERP analysis has the potential to provide us with an exquisitely precise tool for revealing the temporal dynamics of the component processes of reading. Figure 1 is a schematic representation of the locus, size, and reliability of ERP effects associated with different levels of analysis of the VWR (from Barber and Kutas 2007, Fig. 4). Note that P1 and P2 are the first two positive peaks; N400 is a robust negative potential peak occurring typically around 400 ms. LPC represent later positive components such as the P600 (Osterhout et al. 1994). The figures in boxes represent size of effects in milliseconds (i. e. the temporal responsivity of the component amplitude)—darker figures are for effects supported by more than one study. As can be seen from this figure, ERPs can be used to index the temporal stages of the subcomponents of the reading process. Note that it is not just peak amplitude that is the source of information about timing but also the point at which the ERPs diverge as a function of an experimental manipulation. There is still some debate about the precise timing and nature of word identification, since the N400 is rather late compared to timings derived from eye movement behaviour, such as fixation durations (Sereno and Rayner 2003; Kliegl et al. 2006).

Fixation-Related Potentials (FRPs) Although still very much at an exploratory stage, the co-registration of eye movements and EEG has been successfully employed by several research groups (e.g. Baccino et al. 2005; Hutzler et al. 2007; Dambacher and Kliegl 2007; Dimigen et al. 2011).

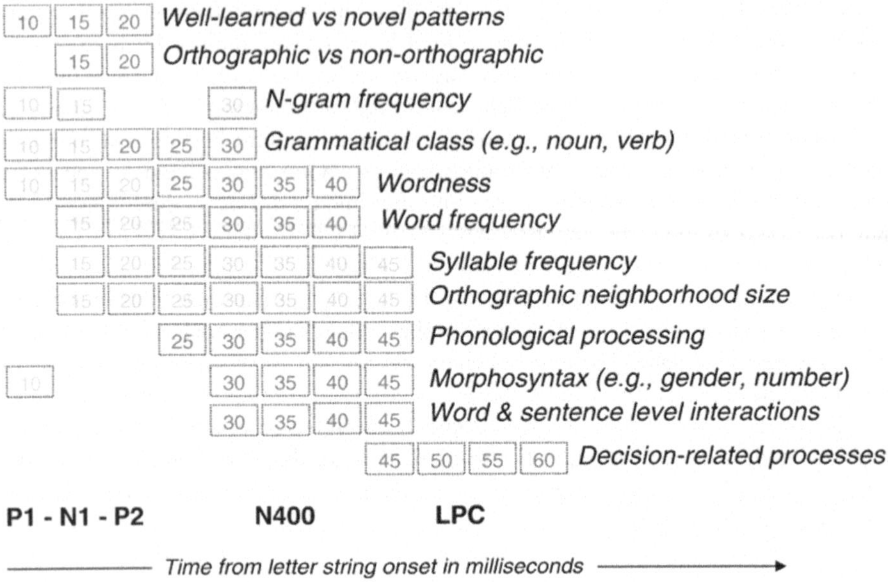

Fig. 1 Time course of effects in visual word recognition. (From Barber and Kutas 2007)

One of the disadvantages of permitting eye movements when recording EEGs is having to deal with artefacts from the movement of the eyes themselves. The eye is, in effect, a large dipole, any movement of which causes significant potential flows when viewed against the background of the much smaller scalp potentials. Nonetheless, the detection and removal of these artefacts is now relatively straightforward. For example, the use of independent component analysis (ICA) has been shown to be quite effective (Hutzler et al. 2007; Tang et al. 2002; Pearlmutter and Jaramillo 2003; Tang and Pearlmutter 2003; Henderson et al. 2013). Moreover, since the time-locking event is now the start of the fixation (hence, the term fixation-related potentials—FRPs—rather than ERPs), from the eye-tracking record we will know precisely when an eye movement has occurred.

The use of a more ecologically valid reading setting has many advantages. For example, in some reading studies involving a lexical decision task where the subject must press one of two buttons to indicate whether a word or non-word is being displayed, one can obtain P300 components that are merely associated with the need to generate a binary response (Kutas and van Petten 1994). These, in turn, can overlap with N400 responses, which are the usual focus of interest in lexical decision experiments. In contrast, by allowing free viewing and using the standard eye movement parameters of fixation duration and saccade extent, we get closer to the real reading process and avoid such procedural artefacts.

Naturally, the analysis of FRPs is not without its own significant challenges. Foremost among these is the problem of spillover of FRP components from one fixation to the next.

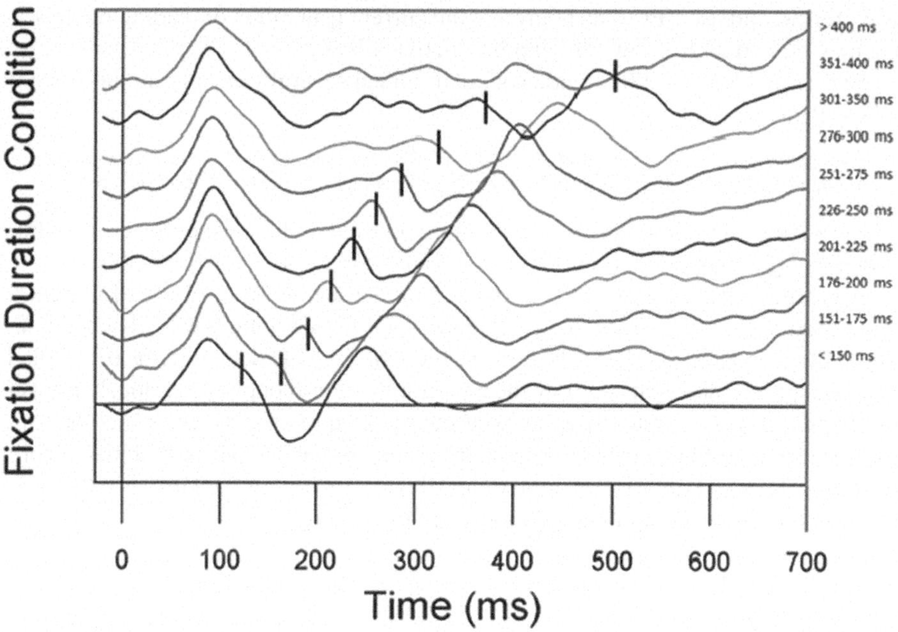

Fig. 2 Electroencephalogram (EEG) waveforms for fixations of different durations from Henderson et al. (2013). The *vertical bars* indicate fixation offset. Note that for short fixations, the P1 peak following fixation offset and associated with the new fixation can occur before some of the later components (e.g. N400) of the preceding fixations

Figure 2 from Henderson et al. (2013) gives a striking example of how spillover can make the interpretation of waveforms from natural reading problematic. In the case of, say, the 151–175-ms waveform in Fig. 2, it is impossible to distinguish the source of the negative inflection following the peak at P1. It could be the expected N1 for the second fixation, or it could have arisen from a possible N400 from the preceding fixation. A possible way to deal with this challenge is to provide context for the interpretation in the form of predictions from a computational model that would allow us to systematically unpack the various contributions to the final waveform.

5 A Synergistic Alliance

Two complementary challenges have been discussed in this chapter: (1) the need to determine which of a number of competing computational accounts of the reading process is the more plausible and (2) how to handle the complex spillover effects we find in co-registered EEG and eye movement data from natural reading experiments. However, by placing one challenge at the service of the other—in other words, use computational models to help disentangle the multiplexed EEG waveforms, and use EEG data to ground the current generation of reading models—we may end up with a powerful and productive alliance.

Barber and Kutas (2007) made a similar appeal several years ago in the context of VWR models (2007):

> We suggest that it is time that computational modelers and neurophysiologists come together in practice and in theory to unravel the mysteries of reading (Barber and Kutas 2007, p. 119).

The pressure to combine forces is even more pressing in the case of the co-registration paradigm in reading, if only to make analysing the data more tractable.

A recent approach to neurally grounded modelling at the level off EEG-generating current flows has been that of dynamic causal modelling (DCM; Kiebel et al. 2008; Stephan et al. 2007). DCM aims to account for EEG data in terms of coupled neuronal groups and analyses how the topography of the couplings variously impact on the brain's response to different experimental conditions. Bayesian methods are then used to select the most likely candidate from among competing patterns of connectivity. This and approaches at a similar level of neural granularity are the next step forward in the computational modelling of reading.

References

Baccino, T., Manunta, Y. (2005). Eye-fixation-related potentials: Insight into parafoveal processing. *Journal of Psychophysiology*, *19*, 204–215.

Barber, H. A., & Kutas, M. (2007). Interplay between computational models and cognitive electrophysiology in visual word recognition. *Brain Research Reviews*, *53*, 98–123.

Dambacher, M., & Kliegl, R. (2007). Synchronizing timelines: Relations between fixation durations and N400 amplitudes during sentence reading. *Brain Research*, *1155*, 147–162.

Davis, C. J., & Bowers, S. J. (2004). What do letter migration errors reveal about letter position coding in visual word recognition? *Journal of Experimental Psychology: Human Perception and Performance*, *30*, 923–941.

Dimigen, O., Sommer, W., Hohlfeld, A., Jacobs, A. M., & Kliegl, R. (2011). Co-Registration of eye movements and EEG in natural reading: Analyses and review. *Journal of Experimental Psychology: General*, *140*, 552–572.

Engbert, R., Nuthmann, A., Richter, E., & Kliegl, R. (2005). SWIFT: A dynamical model of saccade generation during reading. *Psychological Review*, *112*(4), 777–813.

Grainger, J., & Jacobs, A. (1998). *Localist connectionist approaches to human cognition*. Hillsdale: Erlbaum.

Henderson, J. M., Luke, S. G., Schmidt, J., & Richards, J. E. (2013). Co-registration of eye movements and event-related potentials in connected-text paragraph reading. *Frontiers in Systems Neuroscience*, *7*, 28. doi:10.3389/fnsys.2013.00028.

Huey, E. B. (1908). *The psychology and pedagogy of reading*. New York: Macmillan.

Humphreys, G. W., Evett, L. J., & Quinlan, P. T. (1990). Orthographic processing in visual word recognition. *Cognitive Psychology*, *22*, 517–560.

Hutzler, F., Braun, M., Vo, M., Engl, V., Hoffman, M., Dambacher, M., Leder, H., & Jacobs, A. M. (2007). *Brain Research*, *1172*, 124–129.

Kiebel, S. J., Garrido, M. I., Moran, R. J., & Friston, K. J. (2008). Dynamic causal modelling for EEG and MEG. *Cognitive Neurodynamics*, *2*, 121–136.

Kliegl, R., Nuthmann, A., & Engbert, R. (2006). Tracking the mind during reading: The influence of past, present, and future words on fixation durations. *Journal of Experimental Psychology:General*, *136*, 520–529.

Kutas, M., & Van Petten, C. (1994). Psycholinguistics electrified. In M. A. Gernsbacher (Ed.), *Handbook of Psycholinguistics* (pp. 83–143). San Diego: Academic Press.

McClelland, J. L., & Elman, J. L. (1986). The TRACE model of speech perception. *Cognitive psychology, 18*(1), 1–86.

McClelland, J. L., & Rumelhart, D. E. (1981). An interactive activation model of context effects in letter perception: I. An account of basic findings. *Psychological review, 88*(5), 375.

McConkie, G., Kerr, P. W., Reddix, M. D., & Zola, D. (1988). Eye movement control during reading: I. The location of initial eye fixations on words. *Vision Research, 28*, 1107–1118.

Mewhort, D., & Johns, E. E. (1988). Some tests of the interactive activation model for word recognition. *Psychological Research Psychologische Forschung, 50*, 135–147.

Morrison, R. E. (1984). Manipulation of stimulus onset delay in reading: Evidence for parallel programming of saccades. *Journal of Experimental Psychology: Human Perception and Performance, 10*, 667–682.

Norris, D. (1994). Shortlist: A connectionist model of continuous speech recognition. *Cognition, 52*(3), 189–234.

Norris, D. (2005). How do computational models help us develop better theories? In A. Cutler (Ed.), *Twenty-first century psycholinguistics: Four cornerstones* (pp. 331–346). Hillsdale, NJ: LEA.

Nuthmann, A., Engbert, R., & Kliegl, R. (2005). Mislocated fixations during reading and the inverted optimal viewing position effect. *Vision Research, 45*, 2201–2217.

O'Regan, J. K. (1990). Eye movements and reading. In E. Kowler (Ed.), *Reviews of Oculomotor Research, Vol. 4: Eye Movements and their Role in Visual and Cognitive Processes* (pp. 395–453).

Osterhout, L., Holcomb, P. J., & Swinney, D. A. (1994). Brain potentials elicited by garden-path sentences: Evidence of the application of verb information during parsing. *Journal of Experimental Psychology: Learning, Memory, & Cognition, 28*, 786–803.

Pearlmutter, B. A., & Jaramillo, S. (2003). Progress in blind separation of magneto-encephalographic data. In Independent Component Analyses, Wavelets, and Neural Networks, volume 5102 of Procceding of the SPIE, pp. 129–34.

Rayner, K. (1979). Eye guidance in reading: Fixation locations within words. *Perception, 8*, 21–30.

Rayner, K., & Pollatsek, A. (1989). *The psychology of reading*. Boston: Prentice-Hall.

Reichle, E. D., Pollatsek, A., Fisher, D. L., & Rayner, K. (1998). Toward a model of eye movement control in reading. *Psychological Review, 105*, 125–157.

Reichle, E., Rayner, K., & Pollatsek, A. (1999). Eye movement control in reading: Accounting for initial fixation locations and refixations within words. *Vision Research, 39*, 4403–4411.

Reichle, E. D., Rayner, K., & Pollatsek, A. (2003). The E–Z reader model of eye-movement control in reading: Comparisons to other models. *Behavioral and Brain Sciences, 26*, 445–526.

Reilly, R., & Radach, R. (2006). Some empirical tests of an interactive activation model of eye movement control in reading. *Cognitive Systems Research, 7*, 34–55.

Rumelhart, D. E., & McClelland, J. L. (1982). An interactive activation model of context effects in letter perception: II. The contextual enhancement effect and some tests and extensions of the model. *Psychological review, 89*(1), 60.

Sereno, S. C., & Rayner, K. (2003). Measuring word recognition in reading: Eye movements and event-related potentials. *Trends in Cognitive Sciences, 7*(11), 489–493.

Stephan, K. E., Harrison, L. M., Kiebel, S. J., David, O., Penny, W. D., & Friston, K. J. (2007). Dynamic causal models of neural system dynamics: Current state and future extensions. *Journal of Biosciences, 32*, 129–144.

Tang, A. C., & Pearlmutter, B. A. (2003). Independent components of magneto-encephalography: Localization and single-trial response onset detection. In Z.-L. Lu & L. Kaufman (Eds.), *Magnetic source imaging of the human brain* (pp. 159–201). Lawrence Erlbaum Associates.

Tang, A. C., Pearlmutter, B. A., Malaszenko, N. A., Phung, D. B., & Reeb, B. C. (2002). Independent components of magnetoencephalography: Localization. *Neural Computation, 14*(8), 1827–1858.

Oculomotor Control, Brain Potentials, and Timelines of Word Recognition During Natural Reading

Reinhold Kliegl, Michael Dambacher, Olaf Dimigen and Werner Sommer

1 Introduction

Reading is an outstanding achievement of the human brain. The ability to read has substantially formed our history and culture and plays an essential role in our everyday lives. Skilled readers can hardly prevent processing a written stimulus and, most often, they grasp a word's meaning within a fraction of a second. The understanding of the underlying mechanisms as well as the time course of this seemingly effortless skill is the goal of psycholinguistic research.

In general, two major sources of information contribute to language comprehension. First, bottom-up processes transmit neural codes of sensory input to increasingly complex levels. Upon success, the appropriate word representation in long-term memory is activated and semantic information associated with a word becomes available. Second, language-related knowledge and expectancies are expressed in top-down influences that guide the way words are understood. They permit the integration of word meaning into a wider context and hold the potential to bias expectations about upcoming words.

Despite ample evidence for the relevance of bottom-up and top-down processes, their joint role in the timeline of word recognition is insufficiently understood. Clearly, bottom-up processes account for the elaboration of sensory signals and therefore reflect operations giving rise to the retrieval of a word's mental representation, i.e., lexical access. The role of top-down processes, however, is ambiguous. They may be slow and only play a role for mental operations after lexical access;

R. Kliegl (✉)
University of Potsdam, Potsdam, Germany
e-mail: kliegl@uni-potsdam.de

M. Dambacher
University of Konstanz, Konstanz, Germany

O. Dimigen · W. Sommer
Humboldt University, Berlin, Germany

M. Horsley et al. (eds.), *Current Trends in Eye Tracking Research*,
DOI 10.1007/978-3-319-02868-2_10, © Springer International Publishing Switzerland 2014

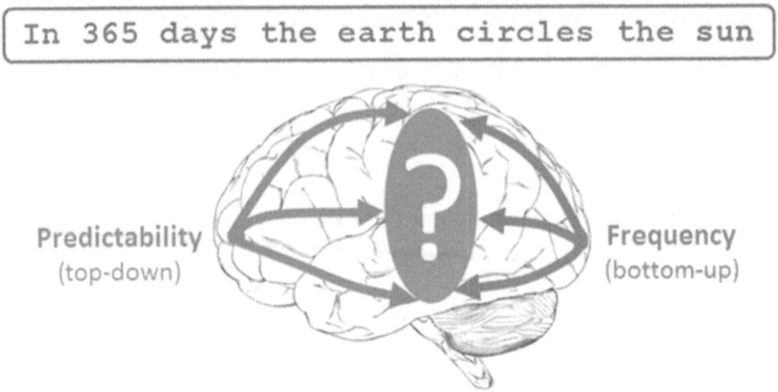

Fig. 1 Illustration of top-down and bottom-up processing in reading: word predictability indexes top-down and word frequency for bottom-up processes

alternatively, they may rapidly impinge on early lexical processes and co-determine the course of word identification.

As illustrated in Fig. 1, when reading the words "In 365 days the earth circles the" the word "sun" will be predicted on the basis of our top-down knowledge. At the same time, when the eyes have landed on the second "the" of this sentence, the perceptual span (i.e., the area from which we process information "bottom-up"; McConkie and Rayner 1975; Rayner 1975) will include the word "sun." Disagreements about bottom-up and top-down processes in reading mainly relate to their dynamics, that is, about the timelines of their interaction and their localization within the brain. The co-registration of eye movements (EMs) and brain potentials is a promising research strategy to test theoretical proposals and resolve some of these disagreements.

2 Word Frequency and Predictability Affect Eye Movements and Brain Potentials

Two major determinants of the speed of word recognition can be considered as proxies of bottom-up and top-down streams in reading, respectively: word frequency and predictability. Normative word frequency is operationalized by counting the occurrences of a word in a large collection of documents. A quantitative index of word predictability is obtained with the cloze task. In this task, subjects are asked to guess the next word, given the previous words of a sentence (Taylor 1953)—the proportion of correct guesses serves as an index of predictability (cloze probability); that is, words are predicted from the prior context in the absence of the visual word form.

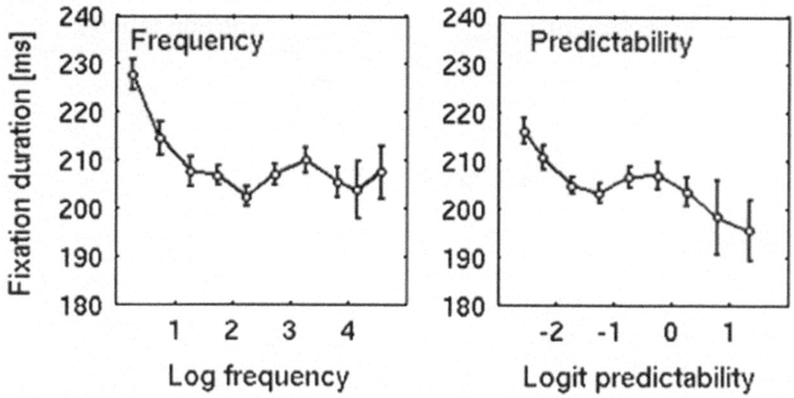

Fig. 2 Frequency and predictability effects on single fixation durations (FDs) during reading of German sentences. (Kliegl et al. 2006)

3 Evidence from Eye Movements (EMs)

There is ample evidence that highly predictable words and highly frequent words require less processing time. Fig. 2 illustrates these effects for single fixation durations (FDs) collected during the reading of German sentences (Kliegl et al. 2006; see also Ehrlich and Rayner 1981, Rayner and Duffy 1989). These effects relate to the fixated word. However, FDs on a given word are also influenced by the frequency and predictability of the neighboring words. Thus, FDs inform us about the preprocessing as well as postprocessing of words, when the eyes have already moved on to the next word. Such results are evidence for distributed processing of words across several fixations. These effects are typically small and are not found in every study (see exchange between Rayner et al. 2007; Kliegl 2007). In our opinion, the question is no longer whether or not such distributed processing occurs; the task is rather to determine which information is used under what conditions.

4 Evidence from Event-Related Brain Potentials

Frequency and predictability effects have also been demonstrated with event-related potentials (ERPs). One of the signature results of neurolinguistic research is the N400-component of the ERP, which is associated with the difficulty of retrieving stored word information (Kutas and Federmeier 2011; Kutas et al. 2006). In response to anomalous or unexpected words in a sentence (e.g., "The pizza was too hot to *drink*"), a pronounced negativity emerges about 200–250 ms after presentation of the anomalous target word (Kutas and Hillyard 1980), reaching a maximum at about 400 ms. The N400-component is also sensitive to variations in cloze probability as they occur while reading normal sentences without semantic violations

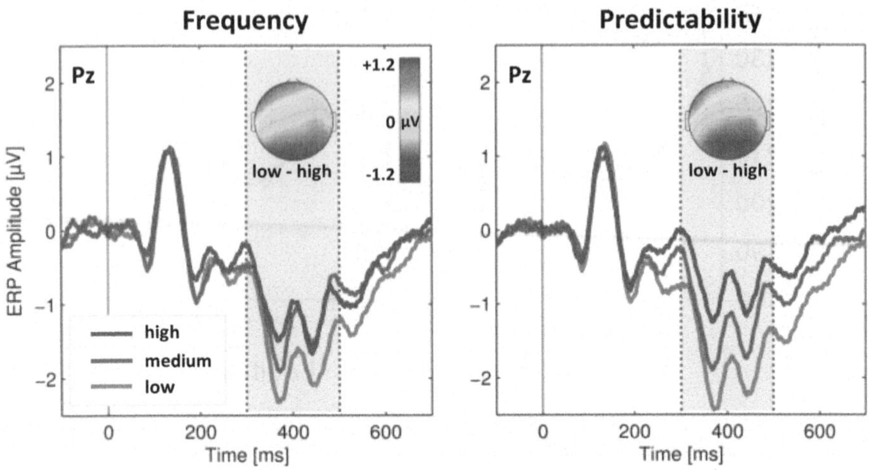

Fig. 3 Frequency and predictability effects on N400-ERP amplitudes during reading German sentences. (Dambacher et al. 2006)

(Dambacher et al. 2006; see right panel in Fig. 3, based on the sentences used by Kliegl et al. 2006).

Dambacher et al. (2006) also reported larger N400 for low-frequency words compared to medium- and high-frequency words (Dambacher et al. 2006; left panel of Fig. 3; see also Rugg 1990; Van Petten and Kutas 1990). Further, there is some evidence that frequency exerts influences on the ERP time course already before 200 ms (Hauk and Pulvermüller 2004; Sereno et al. 1998).

5 Co-Registration of Eye Movements and Brain Potentials

Natural Reading Versus Serial Visual Presentation FDs are strongly affected not only by the properties of the currently fixated word but also by word frequency and predictability of the neighboring words (Kliegl et al. 2006). Dambacher and Kliegl (2007) assembled FDs measured on content words by Kliegl et al. (2006) during natural reading and ERPs measured on the same words by Dambacher et al. (2006) during the "traditional" ERP procedure, where words were presented serially at a fixed rate of 700 ms per word in the center of the display. Using path analysis based on words as units of observations, they were able to show that frequency and predictability of these content words mediated the covariation of FD and N400 amplitudes across successive words in sentences.

Obviously, however, serial visual presentation (SVP) of words differs from natural reading in several important ways. First, fixation times are artificially inflated; a stimulus onset asynchrony (SOA) of 700 ms is about three times the duration

Table 1 Methodological problems of co-recording of eye movements and EEG from Dimigen et al. (2011)

Problem	Solution
Technical	
Data synchronization	Shared trigger pulses or D/A card
50-Hz line noise	Notch filter
Artifacts from head stabilization	Foam cushions or remote eye tracking
Eye-movement artifacts	
Corneoretinal dipole changes	Multiple source eye correction (MSEC)
Extraocular muscles (spike potential)	Independent component analysis, with component selection supported by eye-tracking data
	Constrain analysis to one fixation duration
Overlapping signals	
Temporal overlap with "neighbor" potentials	(Post hoc) matching of similar fixations
	Deconvolution methods
No neutral baseline	Modeling
Low-level neural response variation	
Saccade amplitude	(Post hoc) matching of similar fixations
Luminance around fixated spot	Modeling, e.g., with linear mixed models

of a normal fixation. Second, the SVP paradigm eliminates the need to select the next word target to which to saccade. Consequently, there is also no skipping of upcoming words or regressions to previous words in a sentence. Third, with central presentation of words in one location there is no preview of words to the right of fixation. Although some recent studies attempted to combine SVP with preview of the upcoming word, most SVP paradigms eliminate the core issue of much research on eye-movement control during reading: parafoveal preprocessing.

There are several reasons why most psycholinguistic ERP research uses the SVP paradigm rather than natural reading conditions. First, the SVP paradigm eliminates EMs. This is of paramount concern because EMs generate artifacts due to corneoretinal dipole changes and due to spike potentials (SPs) generated by extraocular muscles. These artifacts are often several magnitudes larger than the neurolinguistic experimental effects one hopes to identify with experimental manipulations of frequency or predictability. Second, the long SOAs reduce the temporal overlap with potentials triggered by the previous or the following word. The left column of Table 1 summarizes these problems. The table lists additional problems and solutions available or under active development. As far as these are not self-explanatory (e.g., the technical problems and their solutions), we refer readers to several recent expositions (e.g., Baccino 2011; Dimigen et al. 2009, 2011, 2012).

Despite the differences between natural and SVP reading, there is no doubt that ERP research has delivered important insights about timelines when various types of lexical and sublexical information become available (see, e.g., Kutas and Federmeier 2011; Lee et al. 2012, for reviews). Nevertheless, since we are interested in eye-movement control during reading with a focus on saccade target selection and parafoveal processing as well as in the timelines uncovered by ERP research, we need to tackle co-registration. Therefore, we adopted a two-pronged approach for our research program. On the one hand, we have started to co-register EMs and brain potentials during natural reading (Dimigen et al. 2011, 2012). On the other hand, we have also stayed with the SVP paradigm, but systematically varied SOA, using both slow presentation rates (SOA of 700 and 490 ms) and a fast presentation rate (280 ms) that approximates the speed of natural reading (Dambacher et al. 2012). The general goal of this concerted effort addresses the important question of how timelines uncovered by traditional ERP research depend on SOA and whether there are task-specific dependencies that need to be taken into account.

Fixation-Related Potentials During Natural Reading The first step was to demonstrate that EEG recordings are feasible during natural reading; in other words, the problems listed in Table 1 are indeed surmountable. Using again the Potsdam sentences, which already featured in Figs. 2 and 3, Dimigen et al. (2011) reported a fairly comprehensive experiment and data analysis. Using the multiple source eye correction method (MSEC; Berg and Scherg 1994; Ille et al. 2002), they were able to remove most of the corneoretinal artifacts. MSEC was only partially successful with the muscle SPs at the saccade onset (Keren et al. 2010). However, due to some topographic overlap with the corneoretinal dipole artifact, the MSEC procedure attenuated the contribution of the SP.

Fixation-Related Potentials In traditional EEG research, the onsets of stimuli serve as the reference events at which EEG segments are aligned and averaged across trials, yielding the ERP. Analogous to this procedure, during natural reading, the onsets of fixations are used as synchronizing events. We refer to the averaged electrical potential that results from aligning EEG epochs with reference to the onsets of critical fixations as fixation-related potential (FRP). In addition, one can also calculate averaged potentials synchronized to the onsets of saccades (saccade-related potentials, SRPs).

Deflections of the FRP The maps at the top of Fig. 4a illustrate scalp topographies of five interpretable FRP deflections: First, the visually evoked lambda response to the previous fixation *N*-1 occurs at around 150 ms prior to the current fixation *N*. Second, the muscular SP emerges at saccade onset. Third, the lambda response evoked by current fixation *N* peaks after around 100 ms at parieto-occipital electrodes. Fourth, we observed an equivalent of the N170 component known from ERP research. Fifth, the N400 component overlapped with the lambda response to the next fixation $N + 1$. In general, the visually evoked FRP potentials show the P1–N1 complex known from ERP research. Before ocular artifact correction, many of the EEG channels were strongly correlated with the participants horizontal gaze position, as measured with the electrically independent eye tracker. After MSEC correction, correlations were nonsignificant or very small. In this way, simultaneous eye tracking can be used as an objective external criterion to evaluate the success of artifact correction.

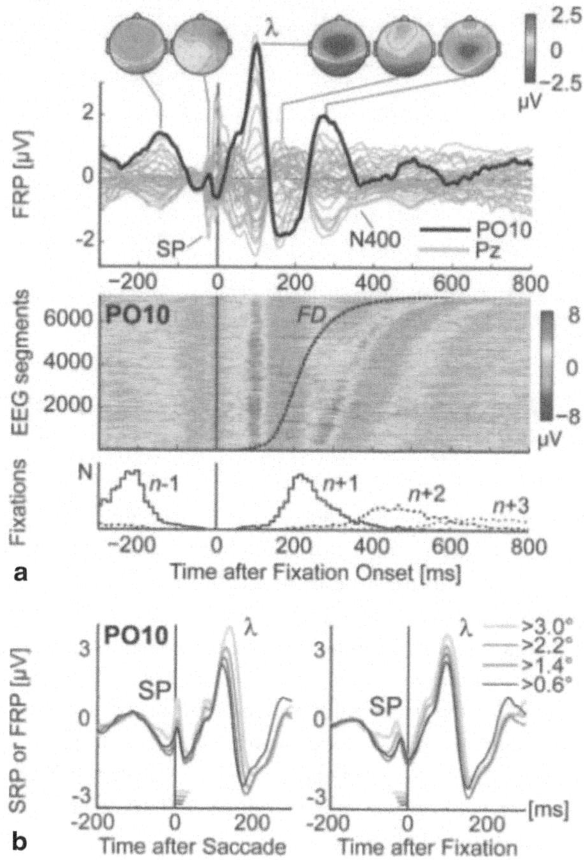

Fig. 4 The fixation-related potential (FRP): illustrations of postsaccadic neural response (for explanation, see text). (Dimigen et al. 2011)

Neural Responses and FDs In the lower panel of Fig. 4a each of the 7,113 EEG segments contributing to the average FRP shown on the top is represented by one horizontal line. The segments are sorted by FD in increasing order. Colors represent EEG amplitudes at every sampling point that were smoothed vertically with a moving average across 50 adjacent segments. The figure shows that the positive peak at around 280 ms in the average FRP is partially affected by the lambda response to fixation $N + 1$. The bottom panel of Fig. 4a displays the distributions of onset latencies of the preceding and succeeding fixations relative to the onset of the current fixation N.

SRPs and Saccade Amplitudes Finally, Fig. 4b shows that the morphology of SRPs and FRPs depend on saccade amplitude: Deflections of both the premotor SP and the visually evoked lambda response increased with saccade size. Because the lambda response largely dominates the shape of FRPs during left-to-right reading, saccade amplitudes can critically affect FRPs even after the correction of EM artifacts.

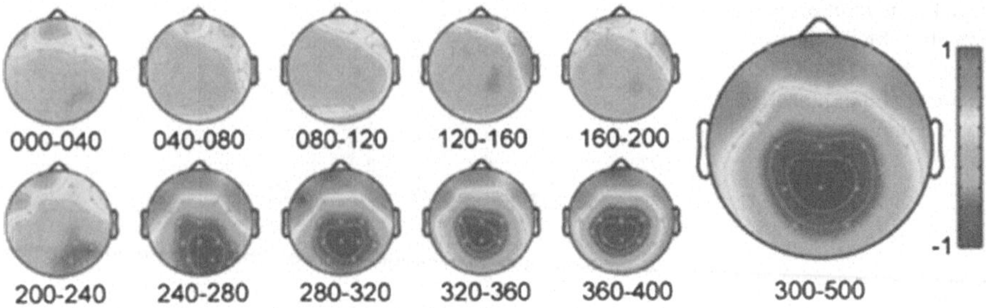

Fig. 5 N400 effects of predictability (*low minus high*) in natural reading start early in fixation-related potentials (FRPs). (Dimigen et al. 2011)

6 The N400 Predictability Effect in FRPs During Natural Reading

The N400 enjoys the status of a benchmark result in psycholinguistic research. Therefore, the first goal of a new methodology, such as the co-registration of EMs during reading, must be to replicate this signature result. As shown in Fig. 5, this goal was indeed reached (see also Marton et al. 1985). The figure displays scalp distributions for differences between low- and high-predictability words for successive 40-ms windows after fixation onset. The large map at the right displays the aggregate for the usual N400 range. After correction for multiple testing, results were significant from 248 ms onwards, which is quite consistent with reports of N400 predictability effects in earlier SVP experiments (Kutas et al. 2006). An inspection of the maps, however, clearly suggests that the effect (a central negativity) emerges already somewhat earlier.

7 Experimental Manipulation of Frequency and Predictability

With the demonstration of a key result of psycholinguistic research, that is, the N400 predictability effect on FRPs, we are in a position to address the question whether the timelines of the effect are different in SVP studies. Thus, we are moving back to ERP research, but with the difference that we are looking at ERP effects at different SOAs (Dambacher et al. 2012). The motivation to examine SOA effects is straightforward: Traditional SVP experiments present one word at a time with relatively long intervals (usually from 500 up to 1000 ms) between successive stimuli. Natural reading, in contrast, takes place at a much faster pace of around 250 ms per word. The SOA manipulation therefore tests whether the rate of incoming words is a relevant factor for the time course of word recognition.

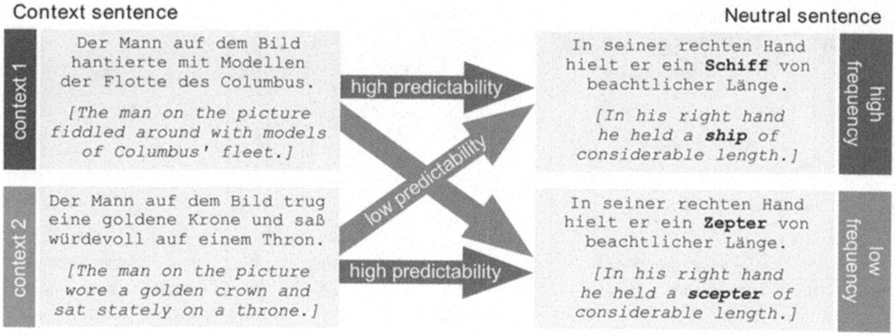

Fig. 6 Experimental paradigm for orthogonal manipulation of frequency and predictability with target words. A set of low- and high-frequency words is presented in high- and low-predictability contexts. (Dambacher et al. 2009)

Moreover, because frequency and predictability are positively correlated in everyday language (cf., Dambacher and Kliegl 2007), we also aimed for uncorrelated effects of predictability and frequency in this research. This required the construction of new experimental material, in which high- and low-frequency target words were embedded in identical neutral sentence frames. Predictability of these targets was manipulated by a preceding context sentence that induced high expectancy either of a high- or a low-frequency target, while its counterpiece was of low predictability (see Fig. 6 for an illustration).

With this corpus, we conducted three SVP experiments with different presentation rates: SOAs of 700 and 490 ms reflected the medium and lower range of presentation times usually used in ERP experiments. The third SOA of 280 ms approximated normal reading speed. In another line of ongoing experiments, this corpus is also used to assess FRPs during normal left-to-right reading.

8 Experimental Manipulation of SOA in the SVP Paradigm

8.1 The ERP Time Course Depends on SOA

The first three experiments were devoted to documenting the SOA dependencies of predictability and frequency effects (Dambacher et al. 2012). SOA effects on the timeline of word recognition are not well established because fast SOAs have been rarely used in psycholinguistic ERP research. The reason is that overlapping components of successive words usually complicate the interpretation of neural responses since they cannot be attributed to individual words. Here, this problem was minimized since the words preceding and succeeding the target are physically

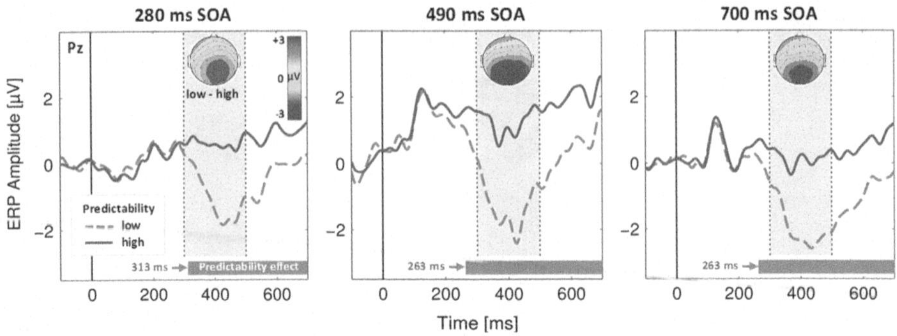

Fig. 7 Onsets of N400 predictability effect for (rapid) serial visual presentations (SVPs) with stimulus onset asynchronies (SOAs) of 700, 490, and 280 ms. (Dambacher et al. 2012)

identical across all conditions and only differ in the expectancy induced by the context sentence.

Predictability Effects Fig. 7 illustrates that SOA affected the N400 predictability effect. Its onset was later and amplitudes were smaller in the SOA of 280 ms compared to the SOAs of 490 and 700 ms. One possible reason for the weaker N400 effect under a reading-like SOA is that the N400 temporally overlapped with the processing of the next word. Thus, two words may have been processed in parallel such that N400-related mechanisms were attenuated and/or decelerated. In fact, the offline comparison of SVP-based N400 amplitudes and FDs from natural reading revealed a cross talk between the currently fixated word and ongoing processing of the previous word (Dambacher and Kliegl 2007). It is also possible, though, that the N400 effect was smaller because the short SOA granted less time to build up expectations about the upcoming word. This limited semantic preactivation may result in a smaller predictability effect (e.g., Federmeier and Kutas 1999; Lau et al. 2008).

At the same time, however, the data of the short SOA revealed evidence that predictability supported lexical processing well before the N400. First, we observed a very early main effect of predictability at around 90 ms poststimulus, which was reliable even within high- and low-frequency words. We suggested that this effect may point to a rapid verification mechanism, i.e., the comparison of the incoming with the expected word (Dambacher et al. 2009). In fact, similar early expectancy effects have been reported with spoken language (Van Berkum et al. 2005). Second, an interaction of frequency and predictability at around 145 ms pointed to an early processing advantage of highly predictable words. Notably, there was no evidence for such early predictability effects with the slower SOAs of 490 and 700 ms.

Frequency Effects Irrespective of SOA, Dambacher et al. (2012) found no evidence for frequency effects in the N400 range. However, frequency effects at earlier intervals indicated that the time course of lexical processing is modulated by SOA. Specifically, frequency effects emerged at 207, 174, and 131 ms with SOAs of 700,

490, and 280 ms, respectively. There is one qualification: At the shortest SOA, the effect was only observed for high-predictable words. These results suggest that lexical processing is accelerated as presentation rate increases. In particular, the shortest SOA rate seems to play a special role for word recognition since only here early lexical processing was supported by predictability. Thus, normal reading speed may indeed reflect the optimal rate of visual word processing since it combines several sources of information to grant rapid lexical access. We want to emphasize that we consistently find frequency effects with robust and highly comparable scalp distributions at the latest around 200 ms. Analyses of frequency effects during natural reading are currently ongoing.

N400 Predictability Effects on FRPs During Natural Reading In experiment 4 of this series, we co-registered EEG during natural reading of these sentences. Our preliminary analysis reveals predictability effects with an N400-like topography beginning as early as 160 ms. Thus, counter to the delay at the shortest relative to longer SOAs, there appears to be a mechanism that supports early retrieval of semantic information of upcoming words in left-to-right reading. One plausible candidate is parafoveal preview of the not-yet-fixated word, which is maintained in natural reading but prevented in SVP settings. In fact, the preview benefit effect, i.e., the reduction of FDs on words that have been visible from the parafovea compared to words that were parafoveally inaccessible, is well established in EM research and the brain correlates of the preview benefit effect has been recently tracked with FRPs (Dimigen et al. 2012).

9 Outlook: Gaze-Contingent Display Change Experiments

The sentence material, comprising an orthogonal manipulation of frequency and predictability of content words, has already proved very useful to unravel SOA effects on timelines of predictability and frequency effects in ERPs during reading in the SVP procedure and, in ongoing research, in FRPs during natural reading. These results strongly suggest that readers make use of properties of words as soon as they are available in parafoveal vision in agreement with eye-movement research about parafoveal processing during reading. This research comprises multivariate analyses of the majority of words in sentences during natural reading (e.g., Kliegl et al. 2006) and the analyses of specific target words for which critical natural correlations have been eliminated by construction, such as the correlation between frequency and predictability (Dambacher et al. 2009; Fig. 6).

In eye-movement research, statistical analyses, covering most words of sentences or a subset of target words, are complemented by experimental manipulations involving gaze-contingent display changes, such as the boundary paradigm (Rayner 1975). In the boundary paradigm, a target word (such as the target words of Fig. 6) is either displayed all the time (like in natural reading) or its location is filled with a random string of letters and the word is displayed first during the saccade to the target word. Obviously, in this latter condition, preview of the target word is prevented just like in SVP. Typically, FDs are 20–50 ms longer compared to natural reading. This difference in FDs is called preview benefit.

Gaze-contingent display changes can also be employed during co-registration of EMs and EEG brain potentials (see Dimigen et al. 2012, for a first application and discussion of feasibility). Indeed, we already carried out two experiments to test predictability- and frequency-related preview effects for the target words used in the last set of experiments. In the first experiment, words were displayed during a saccade to the target word (i.e., the "classic" N + 1-boundary paradigm); in the second experiment, words were displayed already when the eyes approached the word before the target word (i.e., the N + 2-boundary paradigm). Preliminary analyses suggested that preview effects were significant in FRPs for the N + 1-boundary condition; they were not significant for the N + 2-boundary paradigm. Specifically, if word N + 1 is masked, the N400 predictability effect was delayed—similar to the SOA 700 and SOA 490 timelines of Fig. 7; if word N + 2 was masked, the predictability occurred already after 160 ms, that is, at the time when the effect appeared during natural reading. Thus, again, it appears that parafoveal preview exerts effects on timelines of word recognition during reading.

Clear evidence for the preview benefit in FDs and FRPs was also reported for reading a short list of five words with the goal to check for an animal name in the list (Dimigen et al. 2012). Like the sentence-reading studies, this study found that N400 effects are shifted forward during natural reading. As shown in Fig. 8, effects of identity priming from showing repeated words in a list arose 80–120 ms after fixation onset. Similarly, semantic priming effects in the FRP started only 160 ms after fixation onset. Both latencies are significantly shorter than those typically observed for priming effects in SVP reading or in foveal priming studies that do not allow for parafoveal preview (e.g., Rugg 1987). This experiment also tested whether parafoveal preprocessing extends to semantic processing, but did not find any significant evidence in support of this hypothesis. Although semantic priming effects started early, they only began after a semantically related word was directly fixated. Similarly, in other trials of the experiment, which manipulated preview with the boundary paradigm, a semantically related preview word did not lead to shorter FDs or different brain responses than an unrelated preview word. These results stand in contrast to demonstrations of semantic relatedness with the boundary paradigm during reading of sentences (i.e., four experiments in Hohenstein and Kliegl 2012; see also Hohenstein et al. 2010, for evidence with a different, that is, parafoveal fast priming paradigm). Obviously, there are many empirical issues that remain to be sorted out, but we close with the observation that our research program on co-registration of EMs and brain potentials during reading closely recapitulated the development of eye-movement research on reading in a time-compressed manner.

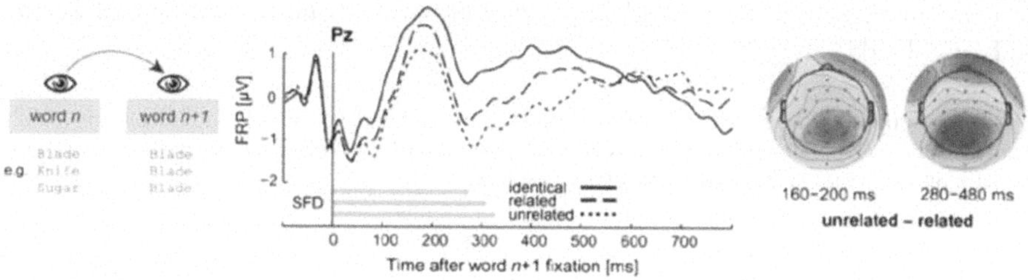

Fig. 8 Priming effects on the N400 component while reading word lists. Within lists of five unrelated German nouns, two neighboring words were either semantically related (e.g., "knife"–"blade"), unrelated (e.g., "sugar"–"blade"), or identical (word repetition, e.g., "blade"–"blade"). The difference between fixating related versus unrelated word pairs modulated the N400 already 160 ms after fixating the second word (see scalp topography). The unusually early onset of semantic priming in fluent reading is most likely explained by the fact that readers had already preprocessed orthographical or phonological properties of the second word parafoveally. Figure adapted from Dimigen et al. (2012).

10 Summary

Our research on top-down predictability and bottom-up frequency effects on eye movements and brain potentials during reading covers experimental paradigms ranging from natural reading to SVP at different SOAs to gaze-contingent display changes. The experimental paradigms are needed to advance our theoretical understanding of the dynamics of word recognition during reading, but they also appear to incur paradigm-specific effects:

- Fixation-related EEG analyses in ecologically valid settings are feasible
- Care is necessary, since FRPs are influenced by a number of potentially confounding factors (e.g., artifacts, overlapping potentials, incoming saccade amplitude)
- N400 topography and amplitude are similar in natural reading and SVP, but there are large differences in timelines
- SOA manipulations in the SVP paradigm and gaze-contingent display-change manipulations (such as the boundary paradigm) show that this is (largely) explained by parafoveal preprocessing; this limits generalizability of ERP data recorded in the SVP paradigm for natural reading

Our understanding of eye-movement control during reading is shaped strongly by computational models of saccade generation during reading (e.g., Engbert and Kliegl 2011; Reichle 2011; Reilly and Radach 2009). Timelines of predictability and frequency effects and their interactions, as revealed by the co-registration of EMs and brain potentials, are important for the further development of these models, because they greatly reduce the degrees of freedom available for simulating this complex human activity. Most importantly, we present this research program as an example that it is possible to overcome this traditional fractionation of research domains in cognitive and biological psychology to the benefit, we hope, of everybody involved.

Acknowledgments This research was supported by DFG Research Unit 868, Project A2.
Address for correspondence: Reinhold Kliegl, Department of Psychology, University of Potsdam, Karl-Liebknecht-Str. 24/24, 14476 Potsdam, Germany.

References

Baccino, T., & Manunta, Y. (2005). Eye-fixation-related potentials: Insight into parafoveal processing. *Journal of Psychophysiology*, *19*, 204–215.

Baccino, T. (2011). Eye movements and concurrent event-related potentials: Eye fixation-related potential investigations in reading. In: S.P. Liversedge, I.D. Gilchrist, & S. Everling (Eds.), *Eye Movements Handbook* (pp. 857–870). Oxford: Oxford University Press.

Berg, P., & Scherg, M. (1994). A multiple source approach to the correction of eye artifacts. *Electroencephalography and Clinical Neurophysiology*, *90*, 229–241.

Dambacher, M. (2010). *Bottom-up and top-down processes in reading*. Potsdam: Universitätsverlag Potsdam.

Dambacher, M., Dimigen, O., Braun, M., Wille, K., Jacobs, A. M., & Kliegl, R. (2012). Stimulus onset asynchrony and the timeline of word recognition: Event-related potentials during sentence reading. *Neuropsychologia*, *50*(8), 1852–70

Dambacher, M., & Kliegl, R. (2007). Synchronizing timelines: Relations between fixation durations and N400 amplitudes during sentence reading. *Brain Research*, *1155*, 147–162.

Dambacher, M., Kliegl, R., Hofmann, M., & Jacobs, A. M. (2006). Frequency and predictability effects on event-related potentials during reading. *Brain Research*, *1084*, 89–103.

Dambacher, M., Rolfs, M., Göllner, K., Kliegl, R., & Jacobs, A. M. (2009). Event-related potentials reveal rapid verification of predicted visual input. *PLoS ONE*, *4*(3), e5047.

Dimigen, O., Kliegl, R., & Sommer, W. (2012). Trans-saccadic parafoveal preview benefits in fluent reading: A study with fixation-related brain potentials. *NeuroImage*, *62*, 381–393.

Dimigen, O., Sommer, W., Hohlfeld, A., Jacobs, A. M., & Kliegl, R. (2011). Co-registration of eye movements and EEG in natural reading: Analyses and review. *Journal of Experimental Psychology: General*, *140*, 552–572.

Dimigen, O., Valsecchi, M., Sommer, W., & Kliegl, R. (2009). Human microsaccade-related visual brain responses. *Journal of Neuroscience*, *29*, 12321–12331.

Engbert, R., & Kliegl, R. (2011). Parallel graded attention models of reading. In: S.P. Liversedge, I.D. Gilchrist, & S. Everling (Eds.), *Eye Movements Handbook* (pp. 787–800). Oxford: Oxford University Press.

Ehrlich, S. F., & Rayner, K. (1981). Contextual effects on word recognition and eye movements during reading. *Journal of Verbal Learning and Verbal Behavior*, *20*, 641–655.

Hauk, O., & Pulvermüller, F. (2004). Effects of word length and frequency on the human event-related potential. *Clinical Neurophysiology*, *115*, 1090–1103.

Hohenstein, S., & Kliegl, R. (2012). Semantic preview benefit during reading. Manuscript submitted for publication.

Hohenstein, S., Laubrock, J., & Kliegl, R. (2010). Semantic preview benefit in eye movements during reading: A parafoveal fast-priming study. *Journal of Experimental Psychology: Learning, Memory, and Cognition*, *36*, 1150–1170.

Ille, N., Berg, P., & Scherg, M. (2002). Artifact correction of the ongoing EEG using spatial filters based on artifact and brain signal topographies. *Journal of Clinical Neurophysiology*, *19*, 113–124.

Keren, A. S., Yuval-Greenberg, S., & Deouell, L. Y. (2010). Saccadic spike potentials in gamma-band EEG: Characterization, detection, and suppression. *NeuroImage*, *49*, 2248–2263.
Kliegl, R. (2007). Toward a perceptual-span theory of distributed processing in reading: A reply to Rayner, Pollatsek, Drieghe, Slattery, and Reichle (2007). *Journal of Experimental Psychology: General*, *136*, 530–537.
Kliegl, R., Dambacher, M., Dimigen, O., Jacobs, A. M., & Sommer, W. (2012). Eye movements and brain electric potentials during reading. *Psychological Research*, *76*, 145–158.
Kliegl, R., Nuthmann, A., & Engbert, R. (2006). Tracking the mind during reading: The influence of past, present, and future words on fixation durations. *Journal of Experimental Psychology: General*, *135*, 12–35.
Kretzschmar, F., Bornkessel-Schlesewsky, I., & Schlesewsky, M. (2009). Parafoveal versus foveal N400s dissociate spreading activation from contextual fit. *Neuroreport*, *20*, 1613–1618.
Kuperman, V., Dambacher, M., Nuthmann, A., Kliegl, R., (2010). The effect of word position on eye-movements in sentence and paragraph reading. *Quarterly Journal of Experimental Psychology*, *63*, 1838–1357.
Kutas, M., & Federmeier, K. D. (2011). Thirty years and counting: Finding meaning in the N400 component of the event-related brain potential (ERP). *Annual Review of Psychology*, *62*, 621–647.
Kutas, M., & Hillyard, S. A. (1980). Reading senseless sentences: Brain potentials reflect semantic incongruity. *Science*, *207*, 203–205.
Kutas, M., Van Petten, C., & Kluender, R. (2006). Psycholinguistics electrified II: 1994–2005. In M. Gernsbacher & M. Traxler (Eds.), *Handbook of psycholinguistics, 2nd edition* (pp. 659–724). New York: Elsevier Press.
Lee, C.-Y., Liu, Y.-N., & Tsai, J.-L. (2012). The time course of contextual effects on visual word recognition. *Frontiers in Psychology*. doi:10.3389/fpsyg.2012.00285.
McConkie, G. W., & Rayner, K. (1975). The span of effective stimulus during a fixation in reading. *Perception & Psychophysics*, *17*, 578–586.
Rayner, K. (1975). The perceptual span and peripheral cues in reading. *Cognitive Psychology*, *7*, 65–81.
Rayner, K., & Duffy, S. (1986) Lexical complexity and fixation times in reading: Effects of word frequency, verb complexity, and lexical ambiguity. *Memory & Cognition*, *14*, 191–201.
Rayner, K., Pollatsek, A., Drieghe, D., Slattery, T., & Reichle, E. D. (2007).
Reichle, E. D. (2011). Serial-attention models in reading. In: S.P. Liversedge, I.D. Gilchrist, & S. Everling (Eds.), *Eye Movements Handbook* (pp. 767–786). Oxford: Oxford University Press.
Reilly, R. G., & Radach, R. (2006). Some empirical tests of an interactive activation model of eye movement control in reading. *Journal of Cognitive Systems Research*, *7*, 34–55.
Rugg, M. D. (1987). Dissociation of semantic priming, word and non-word repetition effects by event-related potentials. *Quarterly Journal of Experimental Psychology*, *39*, 123–148.
Rugg, M. D. (1990). Event-related brain potentials dissociate repetition effects of high- and low-frequency words. *Memory and Cognition*, *18*, 367–379.
Taylor, W. L. (1953). Cloze procedure: A new tool for measuring readability. *Journalism Quarterly*, *30*, 415–433.
Sereno, S. C., & Rayner, K. (2003). Measuring word recognition in reading: Eye movements and event- related potentials. *Trends in Cognitive Science*, *7*, 489–493.
Sereno, S. C., Rayner, K., & Posner, M. I. (1998). Establishing a time-line of word recognition: Evidence from eye movements and event-related potentials. *Neuroreport*, *9*, 2195–2200.
Van Berkum, J. J. A., Brown, C. M., Zwitserlood, P., Kooijman, V., & Hagoort, P. (2005). Anticipating upcoming words in discourse: Evidence from ERPs and reading times. *Journal of Experimental Psychology: Learning, Memory, and Cognition*, *31*, 443–467.
Van Petten, C., & Kutas, M. (1990). Interactions between sentence context and word frequency in event-related brain potentials. *Memory and Cognition*, *18*, 380–393.

Measuring Neuronal Correlates of Reading with Novel Spread-Spectrum Protocols

Ronan G. Reilly

The aim of the research described here is to extend the boundaries of our understanding of the reading process using a combination of eye tracking, electrophysiological measurement, and innovations in signal processing to unpack the millisecond-to-millisecond time course of the underlying neuronal processes. In particular, we are interested in studying the dynamics of attention allocation while the reader fixates a word.

Investigating the neural foundations of reading remains a significant challenge for a number of reasons: (1) reading is an active process, so paradigms that restrict eye movements degrade the process under investigation; (2) reading involves co-ordinated activity in a variety of brain regions from the retina and primary visual cortex to integration areas through to language areas; (3) both inter-subject and intra-subject variabilities is high across various aspects of reading. Despite these difficulties, significant progress has been made using oculomotor recordings, Electroencephalography (EEG), and functional magnetic resonance imaging (fMRI). The primary limitation of fMRI is that the time scale of the blood-oxygen-related dependent (BOLD) effect is orders of magnitude slower than the time scales of interest in the reading brain. EEG, on the other hand, has better temporal characteristics. EEGs are recordings of minute low-frequency electrical potentials on the surface of the skull produced by neural activity within the brain. EEGs on the surface of the scalp reflect the synchronous activity of large populations of cortical neurons of the order of 10,000 or so with similarly aligned current flow. Variations in the magnitude of these potentials in the same spatial location are assumed to reflect underlying cortical processing activity. The primary limitation of EEG is that it is difficult to link the noisy EEG responses to particular aspects of brain function. The goal of this talk is to describe a new technique for establishing this linkage.

Visual-evoked spread-spectrum analysis (VESPA) is a recently developed signal-processing technique that offers the possibility of measuring evoked response potentials (ERPs) during free viewing (Lalor et al. 2007; Lalor et al. 2006).

R. G. Reilly (✉)
Department of Computer Science, National University of Ireland, Maynooth, Ireland
e-mail: ronan.reilly@nuim.ie

M. Horsley et al. (eds.), *Current Trends in Eye Tracking Research*,
DOI 10.1007/978-3-319-02868-2_11, © Springer International Publishing Switzerland 2014

This is achieved by 'tagging' different regions of interest, in our case words, using random luminance variation where these tags can be recovered through a simple convolution process in the subsequent ERP analysis (Lalor et al. 2007). The VESPA amplitude associated with each tag can then, in principle, tell us what word is being attended to at a given point during a fixation.

Neural visual responses are usually measured with EEG using the average response to repeated flashes. These flashes are aversive and disruptive of the oculomotor behaviour characteristic of normal reading. In contrast, the VESPA technique uses a non-aversive smoothly shimmering stimulus. The essence of the approach is to consider the relationship between the brightness of a stimulus, $x(t)$, and the potential measured at an EEG electrode, $y(t)$, as linearly related by convolution with an impulse response function $w(t)$, where $y(t) = w(T) \bullet x(t) + noise$. We can choose a random process that changes continuously but randomly and can then recover $w(t)$, a generalisation of the usual VEP, using numeric deconvolution.

The deployment of covert attention plays a crucial theoretical role in a number of current reading models, in particular the influential E-Z Reader model (Reichle et al. 2003). This model assumes a discrete word-based shift in an attentional spotlight from the currently fixated word to the next one. Alternative accounts argue for an attentional gradient distributed over a small number of words, allowing for parallel processing. So the ability to ascertain the focus of attention during a fixation is an important tool for deciding among competing reading models.

By using the VESPA technique to tag regions of the visual field, we will be able to ascertain when and where attention is allocated. The main question asked is whether there is temporal overlap in the waveforms associated with the processing of consecutive target words, essentially deciding between the alternatives of discrete attention shift and distributed attention gradient. This will directly address one of the most hotly debated issues in the field with direct consequences for model development and reading theory in general (see lnhoff et al. 2005 and Pollatsek et al. 2006 for opposing theoretical positions).

This talk will discuss the methodological details for implementing the VESPA paradigm in a reading study and will present results from a co-registration study involving EEG recording and eye tracking.

References

lnhoff, A. W., Eiter, B. M., & Radach, R. (2005). Time course of linguistic information extraction from consecutive words during eye fixations in reading. *Journal of Experimental Psychology: Human Perception & Performance, 31,* 979–995.

Lalor, E. C., Kelly, S. P., Pearlmutter, B. A., Reilly, R. B., & Foxe, J. J. (2007). Isolating endogenous visuospatial attentional effects using the novel visual-evoked spread spectrum analysis (VESPA) technique. *European Journal of Neuroscience, 26,* 3536–3542.

Lalor, E. C., Pearlmutter, B. A., Foxe, J. J., Reilly, R. B., & McDarby, G. (2006). Spread-spectrum intensity modulation for rapid estimation of a visually evoked potential. *Neurolmage, 32,* 1549–1561.

Pollatsek, A., Reichle, E. D., & Rayner, K. (2006). Serial processing is consistent with the time course of linguistic information extraction from consecutive words during eye fixations in reading: A response To Inhoff, Eiler, and Radach (2005). *Journal of Experimental Psychology: Human Perception & Performance, 32,* 1485–1489.

Reichle, E. D., Rayner, K., & Pollatsek, A. (2003). The E-Z reader model of eye-movement control in reading: Comparisons to other models. *Behavioral and Brain Sciences, 26,* 445–526.

The Quest for Integrating Data in Mixed Research: User Experience Research Revisited

Annika Wiklund-Engblom and Joachim Högväg

1 Introduction

Life is a narrative, and meaningful communication needs to be shaped within the boundaries of this narrative. The implication of this statement is that as a teacher you are a storyteller, as a marketing expert you are a storyteller, as an entertainer you are a storyteller—or as Walter Fisher (1987) says: We are all "homo narrans".

Across cyber space, there is a new grammar of storytelling. Almost all disciplines are touching upon it, but from various perspectives, using diverging terminology, but with one common denominator: The experience of the story is enabled through the use of various digital tools. There is an enormous media convergence taking place evolving our ability to communicate. The well-known media guru Marshall McLuhan would probably consider it as extending our voices as storytellers, or perhaps rather a whole nervous system evolving outside our physical bodies. Think of it as the stories told by the campfires long ago echoing into our time with the help of new tools. Our aim is to understand this media convergence from an interdisciplinary perspective and to learn how to speak the new grammar of digital storytelling. We are embracing it as both content developers and researchers.

Disregarding what your purpose as a digital storyteller is—be it to sell products, to enhance learning, or simply to entertain—the core of this new grammar is about understanding the audience/learner/customer and how you can engage them in your storyworld across media platforms. Therefore, the concept of user experience (UX) is highly relevant, as it places the human in the centre of any kind of design solution, i.e. user-centred design. To take this one step further, Experience Design gives importance to the experience as the guiding star for the design, rather than technology in itself. According to this design genre, the aim of any design should start with a "Why?"—referring to the motives why the end users would want to engage with your product, the so-called be-goals (Hassenzahl 2010; Carver and Scheier 1989). Our emphasis in both research and digital content development lies in experience

A. Wiklund-Engblom (✉) · J. Högväg
MediaCity, Åbo Akademi University, Vaasa, Finland
e-mail: annika.wiklund-engblom@abo.fi

M. Horsley et al. (eds.), *Current Trends in Eye Tracking Research*,
DOI 10.1007/978-3-319-02868-2_12, © Springer International Publishing Switzerland 2014

design, in which technology becomes transparent and human needs and UX guides the design process. It is simply the idea of using technology for all the right reasons (Hassenzahl 2010).

Our aim in this chapter is to describe our notion of UX and audience research related to media content creation and format development. We discuss our mixed research methods in relation to research questions. We further illustrate this using an example how mixed data can be integrated in UX research during iterative design of a website, and how eye tracking is an essential method involved in several steps of analysis.

2 Experience as a Design Aim

Experience design is becoming more widespread among digital content creators. For instance, instructional designers are starting to realize that the job to design an e-learning course is not only about delivering material to be learnt, but also about managing a learning experience. Matthew Moore wrote a post in The eLearning Guild's discussion forum (30 December 2009) on LinkedIn, titled "Do we need to move from instructional design to experience design?" The topic generated 80 comments in 15 days, which, after a quick glance in the forum, is about 99 % more than usual.

An experience is always dependent on many interrelated dimensions and aspects, which differs from person to person (Hassenzahl 2010). We are interested in the dimensions of UX created by encountering various media content—mostly digital, and how we can design media solutions for targeting specific user needs in order to facilitate a great experience and user satisfaction. When testing digital media content on users, we emphasize on the difference between usability and UX. These two aspects of human interaction with any system are two sides of the same coin, whether the focus is on the technological product or the digital media content. Nonetheless, these perspectives raise different questions, require different methods and research instruments, and generate different results. In our research on UX we target both sides of the coin: for instance, users' attributions to product qualities, pragmatic "do-goals", hedonic "be-goals" (Hassenzahl 2010; Carver and Scheier 1989), the affective experience (Watson and Clark 1994; Eilers et al. 1986), needs related to the experience (Sheldon et al. 2001), and mental effort (Zijlstra and van Doorn 1985). We apply several research methods that are specifically chosen to support each other in order to form a holistic picture of the experience. This allows us to capture a broad perspective of UX, and we can validate findings by triangulating data collected by a variety of research instruments. This will be explained in the methods section discussing mixed research.

Hassenzahl (2010) describes experiences as both actual and reflective. Actual experiences are in the moment, objective phenomena dealing with usability, functionality, as well as visceral stimulation. Reflective experiences are subjective

Table 1 Perspectives on the structure of an experience

	Pre-experience	Actual experience	Post-experience
Structure of experiences	Experience as what the designers offer and what the users bring to it Individual difference Fundamental needs Autonomy Competence Relatedness Stimulation Self-esteem Physical WB Security Self-actual Popularity Luxury	Experience as Practical involvement Cognitive involvement Emotional involvement Sensual involvement Experience as Subjective Holistic Situated Dynamic	Experience as interpretation; how we actively construct meaning: Reflective experiences Subjective memories Experienced emotions Benefits felt Stories remembered
Research focus	Preconception of an experience	Process of an experience	Reproducing an experience
Methods	Interview, survey	Eye-tracking, EDA, heart rate, electroencephalography (EEG), concurrent think aloud	Stimulated recall interview retrospective think aloud

memories activated by experienced emotions, benefits felt, as well as stories we remember and tell about our experiences to other people. There are four essential characteristics to an experience: it is always subjective, holistic, situated, and dynamic. Given these characteristics, experiences can be shaped. This fact is something that the field of human–computer interaction still needs to recognize on a wider scale (Hassenzahl 2010), and it is something that we find of great importance in our research on UX and media content development.

Experience can be regarded as meaning making. There are four concepts that can guide us in the design of technology and digital content, in order to support the end-user experience: (1) experience in interaction involves the dynamic process, (2) experience in interpretation involves how we actively construct meaning, (3) experience as what the designers offer and what the users bring to it, and (4) involves four dimensions of experience, which are inseparable, interrelated, and experienced as a coherent whole, i.e. practical involvement, cognitive involvement, emotional involvement, and sensual involvement (Vyas and van der Veer 2006).

Table 1 illustrates categorization of experience according to Hassenzahl (2010), as well as Vyas and van der Veer (2006). We can look at an experience from three perspectives based on time: before, during, and after—or in other words, pre-experience, actual experience, and post-experience. This has implications for the methods we chose for looking at different aspects of an experience.

3 Needs as Determinants of Experiences

The ultimate goal of the digital media content and products we are designing and developing is to invoke a positive UX. One approach to achieve this endeavour is to look at the role of human fundamental needs for the content of our experiences. We suggest that the source of UX lies in the needs that drive our actions and choices (Hassenzahl 2010; Wiklund-Engblom et al. 2009). Research on needs address the content of experiences, rather than focusing on the structure of experience. Where, for instance, McCarthy and Wright's (2004) experience framework, stop with the observation that experiences have an emotional–motivational and a meaning-making thread, the needs perspective clarifies where the emotion, motivation, and meaning comes from, as well as the essence of it. By that, it becomes much easier to address experiences in the context of product and media content design (Wiklund-Engblom et al. 2009).

According to Sheldon et al. (2001), fundamental needs are limited in number. By looking at various theories, they found ten needs that seemed to represent the most important categories. This summary of needs builds upon well-known theories, such as Ryan and Deci's self-determination theory, Maslow's theory of personality, Epstein's cognitive–experiential self-theory, and Derber's lay theory of human needs. The ten needs are: *self-esteem, autonomy, competence, relatedness, pleasure-stimulation, physical well-being, self-actualization-meaning, security, popularity-influence,* and *money-luxury*.

Sheldon et al. (2001) wanted to know which basic human needs were mostly related to satisfying events. They developed a Likert-scale-based questionnaire including three items for each of the ten needs. In a number of studies, college students were asked to think of the most satisfying event they remember from the last month, and then fill in the questionnaire while thinking of this event. These studies continuously showed that the four most important needs that people related to a satisfying event were self-esteem, relatedness, competence, and autonomy. We are interested in creating positive experiences from digital media content. The needs approach is one way of defining the aimed experience. This approach was, for example, applied in the pre-design phase of a multiplatform music show *The Mill Sessions*. The ten fundamental needs were used as a base for a survey to capture peoples' conceptions of what a great music experience is in relation to the ten needs (Esch et al. 2011). The aim of the study was to confirm the design idea of the show.

4 Research Questions

Our questions involve looking at UX from (1) an ontological perspective; what a great UX is in specific contexts, (2) a methodological perspective; which methods should be applied in various studies in order to capture end-users' experiences, as well as (3) a content design perspective; exploring how to design media content for a great UX.

The media convergence culture is upon us, and new generations are more or less expecting it. The convergence involves "the flow of content across multiple media platforms, the cooperation between multiple media industries, and the migratory behavior of media audiences who will go almost anywhere in search of the kinds of entertainment experiences they want". (Jenkins 2006, p. 2). To experience the flow of a story is often the aim with digital content. When the story flows between different media platforms it becomes transmedia storytelling. One of the most obvious questions designers have when creating transmedia solutions is the notion of keeping the audience with the story flow between options of content and discontinuities between technological platforms. The experience of the story (transmedia content) is intended to evoke feelings and thoughts by allowing for identification and participation. The question is how this transmedia UX is optimized—especially at the trigger points where users need to be motivated to move with the flow of the story. This is where UX research becomes critical in order to spot flaws of the design, as well as to explore potential design choices. To design for a transmedia experience, we have to account for media discontinuity as part of the experience, and this is not necessarily a negative thing. However, it is not only about delivering content, it is also about participation—letting the audience become part of a story in order to enable a deeper level of identification.

For transmedia learning, the UX research further involves the dimensions of the learning process with its specific aims for learning, such as motivating self-regulation, reflection, and higher order thinking. From the user's perspective, transmedia storybuilding is the process of taking in a story told across platforms and creating your own version of it through your own choices and your own experience of the content (Wiklund-Engblom et al. 2013). The transmedia seams across media discontinuities represent the choices you make for your own learning to move with the flow in co-constructing the content. The totality of your experience becomes your own storybuilding and transmedia learning process; what you have learned. The what, how, and why of your choices for constructing your own story reveal important aspects of your UX. These are, for instance, the physical, the sensory, the cognitive, and the affective influences on the experience. Other aspects are expectations, prior knowledge, prior experiences, social context, etc.

5 Mixed Methods

UX of media users is in many ways an abstract notion. How to best measure it is a debatable question. For instance, multidimensionality is an essential factor of UX, but still, not many studies explore the interrelations of dimensions (Bargas-Avila and Hornbæk 2011). Over the years, we have used many different methods to collect both quantitative and qualitative data targeting the multidimensionality of UX, for instance, users' observable behaviour, a variety of physiological reactions, product evaluations and attitudes, judgements of affect and effort, as well as users' needs and reflections.

"Mixed methods" is the most common term for blending of research methods and methodological paradigms in a study. Broader terms are mixed research or integrative research, which according to Johnson and Onwuegbuzie (2004) would provide a more inclusive meaning. Creswell and Plano Clark (2007, p. 5) refer to mixed methods research "as a research design with philosophical assumptions as well as quantitative and qualitative methods". Others argue that the minimum criterion for defining a research design as mixed methods is that you have an "interdependence of component approaches during the analytic writing process" (Bazeley and Kemp 2012, p. 70). The key is the integration of data at some phase of the analysing process (Creswell et al. 2003; Bazeley and Kemp 2012).

The purpose of using multiple methods in our UX research has been to capture the phenomenon as closely as possible, not being restricted by any methodological paradigm, but rather to explore how the use of methods can be expanded. Bazeley and Kemp (2012, p. 16) suggest that researchers should take "every opportunity to fully exploit the integrative potential of mixed data sources and analysis methods" if it benefits the research by resulting in better validity and stronger conclusions. Our goal is to find out the best ways of integrating mixed data in order to answer research questions. Our approach to research is, therefore, pragmatic. This is understandable, as we are looking at different media solutions in various media contexts, and these are forever changing into innovative new forms. Similarly, we need to be constantly re-evaluating what the best ways to answer research questions are. For instance, new technological solutions and software are providing new possibilities for analysing larger amounts of mixed data, which was not an option before because of the time it demanded.

Although mixed methods research has been carried out in various forms since the 1950s, it is only in the recent years that it has become a more widespread acknowledged research approach (Creswell and Plano Clark 2007). Historically, there have been two base camps of purists when it comes to methodological approaches: the quantitative vs. the qualitative advocates. These often argue for the incompatibility thesis that the two paradigms never should or even could be mixed. However, nowadays there is a new set of advocates suggesting that focus should rather be on the many similarities between the two approaches. In this debate, the mixed methods research is seen as a third paradigm drawing on the strengths and counteracting the weaknesses of each of the first two methodological research approaches. As such, qualitative and quantitative research are seen as end points on a continuum where mixed methods are represented by the large middle area (Johnson and Onwuegbuzie 2004).

5.1 Differentiating Methods

We explore methods to discover the added value of combining instruments and methods for collecting data. This methodological research process in itself is one of our areas of interest. We are using a number of laboratory-based methods to collect both quantitative and qualitative data targeting the multidimensionality of UX.

Table 2 Methods matrix illustrating types of mixed data

Types of Data	*Formative (Qualitative) Data*	*Summative (Quantitative) Data*
Objective Data	Q1: What did they do? • Observed Actions • Observed Focal Points	Q3: How much reaction and action? • Skin Conductance • Eye Movements • EEG
Subjective Data	Q2: Why did they do/think? • Stimulated Recall Interview • Concurrent Think Aloud	Q4: Who are they: Identity & Attributes? • Demographics • Questionnaires

In our laboratory for UX, we have equipment such as eye trackers, electroencephalography (EEG) helmets, and various psychophysiological instruments for measuring skin response and heart rate. In combination with these methods, we use stimulated recall interviews, think-aloud protocols, as well as traditional questionnaires for measuring emotional experiences, needs, and evaluations targeting hedonic and pragmatic qualities of media solutions.

In order to better understand the varieties of data we collect, we visualize it in Table 2. The table is a matrix of the methods used, categorizing them according to type of data generated. The matrix is organized by four categories of data characteristics: objective, subjective, formative (qualitative), and summative (quantitative). Which methods we use in a study is, of course, determined by the particular research questions at hand—usually generated by the phase of development during the iterative design of the digital content.

The table further illustrates how the differing data target a variety of research questions. In the following, the four questions heading each of the four centre cells will be described in relation to the variety of methods we employ.

Question 1: What did they do? Observations of user activities and eye movements recorded on the screen generate *objective formative* data. User activities on the screen are recorded by video and audio using the functions of the TobiiStudio recording software. The qualitative eye-tracking data, i.e. the eye movements or scan paths, which are synchronized with the screen recording, can be imported as video into, for instance, the QSR Nvivo analysing software. Eye-tracking data may be used as a base for a narrative description of the event, or for describing and categorizing activities and task performance. Participants are also filmed during the session. These video recordings are used as validity check for psychophysiological reactions, which are sensitive to body movements and other disturbances.

Question 2: Why did they do/think what they did? Interviews and concurrent think aloud by participants during task performance generate *subjective formative* data.

Instant recall interview questions relating to usability and UX are often asked after each task during a test session. Screen recordings, including eye tracking, are used as a stimulus for the stimulated recall interviews. Furthermore, the participants are often encouraged to think aloud during task performance, in order to visualize their thought process and motivations for actions. This is referred to as concurrent think aloud in contrast to retrospective think aloud where participants reflect on their actions while watching their eye movements after the actual task performance is over (Tobii 2009). This is an example of the difference between researching UX as a process or as a memory in retrospect. Interviews and the concurrent think aloud are usually transcribed and may be imported into, for instance, the QSR Nvivo software for qualitative categorization and/or mixed analyses in which it is compared and correlated to quantitative data. The subjective formative data are in many ways the most critical in determining the essence of the UX, and also for validating objective data analyses.

Question 3: How much reaction and action was measured? *Objective summative* data are generated from measuring psychophysiological reactions and eye movements. Skin response data are measured with the Affectiva Q-sensor or clip-on sensors on the index finger. The sensor registers changes in the electrical conductance of the skin that are driven by sympathetic or parasympathetic nervous system activation controlled by the brain. The sympathetic nervous system is activated when we have an emotional response or a stress response to stimuli. As opposed to this, parasympathetic activation means the subject is calm and has low stress levels. High skin conductance levels, or electrodermal activation (EDA) levels, mean sympathetic activation and emotional response; low EDA levels mean parasympathetic activation and low emotional stress levels (Poh et al. 2010). The Affectiva Q-sensor is placed on the wrist and is unobtrusive to the test persons. EDA data are transferred from the sensor to a computer and synchronized with eye-tracking data and video recordings for the analysis. By defining areas of interest (AOI) in the eye-tracking data, we can correlate the user's attention to these AOIs with psychophysiological reactions during this defined attention span.

Participants' eye movements are recorded using TobiiX120 and the recording software TobiiStudio. Our eye movements can be categorized in two ways: the fixation on something we are looking at, and the movements in between fixations, which are called saccades. The eye tracker records both fixations and saccades (Tobii 2009). This enables us to measure, for instance, time spent in areas of interest, initial perception, and order of actions taken and fixations made within the media environment. An analysis of the quantitative eye-tracking data would include measuring fixations and saccades in AOIs, mouse clicks, time, etc. The quantitative eye-tracking data, i.e. number and length of fixations within pre-defined AOI and areas of non-interest (AOINS), may be exported as an Excel file, which enables import into, for instance, Statistical Package for the Social Sciences (SPSS) or the QSR Nvivo software.

Brain waves (EEG data) are registered with the Emotiv EPOC, including 14 sensors to be placed on the scalp. The sensors detect a person's emotional experience by tuning into the electrical signals the brain is producing. In addition to raw

EEG data, the EPOC uses an algorithm to estimate a person's levels of frustration, engagement, and meditation related to the experience subjected to study.

Question 4: Who are they: identity and attributes? Demographic data about the participants, as well as questionnaire data covering attitudes, needs, preferences, etc., can be used for describing the users in relation to the experience they had during their activities in the session. These kinds of instruments generate subjective summative data. As a post-questionnaire, we often use the abridged version of the AttrakDiff2 (Hassenzahl and Monk 2010), developed for evaluating interactive products. It consists of ten seven-point semantic differential items, including four items measuring pragmatic quality (confusing–structured, impractical–practical, unpredictable–predictable, complicated–simple), four items measuring hedonic quality (dull–captivating, tacky–stylish, cheap–premium, unimaginative–imaginative), and two items measuring general product evaluation, i.e. goodness (good–bad) and beauty (beautiful–ugly). Traditionally, subjective summative data are mostly used for studies aiming for generalizations based on large data samples. However, we use it as indications for product evaluations, in order to see if the targeted object of the evaluation is up to the standards aimed for in the product development. These quantitative data can further be used in mixed analyses in correlations with qualitative data, or with objective summative data, such as eye-tracking measures. This will be further explained in the example below.

For the non-verbal measuring of mental effort, positive/negative affect, and enthusiasm/serenity, we use the *Subjective Mental Effort Questionnaire* (SMEQ) and the *Self-Assessment Manikin* (SAM). The SMEQ is a self-report questionnaire measuring the amount of mental effort invested during task performance. There is a numeric scale parallel to a verbal scale, which ranges from 0 (no effort at all) to 220 (very much effort). Thus, the participant can judge his/her mental effort according to both scales (Eilers et al. 1986). The SAM scale, developed by Lang (1980), is a non-verbal, pictorial questionnaire measuring general emotional states, including *valence*, *arousal*, and *dominance*. (The dominance scale is not included in our research at this point.)

There are a variety of reasons for the addition of these questionnaires in our hands-on system testing. First of all, the questionnaires are applicable to studies conducted in different contexts, on different target groups and on different media content as well as media technology. As the scales are short and easy to fill in, it is possible to measure test participants' experiences several times during a media encounter.

5.2 Example of Mixed Research for Measuring UX

The integration analysis, i.e. mixed research using both qualitative and quantitative data, should be guided by key issues related to the research questions, rather than separating the analysis by method (Bazeley and Kemp 2012). The following example of an integration analysis is guided by the aim to explore usability and UX of a website (Table 3).

Table 3 Type of data, methods, and research targets in relation to UX key issue

Question 1: What did they do? (Objective Formative Data)	
Method	Observing user activities and eye movements
Research Targets	• Initial perception on the web site
Question 2: Why did they do/think what they did? (Subjective Formative Data)	
Method	Instant Recall Interview with open-ended questions
Research Targets	• Post-comments on first impression of the site • Post-comments on perceived purpose of the site • Post-comments on motivation to interact
Method	Concurrent think aloud during activities on the web site
Research Targets	• Describing thought process of the first impression of the site • Describing motivations for actions in the moment
Question 3: How much reaction and action was measured? (Objective Summative Data)	
Method	Measuring Electrodermal Activation Levels (EDA)
Research Targets	• EDA for initial task on the web site • High EDA versus low EDA triggers in relation to content • Number of high EDA moments
Method	Tracking eye movements (statistics)
Research Targets	• Visual attention to, measured as number and time of fixations to the heading of the information about aim and function of the site. • Time to first fixation
Question 4: Who are they: Identity & Attributes? (Subjective Summative Data)	
Method	Post-questionnaire (Attrakdif2)
Research Targets	• Perceived relevance and value

Key issues concern, for instance, first impressions, perceived purpose, and motivation. In order to illustrate the potentials of integration of mixed methods in analysis, we will present one key issue targeting the relation between first impression, perceived purpose, and motivation.

The first impression we get of something affects our experience. Therefore, participants' first impressions are crucial in determining UX. It is further of importance for determining usability, as the functions of the website need to be designed to guide the user in a positive direction in using the site from the first point of entry. By using mixed methods, we can look at this key issue from several perspectives:

- Users' initial perceptions are recorded from their eye movements
- Their first thoughts while using the site are recorded from their concurrent think aloud
- Their physiological reactions (EDA) are recorded, from which the first encounter and first task can be extracted
- Time to their first fixation on the screen can be measured from the eye-tracking statistics
- Their impressions of the site are discussed during the stimulated recall interview

Table 4 lists the different methods we can use to approach this key issue. It is an example of how multiple methods are employed in order to answer research questions

Table 4 Matrix illustrating integration in analysis using mixed methods

<table>
<tr><th rowspan="2">Matrix</th><th colspan="2">Motivation</th><th colspan="3">First Impression</th></tr>
<tr><th>Intrinsic</th><th>Extrinsic</th><th>Positive</th><th>Negative</th><th>Neutral</th></tr>
<tr><td>Fast first fixation</td><td></td><td></td><td></td><td></td><td></td></tr>
<tr><td>Slow first fixation</td><td rowspan="4" colspan="5">These cells contain qualitative text sources related to categories presented in the upper matrix row sorted by the quantitative categories in the left column</td></tr>
<tr><td>High EDA for first fixation</td></tr>
<tr><td>Low EDA for first fixation</td></tr>
<tr><td>High perceived purpose</td></tr>
<tr><td>Low perceived purpose</td><td></td><td></td><td></td><td></td><td></td></tr>
</table>

related to UX. In the table, research targets are separated by the kind of data generated (as described earlier in Table 2).

6 Iterations of Analysis

NVivo nodes (categories) are created for important aspects related to the first impression using qualitative data. Participants are clustered (fast vs. slow) by how fast they make their first fixation on the website using eye-tracking data. They are further clustered by their physiological reactions for the first fixation in which EDA data and eye-tracking data are synchronized. These clusters are imported into NVivo as case attributes linked to each participant. All case attributes can be cross-tabulated in a matrix in order to explore the qualitative data, which will be sorted according to matrix cross-tabulations as shown in Table 4.

Furthermore, our hypothesis is that a person's motivation for and perceived purpose and value of a website will influence his/her first impression. The scaled questionnaire variables related to perceived purpose are dichotomized (high vs. low), and participant clusters are made based on these. Thereafter, the clusters are imported into NVivo as case attributes. The interview discussions related to motivation are categorized and coded. NVivo nodes for motivation are then used in a matrix, as presented in Table 3, in the same manner as described above.

7 Software for Data Manipulation and Synchronization

The rich data of varied types that are produced during investigations cause concern as it is very time consuming to do analyses using mixed data of these kinds. Therefore, we are further developing our software program, eValu8, to assist in the analysing process of data collected with the above-mentioned instruments. Important steps of analysis according to the mixed research process model are: data reduction, data display, data synchronization, data transformation, data comparison,

Table 5 Examples of qualitative and quantitative data manipulation and synchronization in mixed research

Terminology	*Qualitative Data; examples*	*Quantitative Data; examples*
Data Reduction	Reducing the dimensionality of data.	
	E.g., exploratory thematic analysis, memoing, bookmarking, and defining Areas of Interest.	E.g., descriptive statistics, exploratory factor analysis, clustering, bookmarking, and defining Areas of Interest.
Data Display	Describing data visually.	
	E.g., matrices, charts, graphs, networks, lists, rubrics, and Venn diagrams.	E.g., tables, graphs, visualisations of EEG and EDA data.
Data Transformation	Transforming one kind of data into another.	
	E.g., qualitative data are converted into numerical codes that can be represented statistically.	E.g., quantitative data are converted into narrative data that can be analyzed qualitatively.
Data Correlation	Qualitative data are correlated with quantitative data, or vise versa.	
	E.g., by correlating quantitative questionnaire variables with qualitative interview categories, which are sorted into a matrix in accordance with the quantitative categories.	
Data Consolidation	Mixed data are combined to create new or consolidated variables or data sets from categorizations of both quantitative and qualitative data.	
	E.g., clusters based on questionnaire variables are correlated with interview data, and new qualitative categories are defined by consolidating data.	
Data Comparison	Comparing data from qualitative and quantitative data sources.	
	E.g., comparing qualitative observational data to various physiological reactions (EEG, EDA, Heart rate, Eye tracking measures), or questionnaire variables.	
Data Integration	Both quantitative and qualitative data are integrated into either a coherent whole or two separate sets (i.e., qualitative and quantitative) of coherent wholes.	
	An example of integrating multiple qualitative data are, for instance, to integrate stimulated instant recall interview data with observational data into a coherent narrative.	An example of integrating multiple quantitative data are, for instance, integrating EEG, EDA,and Eye tracking measures into coherent visual displays based on areas of interest defined by eye tracking data.

data correlation, data consolidation, as well as data integration. Our aim is to be able to develop the eValu8 software as a useful tool to assist these steps of the analysing process. It is also important that the eValu8 can be synchronized with other analysing tools such as SPSS and the QSR Nvivo software, as these are inevitable in for instance the data reduction process (Table 5).

The core of mixed methods research is legitimation, assessing the trustworthiness of data. The mixed data are used for validating findings of both the qualitative and quantitative data and subsequent interpretations. The legitimation process might include additional data collection, data analysis, and/or data interpretation until as many rival explanations as possible have been reduced or eliminated. In Table 5, we are using terminology from mixed methods research (Johnson and

Onwuegbuzie 2004) to describe the functions we aspire for in developing the data synchronization tool eValu8.

8 Concluding Thoughts on Mixed Methods in UX Research

When exploring UX in, for instance, a computer-based environment, as in the above-mentioned example, for the purpose of improving the user interface, the research questions are not always obvious or even possible to pinpoint on a specific level. Rather, the researcher must be able to shift focus whenever something of interest arises. The research situation needs to remain flexible to the dynamics of human experiences targeting a specific event or process of activities.

The added value of using mixed methods within the field of UX research is that it provides the potential for diving deeper into the analyses, and we are able to ask more intricate questions of the data. Instead of separating methods as incompatible as they generate different types of data, we would rather like to see them as pieces of a puzzle, in need of a little twisting in order to fit the larger picture. However, analysing mixed methods data is messy. It is also very much experimental. As a mixed methods researcher you need to have "flexibility and pragmatism about design, openness to data, and a touch of inventiveness in approach to analysis" (Bazeley 2012, p. 825). One aggravating circumstance is that several iterations of analyses need to be made whenever a study includes rich sources of mixed data. However, this is also positive as the researcher builds deep levels of meaning through an iterative analysing process (di Gregorio and Davidson 2008).

One thing we are certain about is that an experience of a media user needs to be explored as an event in which both objective and subjective data illuminate important parts of the puzzle as a whole. Another methodological reflection is that research methods must be flexible enough to explore an unlimited number of experience dimensions. For us, this is especially interesting as we are involved in both development and content testing research in the field of crossmedia and transmedia creation and innovation, in which the essence of the UX makes the difference between failure and success (Hassenzahl 2010). This kind of exploration of UX is limited if the research remains restricted by paradigmatic boxes and divisional thinking. Our challenge lies in looking beyond the norm of what UX is and how it can be measured, observed, and explored. Therefore, we are advocates of mixing methods, team building, and collaboration across all kinds of borders, be it normative, paradigmatic, scientific, or merely psychological.

There are several challenges of UX research as part of a creative development process. These challenges can shortly be referred to as time, money, and asking the right questions at the right phase of development. Every project presents its own challenges, and for the UX researcher this is a never-ending, but nonetheless exciting learning process.

References

Bargas-Avila, J. A., & Hornbæk, K. (2011). Old wine in new bottles or nov9l challenges? A critical analysis of empirical studies of user experience. Proceedings of CHI 2011, May 7–12, Vancouver, Canada.
Bazeley, P. (2012). Integrative strategies for mixed data sources. *American Behavioral Scientist, 56*(6), 814–828.
Bazeley, P., & Kemp, L. (2012). Mosaics, triangles, and DNA: Metaphors for integrated analysis in mixed methods research. *Journal of Mixed Methods Research, 6*(1), 55–72.
Carver, C. S., & Scheier, M. F. (1989). *On the self-regulation of behavior*. New York: Cambridge University Press.
Creswell, J. W., & Plano Clark, V. L. (2007). *Designing and conducting mixed methods research*. Thousand Oaks: Sage.
Creswell, J. W., Plano Clark, V. L., Gutmann, M. L., & Hanson, W. E. (2003). Advanced mixed methods research designs. In A. Tashakkori & C. Teddlie (Eds.), *Handbook of mixed methods in social and behavioral research* (pp. 209–240). Thousand Oaks: Sage.
di Gregorio, S., & Davidson, J. (2008). *Qualitative research design for software users*. London: Open University Press/McGraw-Hill.
Eilers, K., Nachreiner, F., & Hänecke, K. (1986). Entwicklungund Über prüfung einer Skala zur Erfassung subjektiv erlebter Antrengung [Development and evaluation of a questionnaire to assess subjectively experienced effort]. *Zeitschrift für Arbeitswissenschaft, 40,* 215–224
Esch, M., Wiklund-Engblom, A., & Staffans, S. (2011). Experience as a Starting Point of Designing Transmedia Content. Proceedings of the EuroITV 2011 conference. Available online at the ACM Digital Library.
Fisher, W. R. (1987). *Human communication as narration: Toward a philosophy of reason, value, and action*. Columbia: University of South Carolina Press.
Hassenzahl, M. (2010). Experience design: Technology for all the right reasons. *Synthesis Lectures on Human-Centered Informatics, 3*(1), 1–95.
Hassenzahl, M., & Monk, A. (2010). The inference of perceived usability from beauty. *Human-Computer Interaction, 25*(3), 235–260.
Jenkins, H. (2006). *Convergence culture: Where old and new media collide*. New York: New York University Press.
Johnson, R. B., & Onwuegbuzie, A. J. (2004). Mixed methods research: A research paradigm whose time has come. *Educational Researcher, 33*(14).
McCarthy, J., & Wright, P. C. (2004). *Technology as Experience*. Cambridge: MIT Press.
Moore, M. (2009). *'Do we need to move from instructional design to experience design?'* The eLearning Guild's discussion forum on LinkedIn, December 30th, 2009
Poh, M.-Z., Swenson, N. C., & Picard, R. W. (2010). A wearable sensor for unobtrusive, long-term assessment of electrodermal activity. *IEEE Transactions on Biomedical Engineering, 57*(5)
Sheldon, K. M., Elliot, A. J., Kim, Y., & Kasser, T. (2001). What is satisfying about satisfying needs? Testing 10 candidate psychological needs. *Journal of Personality and Social Psychology, 80,* 325–339
Tobii (2009). *Retrospective Think Aloud and Eye Tracking*. Whitepaper by Tobii Technology. http://www.scribd.com/doc/20155331/Tobii-RTA-Whitepaper
Vyas, D., & van der Veer, G. (2006). Experience as meaning: some underlying concepts and implications for design. ECCE '06 Proceedings of the 13th Eurpoean conference on Cognitive ergonomics: trust and control in complex socio-technical systems. doi:10.1145/1274892.1274906.
Watson, D., & Clark, L. A. (1994). *The PANAS-X: Manual for the Positive and Negative Affect Schedule—Expanded Form*. Iowa City: University of Iowa.
Wiklund-Engblom, A., Hassenzahl, M., Bengs, A., & Sperring, S. (2009). What needs tell us about user experience. In T. Gross, J. Gulliksen, P. Kotzé, L. Oestreicher, P. Palanque, R. O. Prates, M. Winckler (Eds.) Proceedings of INTERACT 2009, 12th IFIP TC13 Conference on Human-Computer Interaction.

Wiklund-Engblom, A., Hiltunen, K., Hartvik, J., & Porko-Hudd, M. (2013). *Transmedia storybuilding in sloyd.* Proceedings of IADIS Mobile Learning 2013.
Zijlstra, R., & van Doorn, L. (1985). *The Construction of a Scale to Measure Subjective Effort. Delft.* The Netherlands: Delft University of Technology, Department of Philosophy and Social Sciences.

Part III
Eye Tracking and Marketing and Social Applications

Eye Tracking as a Research Method in Social and Marketing Applications

Mike Horsley

Human behaviour is incredibly complex and astoundingly broad. As a result, many disciplines have been developed to study it in great detail. Each discipline develops its own unique shared understandings and common discourse and professional community of researchers. Each discipline also develops various research paradigms and assumptions which create boundaries between other disciplines in exploring and investigating human and physical phenomenon (Kuhn 1962). Central to developing research paradigms are research methodologies—the accepted and foundational ways that researchers in a discipline support—to more deeply understand and investigate the world.

Although individual disciplines develop unique and different research methodologies and processes that pertain only to their own discipline, many research methodologies are shared across disciplines. This sharing is an increasing phenomenon in cross-disciplinary and multi-disciplinary studies. Eye-tracking research methodologies have existed for a considerable time (Wade and Tatler 2011). However, spurred on by developments in eye-tracking technologies and analysis software, and increasingly mobile and low-cost eye tracking, the applications of eye tracking are continuing to spread from discipline to discipline.

Eye movements are at the core of many human behaviours (Wade and Tatler 2011). So examining eye movements during behaviour is providing a new and powerful opportunity to develop a new research methodology across disciplines. This is the focus of this section, to present a series of research studies that utilise eye-tracking technologies and research methods in a wide range of social applications—from investigating forged signatures (Dyer, Found, Merlino, Pepe and Sita 2013) to using saccades to assess cognitive decline in ageing (Bowling and Draper 2013). The centrality of eye movement to so many human behaviours means that the application of eye-tracking research approaches to social and marketing research problems will continue to both expand and deepen.

M. Horsley (✉)
Central Queensland University, Noosaville, Australia
e-mail: m.horsley@cqu.edu.au

M. Horsley et al. (eds.), *Current Trends in Eye Tracking Research,*
DOI 10.1007/978-3-319-02868-2_13, © Springer International Publishing Switzerland 2014

1 Core Eye-Tracking Research Methods in Social and Marketing Applications

Concomitant with the increasing application of eye-tracking methodologies to a range of social and marketing research is the development of some core research processes in investigating eye movements in these applications. Increasing use of fixation and saccade analysis through heat maps, AOI, and descriptive and inferential statistical procedures provide new ways of approaching the measurement of cognitive processes and information requirements during behaviour. The development of benchmarks and measures, comparison of different groups in relation to their eye movements and explanations of the relationship between the visual and underlying cognitive and physiological processes are beginning to constitute a new research paradigm (Kuhn 1962). This paradigm has a common core of eye-tracking research elements. This common core of research elements is increasingly being deployed across myriad social behaviours. This can be seen in Table 1.

Table 1 illustrates a core set of research methodologies, based on eye tracking that are increasingly seen at the core of social and marketing applications.

2 Area of Interest Analysis in Social and Marketing Applications

Area of interest (AOI) analysis involves the use of eye-tracking software to explore fixation duration, frequency and return between the different elements, parts or components in any visual scene or visual display. AOI analyses are increasingly used in social and marketing applications to investigate the differences between various groups. As a result, AOI analysis is present in three of the chapters presented in Part 3. AOI analysis is a key research procedure in use when researchers wish to examine and investigate the differences between novice and expert groups. The use of AOI methods and novice and expert comparisons reveals rich insights into learning and development. AOI and expert and novice analysis can also be useful in understanding the nature and visual structure of expertise.

3 Gaze Plots, Heat Maps and Scan Plots in Social and Marketing Applications

Gaze plot, heat map and fixation analysis explore the structure and development of gaze. Fixation analysis allows researchers to explore and investigate the structure of attention and distraction in scenes and visual displays. Gaze and scan path analysis provide a new ways of investigating and exploring the actual structure of

Table 1 Disciplines, research problems and eye-tracking research methods used in research applications given in Part 3

Authors	Discipline	Research area/problem	Eye-tracking research methods
Grigg, Griffin	Accounting	Understanding what readers see in financial report	AOI analysis, gaze plot analysis, novice and expert comparisons
Harwood, Jones	Marketing	Consumers visual attention in a retail environment	Content analysis, coded fixations, saliency analysis, groups comparisons
Li, Breeze, Horsley, Brierly	Marketing	Efficacy of retail marketing tools, reading direction and efficacy of brand motions	Stimulated recall, AOI analysis, fixation analysis: – counts duration
Dyer, Found, Pepe, Merlino, Rogers, Sita	Writing	Evaluating signature forgeries	AOI analysis scan path and gaze plot analysis novice and expert comparisons
Bowling, Draper	Aging	Impact of ageing and eye movements	Saccade analysis, older and younger comparisons
Chen	Social anxiety disorders/mental illness	Psychosocial stress simulation	Fixation analysis, fixation detection algorithms

looking and eye movements. Eye movements are not conscious and are extremely rapid (Wade and Tatler 2011). In addition, the visual system is highly automated (Kahneman 2011), and attention requires significant cognitive energy. How vision is structured through eye movements over scenes and displays can be investigated through gaze plot and scan path analysis, saccade data and algorithms and forms of saliency analysis. These types of eye-tracking research methods form the basis of three of the research projects and studies in three of the chapters in Part 3.

4 Stimulated Recall in Social and Marketing Applications

The inclusion of video playback for subjects being eye tracked and the ability for subjects to observe their own gaze plots is now a feature of new eye-tracking technologies and software. Sometimes called think aloud or stimulated recall, this feature of recent eye tracking provides a new way for researchers to investigate social and marketing phenomena by linking the thoughts of subjects to the underlying processes guiding the structure of their eye movements and gaze plots. A number of new studies are beginning to use the power of the simulated recall methodology to investigate the social and cognitive process underlying eye movements. This section includes one chapter that employs this important research method.

References

Kuhn, T. (1962). *The structure of scientific revolutions*. Chicago: University of Chicago Press.
Wade, N. J., & Tatler, B. W. (2011). Origins and applications of eye movement research. In S. Liversedge, I. D. Gilchrist & S. Everling (Eds.), *Oxford handbook on eye movements* (pp. 17–46). Oxford: Oxford University Press.
Kahneman, D. (2011). *Thinking, fast and slow*. Macmillan

Mobile Eye-Tracking in Retail Research

Tracy Harwood and Martin Jones

1 Introduction

Retail has become an increasingly competitive environment, squeezed by online offers that emphasise convenience and cost, while offline, the proliferation of stores results in low perceived differences between physical stores and brand environments. Thus, there is renewed emphasis on the importance of understanding consumer behaviour in stores in order to build profitable propositions, yet little research has been undertaken that considers the holistic environment from the consumer's perspective. This is primarily because of the development of research methods that have, until recently, been limited to mainly quantitative studies using questionnaires to assess attitude, perception and recall.

With the advent of new classes of observational technologies such as mobile eye tracking that captures consumers' audio-visual attention, it is now possible to explore environments using first-person perspectives, with opportunities to provide new insight into naturalistic shopping behaviour.

This chapter reports on an empirical investigation into consumer behaviour in a UK branded retail environment using mobile eye-tracking technology. The chapter begins with a review of relevant literature in the domain of store environments, atmospherics and typical consumer behaviour. Subsequently, a review of eye-tracking literature in the marketing domain is presented. Thereafter, the chapter discusses the research design adopted for the study, which takes a qualitative dominant-less dominant mixed methods approach using content analysis of consumers' gaze fixations, supported by pre- and post-tracking questionnaires and critical incident analysis to assess key elements within the first-person observations. Findings are then presented and discussed.

T. Harwood (✉)
Institute of Creative Technologies and Usability Lab, De Montfort University, Leicester, UK
e-mail: tharwood@dmu.ac.uk

M. Jones
Retail Lab, De Montfort University, Leicester, UK

M. Horsley et al. (eds.), *Current Trends in Eye Tracking Research*,
DOI 10.1007/978-3-319-02868-2_14, © Springer International Publishing Switzerland 2014

2 Literature Review

Retail stores are complex and visually rich environments that support both hedonic and utilitarian consumer experiences by managing architectural features, layout, product presentation, design, lighting, scent, etc. (Kotler 1974; Spanenburg et al. 1996; Turley and Milliman 2000; Custers et al. 2010). Navigation and store layout are linked to how products are presented to consumers—walking pace, range (distance) and degree of viewing angle are all aspects which interior designers incorporate into retail environments through the creation of immersive 'theatre' (e.g. Davis 1991), where architecture becomes advertising (Cairns 2010). The resultant store atmosphere influences the emotional state of the consumer, and this influences arousal levels that, if positive, may increase time spent in the store and likelihood of purchase (Mehrabian and Russell 1974; Donovan and Rossiter 1982; Bitner 1992; Rook 1987; Bagozzi and Dholakia 1999). Important elements have been found to be the sense of space and spacing of products which stimulate consumer senses; aesthetic surroundings which influence perceptions and attitudes; product presentation which influences consumer reactions; social surroundings which influence consumption patterns (e.g. conformity); and complexity, i.e. the number of components within the environment, which influences arousal leading to preference (Markin et al. 1976; Belk 1974; Nasar 1987; Uhrich and Benkenstein 2010). Consumer response is subject to intrinsic and extrinsic motivational influences, including time available (Park et al. 1989), mood state (Lee and Sternthal 1999), familiarity with an environment (Hoch and Deighton 1989; Alba and Chattopadhyay 1986), attentional priorities and need (Kent and Allen 1994). These areas have been categorised into three broad domains in the literature: atmosphere, ambience and relationships; yet, little research has considered the interrelationships between them at a holistic level (Ballantyne et al. 2010). The range and arrangement of positive atmosphere building cues remain, however, rather a 'black art', dependent upon retailer brand strategy and implementation and management preferences (e.g. Gauri et al. 2008).

Whilst there is a large body of literature into the components of stores and their impact on different aspects of consumer behaviour, this is primarily examined through evaluation of stated intent, perception and third-person observation techniques. There is limited research on actual browsing and purchasing behaviour and the elements that may influence behaviour in stores using first-person observational techniques (Suhur and Sorenson 2010). Third-person observation via remote forms of tracking, whilst popular in industry, is especially difficult to interpret because often only partial, say photographic or moving image, is used which is then subject to interpretation (e.g. Basil 2011). Consumer behaviour viewed from a first-person perspective is, however, important in generating insight to develop retail 'experience environments' that engage and retain consumers from both design and marketing perspectives (Venkatesh et al. 2011). Of particular interest are the ways in which the complexity of the store environment influences consumers' shopping behaviour. New technologies such as mobile eye-tracking are capable of providing such insight 'in action' rather than reflection of some prior activity ('on action').

2.1 *Eye-Tracking in Marketing*

This research uses mobile eye-tracking to evaluate consumer responses at a holistic level. This type of observational technology provides a first-person perspective into embodied behaviour reflecting goal orientation that third-person techniques cannot capture (e.g. Paletta and Rome 2007; Istance 2008). Thus far within marketing, eye-tracking technologies have been used to understand consumer cognitive and emotional responses to advertising communications, focussing on impact of branding, images and text (Pieters et al. 2007; Wedel and Pieters 2008a, b) in media, such as print and feature advertisements (e.g. Aribarg et al. 2010; Zhang et al. 2009), billboards (e.g. Dreze and Hussherr 2003), product labelling (e.g. Fox et al. 1998), TV commercials (e.g. Janiszewski 1998) and supermarket shelving (e.g. Chandon et al. 2007; Van der Lans et al. 2008). Researchers have found correlations between visual attention, i.e. frequency and duration of fixations and product preferences (Maughan et al. 2007) where positive attention results in more and longer fixations, albeit this may be a function of gender, age, personality (Rosler et al. 2005; Isaacowitz 2005) and familiarity with brands (Russo and Leclerc 1994). Of interest to marketers is how well respondents remember what they have seen, for example, firms invest heavily in differentiating brands predicated upon their distinctive and memorable features. Wedel and Pieters (2008a) have also proposed that visual attention in marketing is influenced by both motivational (intrinsic) and environmental (extrinsic) factors. This is, however, a controversial claim within the neuroscience research domain where opposing views suggest that visual attention is a function of purely intrinsic motivations (e.g. Theeuwes 1992) or purely extrinsic stimuli (e.g. Folk and Remington 1998) although these features may be processed in different parts of the brain, and visual attention itself is more a function of the proximity of an individual to relevant features (Becker et al. 2010; Becker 2012).

Thus, this research integrates the use of mobile eye-tracking to assess visual attention in a retail environment in order to explore the types of influences, both intrinsic and extrinsic, upon consumer behaviour at a holistic level. The development of the research methodology is discussed in the next section.

3 Methodology

Wedel and Pieters' (2008a) conceptualisation of visual attention was used as the basis for the research design that aimed to evaluate the naturalistic and holistic visual attention of consumers in a retail environment. A qualitative dominant-less dominant mixed method approach was used for the research design to explore visual attention. Mobile eye-tracking was selected as the dominant method as it provides a flexible means to capture first-person observational data (Duchowski 2007). Wedel and Pieters (2008a) suggest visual attention is both a function of consumer goals that inform what and where to look (intrinsic factors) and saliency of marketing stimuli. Saliency in the context of marketing is defined as the prominence of objects discerned by consumers within the scene (extrinsic factors) because of their relative luminosity, sharpness, brightness, contrast and colour (e.g. Summers and Herbert 2001; Custers et al. 2010). Intrinsic and extrinsic factors combine to produce attentional priorities and simultaneously suppress non-target perceptual features. Consumers may, however, only partially recall aspects of the goal such as characteristics of a brand (colour, shape, distinctive feature) and this is further influenced by the complexity of the scene (Pieters et al. 2010; Clement 2007). Thus, within a retail environment, visual attention is likely to be dominated by intrinsic relevancy of information to the individual ('informativeness') with the extrinsic effectiveness of marketing stimuli varying according to specific goals (Pieters and Wedel 2008a).

The study was conducted in a department store combining home and garden with men's and women's fashions and incorporating a café bar in the South of England, part of a UK-wide high street chain. Intrinsic factors that were anticipated as being relevant were the specific purchase goals and intentions of consumers coupled with the familiarity (previous experience) with the store layout. Pre- and post-tracking administered questionnaires (Eger et al. 2007) were used to understand these aspects of shoppers' journeys and to identify primary search behaviour and attention within the store environment, including, for example, any pre-planned and impulse purchase and browsing behaviours. Market Research Society ethics were adopted for the conduct of the study.

The technology used was Tobii Mobile™ glasses (manufactured 2010, recording resolution 640×480 at 30 Hz) in conjunction with Tobii Studio (version 2.0) and Noldus Observer XT (version 10.0), which enabled capture and analysis of both visual and audio data from research participants. Thus, the research was able to consider the role of any verbal data in assessing consumer response and together with fixations recorded by the mobile tracker (visual attention) enabling the researchers to identify behavioural patterns in the shoppers' journeys using content analysis and critical incident analysis identified in the videos to give rich insight into observed behaviours (Yin 1984; Miles and Huberman 1994). The technology, therefore, enabled the researchers to examine not only consumers' cognitive and affective behaviour but also conative (i.e. apparently instinctive) behaviour in the environment. Eye-tracking data were, firstly, content analysed and this subsequently informed critical incident analysis using an ethnographic approach (Arnould and Wallendorf 1994; Chong 2010) by coding, troping and representing behaviour constellations (Arnould and Wallendorf 1994) using both fixations and 'scanpaths' of the direction of visual attention (Duchowski 2007).

Firstly, content analysis was used to classify and analyse the frequency of fixations of participants in the store generated using the technology (visual attention). Given that proximity of consumers to extrinsic stimuli varied as they passed through the environment, necessitating a degree of flexibility in interpreting the focus of attention (e.g. near sight of, say, a meter and far sight of several meters, depending on the scene and apparent activity of the participant such as browsing products, walking through the store environment, etc.) it was not deemed possible to use the automated functionality provided by Tobii Studio to analyse locus of attention

Table 1 Study participants

Age	Male	Female	All
18–30 years	2	0	2
31–45 years	2	3	5
46–55 years	0	6	6
55+	0	3	3
Total	4	12	16

and draw meaningful conclusions. Therefore, eye-tracking videos were exported to Noldus Observer for content analysis by hand using fixation data to deduce visual attention. From the initial review of the eye-tracking data collected, the researchers, working with four coders, developed and described mutually exclusive categories (codes) for the content analysis in a qualitative mode (Haney et al. 1998). Extrinsic factors (codes) were identified as the actual store layout (space and product), navigational and section signage, promotional merchandising, product display, in-store offers, sales assistants/store staff and roles played by other consumers that facilitate search and goal-related behaviour.

Subsequently eye-tracking videos were sequentially coded to generate frequency (of occurrence) data for content analysis in a quantitative mode (Berelson 1952; Weber 1990; Krippendorff 2004). Each substantive change of focal attention (fixation) was noted and recorded for each participant, e.g. product A to product X generated two separate counts for the 'product' category. This is a method of event-based coding. A table of data was subsequently extracted and used as the basis for further evaluation. The method of content analysis does not represent a comprehensive analysis of all fixation points produced by the Tobii Mobile^{TM} equipment; however, it facilitated coding for data reduction and subsequent evaluation of critical incidents. A valence for each of the store departments was applied in order to assess focal attention (fixations) within the different areas of the store. Cohen's (1960) kappa coefficient was used to test inter- and intra-coder reliability of content analysis with one recording used as the basis for testing the level of agreement among all coders.

Within the retail environment, 7 h 16 min of eye-tracking data were collected at the store location from 16 adult visiting consumers (see Table 1), 10 of whom were accompanied by other adults (family or friend). Participants were asked not to change their intended shopping behaviour during the data collection process (data suggest that visual attention was naturalistic as it did not appear to alter stated consumer intentions). Non-response to participation was noted to be due to time constraints or unwillingness to be recorded: it was evident the technology was an attractor for some and repellent for others. The constraints of the technology precluded participants wearing prescription glasses, and the researchers also selected out adults with small children (because their focus would be likely to be on the child rather than the shopping experience). Sampling was, therefore, convenience based and bias may be present in the data from both self-selection and unknown/unanticipated factors. Nonetheless, all consumers had approached the store voluntarily before being asked to participate in the study. The preliminary recruitment phase included calibrating the equipment to each participant, followed by the pre-tracking questionnaire (see Table 2)

Table 2 Primary reason for visit compared to familiarity with store

Primary reason for store visit	Familiarity with store			Total
	First-time visit	Been twice before	Been 3+ times before	
Buy a particular product	1	3	1	5 (31.3 %)
Compare specific products	1	0	0	1 (6.3 %)
Browse products generally	4	1	3	8 (50.0 %)
Coffee	1	0	1	2 (12.5 %)
Total	7 (43.8 %)	4 (25.0 %)	5 (31.3 %)	16

Table 3 General satisfaction with store visit

Agreement with statement of satisfaction	Mean score (*n*=16) (1= disagree, 2= disagree slightly, 3=agree slightly, 4=agree)
The atmosphere in the store was appropriate for me (lighting, ventilation and heating)	4.0
I found what I was looking for	3.8
The store layout made it easy for me to find what I was looking for	3.8
The store layout made it easy to browse/find other interesting products	3.8
Information I used in the store was helpful	3.7
The information about the products I was interested in was clear	3.6
The sight lines in the store made it easy to navigate around	3.4
The signs for store departments/sections were clear	3.3

(see http://www.tobii.com/en/eye-tracking-research/global/library/manuals/ for a full description of this process). Consumers then shopped the store in their own time and subsequently returned to the researchers where the technology was removed and a post-tracking questionnaire completed the process (see Table 3). Given the qualitative sample used, it is not possible to apply statistics to produce more than a descriptive overview. Triangulation of methods and investigators (data coders) were primary modes of validation (Shapiro and Markoff 1997).

Table 4 summarises the frequency (count) and percentage of data coded for each category, i.e. coded fixations representing visual attention (total number of observations coded=44,625). Table 5 identifies the locations in the store browsed/shopped by the participants. Data indicate that participants primarily browsed/shopped within the home (52.3 % coded behaviours), women's/children's (28 %) and garden (7 %) departments of the store.

The findings show each area of the store elicits different visual attention patterns alluding to the interplay between the design, ambience and social-relational features and characteristics of the environment (Allen 2005). Of note are the roles of shopping partners and staff, who appear to play a greater role in the experience in home than observed in women's, where labelling played a greater role than in home

Table 4 Content analysis of eye-tracking data

Category	Category description	Frequency of coded fixations (n=44,625)	% coded behaviours to all (mean)
Product	All products in store	32,342	72.5
Shopping partner	A person in the store identified as shopping with the glasses wearer	2,868	6.4
Product label/price	All labels that identify product details or pricing	2,615	5.9
Staff	A person in the store identified as a member of staff (with or without an apron)	1,257	2.8
Customer	A person in the store shopping, excluding a shopping partner or member of staff	878	2.0
Brochure/leaflet	A loose brochure or leaflet about products, services, offers, prices, etc.	366	0.8
Set offer	All signs that identify a promotional offer located within a home design style set	179	0.4
Navigational sign	All signs that identify location or direction of a store section	159	0.4
Other	Looking at floor/ceiling/wall; no obvious focus on another defined category	3,961	8.9

Table 5 Content analysis of eye-tracking data by store department

Category	Home n=23,361 (%) (rank order)	Women's/ children's fashion n=12,488 (%) (rank order)	Garden n=3,142 (%) (rank order)
Product	74.4 (1)	82.7 (1)	87.6 (1)
Shopping partner	8.9 (2)	3.3 (4)	3.9 (2)
Product label/price	4.3 (4)	6.0 (2)	3.5 (3)
Staff	2.2 (5)	0.9 (6)	1.9 (4)
Customer	1.8 (6)	1.7 (5)	0.9 (6)
Brochure/leaflet	1.1 (7)	0.6 (7)	0.2 (7)
Set offer	0.7 (8)	0 (9)	0.1 (8)
Navigational sign	0.5 (9)	0.2 (8)	0 (9)
Other	6.1 (3)	4.6 (3)	1.8 (5)

and garden. This is broadly consistent with findings of research by Hu and Jasper (2006) into the positive impact of attentive store staff on consumer experiences. In the present study, the differences observed were felt likely to be due to the more specialized and technical nature of the product offer in home and garden and a greater familiarity with the purchase environment in fashion. It may also be a coincidental

Table 6 Summary of components used by consumers in store (excluding signage)

Component	Description
Light contrast	A product or visual component of the scene that is brightly illuminated or is in light contrast to its immediate vicinity (may be reflecting light, a feature light/product or natural day light)
Colour contrast	A product or visual component of the scene that is prominent by its colour (either neutral tones in a dark-colour scene or dark tones in a light-colour scene)
Vertical sight line (VSL)	A view in the shopping expedition where the consumer makes a directional decision and the decision scene is predominantly made up of vertical lines
Horizontal sight line (HSL)	A view in the shopping expedition where the consumer makes a directional decision and the decision scene is predominantly made up of horizontal lines ('panorama')
Socialisation	'Active' consumption experience—interactive elements that engaged consumers in experimenting with products
Flow	Immersion in the store layout enabling associations between product sections and departments

observation as a consequence of uneven distribution of staff in the departments across the store, with more staff evident in the technical store areas than elsewhere. Overall, content analysis of visual attention illustrates that participants focus most on products, whilst relatively little attention is given to signage, particularly navigational signage, in all areas of the store. From a marketing management perspective, this finding is especially important—marketing and merchandising materials (signage and display information) that support the consumer experience in store are considerable areas of spend for organizations but our data suggest they receive relatively little attention in this type of environment. Thus, ethnography was used to provide insight into critical incidents in the shopping episodes of consumers.

Critical incidents were identified as being those points in the store visit where decisions on what do, where to go or who to speak to were important considerations for the consumers, i.e. key decision points. The researchers aimed to identify elements that were intrinsic to the consumer and extrinsic in relation to the retail environment. Findings highlighted a range of elements that appeared to influence consumer behaviour in store (see Table 6). The intrinsic cues, comprising the products of relevance to an individual consumer, were most clearly evident in those observations where consumers were accompanied by partners through verbal discussion or sought assistance from a member of staff. The analysis identified behaviour patterns in the observations that were replicated across the data set. Patterns are noted in response to sight lines (vertical and horizontal) and saliency achieved through lighting and colour. Saliency (contrast) was found to be the dominant cue in the environment used by consumers to navigate the store. Lighting also supported socialising and flow in the environment. The patterns observed in relation to these findings are now discussed in turn, using video frames from participant eye tracking (fixations and scanpaths removed) to illustrate.

Fig. 1 Vertical sight line (VSL) towards back of store (*left*)

3.1 Sight lines and saliency

On entry to the store, participants paused a few seconds to look at an early feature in the store, low tables filled with highly salient products, and then towards either the back of the store (Fig. 1) or to the right using a vertical sight line (VSL). A less evident VSL was apparent to the left but not selected—research participants clearly preferred the more constrained aisle spaces offered. Thus, the VSL presentation appeared to facilitate the directional decision-making process which may also have been influenced by the highly salient product presentation evident along the aisles. Sight lines are an explicit component of store design that serve both functional and aesthetic needs. Functionally, they enhance store merchandise security by providing increased visibility to products and consumers; aesthetically, sight lines support the organization's aims and objectives for consumption activities within the space that underpin the creation of store ambience (Cohen 1998).

Whichever aisle participants moved into or towards, they initially focussed on feature-lit salient points in the scene—typically products (Fig. 2). They followed the VSL looking left and right, drawn quickly through the environment by the most salient features that their walking speed allowed. Whilst in an aisle, participants found something of more specific (intrinsic) interest that often appeared to be a neutral-coloured object, most likely located in a salient part of the scene. Aisles were filled with products and related information; thus, the focus on relatively neutral objects appears to reflect the idiosyncratic curiosity and preferences of research participants (Berlyne 1960). This may be because the 'noisiness' of the visual cues in the environment drives consumers to consider non-conforming stimuli but may also be because they have a directional aim and merely seek variety as they continue to their goal (e.g. Howard and Sheth 1969; Steenkamp and Baumgartner 1992).

Fig. 2 Feature lit salient product (*right*)

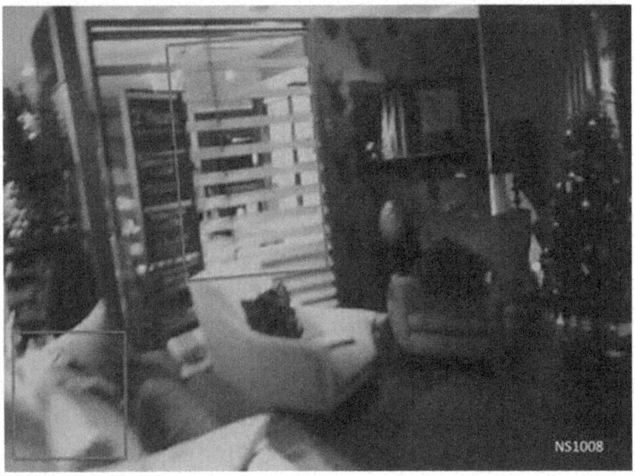

Fig. 3 Colour contrast saliency (neutral in dark tones, *left*)

Further investigation of browsing using saliency identified it was influenced also by contrast emphasised by colour blocks (Fig. 3). Colour blocking is likely to be both an extrinsic and intrinsic cue—salient within a scene but relevant to the participant in some way; for example, there is evidence that colour preference is used in consumers' search patterns evoking empathy with the scene in some way (Klonk 2005) such as red-themed store offers which were dominant for some participants (Clynes 1977). Colour has a long association with designed environments (Fortune 1930), influencing consumer behaviour by enhancing ambience (Nayar 2008) and influencing mood states in stores (Martineau 1958; Bellizzi et al. 1983).

Horizontal sight lines (HSLs), where the scene was presented as a 'panorama' (Fig. 4), were evident in a few parts of the store and these elicited different visual search behaviour to VSLs. Visual behaviour at HSLs involved pausing for a longer

Fig. 4 Horizontal sight line (HSL, *right*)

period of time than VSLs and scanning the scene rapidly far left to far right, fixating on feature-lit objects across the scene. The decision to move appeared to follow some apparent indecision to return to an aisle, i.e. turning around and looking back. Feature lighting and colour-blocked objects acted as draws in the 'panorama'; and once in the scene, participants seemed to revert to rapid browsing behaviour as well as scanning their visual horizon to navigate through the space, until they found a product or feature they may be more intrinsically interested in. Browsing behaviour is most evident when consumers are in scenes presented as VSLs.

Whereas HSLs support store management, because there are fewer visual barriers for security cameras, the findings suggest these areas within the store discourage browsing behaviour. The most likely reason is that consumers are simply too far away from products to investigate them in sufficient detail; however, another reason may be the holistic impact generated by the store ambience. Whereas previous research has found space to positively influence consumer behaviour (Ballantyne et al. 2010), this appears not to be the case in this research. Consumers clearly found the store engaging as a social space, evidenced through the ways in which they interacted with products, staff and other consumers as they shopped. The enforced distance between the consumer and the merchandise apparently presented through an HSL effectively decreased proximity (immediacy) of staff, other consumers (Janakiraman and Niraj 2011; Xu et al. 2012) and products, which the consumers participating in the study evidently found discomforting. Social interaction theory has been found to explain the affective influence of the presence of people on consumer behaviour in shopping environments (Argo et al. 2005), but the impact of store ambience and the ways in which it may change over the store context is not well understood despite a number of researchers investigating this from the stimulus–organism–response (Turley and Milliman 2000; Baker et al. 2002; Hyde 2006) or hedonic-consumption (combining attractive and facilitating stimuli) perspectives (Kotler 1974; Hirschman and Holbrook 1982; Ballantyne et al. 2010).

3.2 Socialisation

Socialisation is more than just the interactions between consumers and staff; it encompasses their interactions with the whole environment—in our evaluation, social consumption is central to the ambience in the store that has been intentionally incorporated into the design of the space, including the presentation and merchandising of products and inclusion of a café concession. A purchase decision from a design (product) set, which was used by the store to present connections between the products in interactive displays, typically involved a staff member, either in locating a particular product on a shelf display or in completing an order process for a product. Elsewhere in the store, products on shelves, or otherwise grouped together into a display, were selected from the display and taken to a check out. Store staff were an active part of many scenes, appeared to be easily approachable and drawn into both browsing and purchase episodes. Participants also actively engaged with design sets, such as room layouts in the store, 'trying out' products by enacting some imagined behaviour from their personal lives and involving others in the consumption experience, e.g. putting feet up on a foot stool, laying back on a sofa, sitting at a dining table, etc. Interesting products placed in design sets were subsequently sought out from shelves (e.g. cushions, bed linen, ornaments), but areas where products were displayed on shelves were of less focal interest to participants than products in sets (which was consistent with stated browsing intentions). Thus, the creative display of content appealed to participants, giving them an opportunity to browse, relax and enjoy the physical and visual stimulation provided by the environment. This was reinforced by the café, where consumers paused in their shopping trip for refreshments, discussion with their shopping partners or just to sit and contemplate the store environment.

The ambience generated by the socialisation within the environment was therefore co-created by consumers (Vargo and Lusch 2004). This finding is consistent with research into hedonic consumption but also highlights the store's intentions in creating experiences through its designed layout and merchandise presentation. Crilly (2011) uses visual rhetoric to consider the role of design in persuading consumers to engage and affect behavioural patterns, albeit an underdeveloped concept. He states there is a tension between informing and persuading through the design process that only the consumer experiences in order to infer 'persuasive intention', depending on their motivations for participation. Considering such an approach, consumers may infer from the presentation of products and presence of a café in the store that interaction and role play are intended to be part of the consumption experience in addition to their purchase behaviour.

3.3 Flow

Flow is the sense of immersion in an experience, a mood state between arousal and relaxation wherein the individual is in some optimum state of intrinsic motivation (Csíkszentmihályi 1990). Within a store, antecedents may be the challenge and skills consumers have in engaging in the experience environment (Wang and Hsiao 2012), such as the nature of the products and number of choices (Swaminathan 2003) or familiarity with the store (Wang and Hsiao 2012). Evident in the discussion above in the ways in which consumers engaged with design sets, flow was apparent when the store layout and navigational signage became frustrating and consumers could not find something they were specifically interested in, say that they had identified in a design set or possibly from a previous store visit, or the content of the section did not appear to 'make sense'. Findings suggest participants were using a narrative approach to shop the store that was more intrinsically directed than random or directed by the store navigation system. For example, participants appeared to correlate content between sections and subsections which may psychologically influence flow behaviour: there were links made between living room and decorative items; kitchens and utilitarian items; lighting, mirrors and wall paints; women's clothing and shoes; etc. The store presentation clearly reinforced the interconnections but large sections of the store were not visited by participants which may represent a disconnect in the narrative they pursued—the sections may repel the consumers because there are no clear associations being made between product categories as presented in the sections and subsections.

4 Conclusions

Our findings provide general support for Wedel and Pieters' (2008) theory of visual attention, i.e. consumer goals (intrinsic motivation) and saliency (prominence) of marketing stimuli (extrinsic stimuli) appear to have combined to establish attentional priorities within the store environment: visual attention is dominated by intrinsic relevancy of information to the individual ('informativeness') with the extrinsic effectiveness of marketing stimuli (Pieters and Wedel 2008a) driving both goal/task-related and browsing behaviour. Findings also highlight that attention is influenced by proximity to relevant features (Becker et al. 2010; Becker 2012). Within the store, signage was relatively little used compared to other components of visual scenes such as products and people. Therefore, although both intrinsic and extrinsic motivators appear to play a role, key elements consumers used in the store were lighting, colour blocks, sight lines and product. These combined and were presented through design sets that, in turn, promoted socialising behaviour and flow, where consumers engaged in 'active consumption' of the store environment. This pattern appears to contribute to store ambience, but more needs to be understood about the dynamic interplay between the elements and the ways in which they influence behaviour during a shopping episode.

Saliency (prominence) is highlighted as a dominant aspect that impacts on consumers' attentional priorities. Within the store, used in this study, saliency was primarily achieved with lighting, which was evidently used by consumers as a stimulus that encourages exploration of the departments of the store. Whereas previous studies have tended to associate saliency, using contrast and colour, with recall (Alba and Chattopadhyay 1986; Lynch and Srull 1982), this finding is particularly interesting because it adds a further dimension to our understanding of store environments. For example, the store clearly used lighting to highlight details and features of products or product areas but in so doing it has created contrast that added to the 'drama' into the store experience that appears to influence consumers to socialise and induce flow (Csikszentmihalyi 1990). The dominance of the lighting in the store appears to have impacted the hierarchy of effects of other components such as marketing messages and navigational aids. These findings are broadly consistent with design principles (e.g. Israel 1994); yet, this is an aspect that is apparently ignored in the marketing literature (Ballantyne et al. 2010). Much more needs to be understood about the role of lighting in influencing consumer response patterns, especially in relation to store navigation, because of the potential impact it has on the use of traditional (and expensive) promotional materials.

The research undertaken has obvious limitations that relate to the use of small samples and qualitative methodology. The research reported on is, however, an exploratory study that sought to evaluate naturalistic behaviour at a holistic level and its design was intended to consider visual attention from a first-person perspective. Whilst the size of samples used in the research limits the generalizability of findings, the external validity of findings reported on may, nonetheless, be used to develop hypotheses for more quantitative focussed investigations in the future.

References

Alba, J., & Chattopadhyay, A. (1986). Salience effects in brand recall. *Journal of Marketing Research*, *23*(Nov), 363–369

Anon. (1930). Color in industry. *Fortune,* Feb, pp 85–94.

Argo, J. J., Dahl, D. W., & Manchanda, R. V. (2005). The influence of a mere social presence in a retail context. *Journal of Consumer Research, 32*(2), 207–212.

Aribarg, A. R., Pieters, R., & Wedel, M. (2010). Raising the BAR: Bias adjustment of recognition tests in advertising. *Journal of Marketing Research, 47*(3), 387–400.

Arnould, E. J., & Wallendorf, M. (1994). Market-oriented ethnography: Interpretation building and marketing strategy formulation. *Journal of Marketing Research*, *XXXI* (Nov), 484–504

Bagozzi, R. P., & Dholakia, U. (1999). Goal setting and goal striving in consumer behavior. *Journal of Marketing*, *63*(Special Issue), 19–32

Baker, J., Parasuraman, A., Grewal, D., & Voss, G. (2002). The influence of multiple store environment cues on perceived merchandise value and patronage intentions. *Journal of Marketing, 66*(2), 120–141.

Ballantyne, P. W., Jack, R., & Parsons, A. G. (2010). Atmospheric cues and their effect on the hedonic retail experience. *International Journal Of Retail & Distribution Management, 38*(8), 641–653.

Basil, M. (2011). Use of photography and video in observational research. *Qualitative Market Research: An International Journal, 14*(3), 246–257.

Becker, S. I. (2012). Keynote: Transient versus sustained attention are based on different mechanisms: An eye movement study. Eye-Track Australia 2012 Conference, 31 May to 1 June, University of Central Queensland, Australia

Becker, S. I., Folk, C. L., & Remington, R. W. (2010). The role of relational information in contingent capture. *Journal of Experimental Psychology: Human Perception and Performance, 36,* 1460–1476.

Belk, R. (1974). An exploratory assessment of situational effects in buyer behaviour. *Journal of Marketing Research, 11*(May), 160

Bellizzi, J. A., Crowley, A. E., & Hasty, R. W. (1983). The effects of color in store design. *Journal of Retailing, 59*(1), 21.

Berelson, B. (1952). *Content analysis in communication research.* Glencoe, Ill: Free Press.

Berlyne, D. E. (1960). *Conflict, arousal and curiosity.* London: McGraw Hill.

Bitner, M. J. (1992). Servicescapes: The impact of physical surroundings on customers and employees. *Journal of Marketing, 56,* 57–71.

Chandon, P. J., Hutchinson, W., Bradlow, E. T., & Young, S. (2007). Measuring the value of point-of-purchase marketing with commercial eye-tracking data. In M. Wedel & R. Pieters (Eds.), *Visual marketing: From attention to action (pp. 225–258),* Mahwah, NJ: Lawrence Erlbaum Associates.

Chong, J. L. L. (2010). Evaluating the impact of Arnould and Wallendorf's (1994) market-oriented ethnography. *Journal of Business Research, 63,* 12(Dec), 1295–1300.

Clement, J. (2007). Visual influence on in-store buying decisions: An eye-track experiment on the visual influence of packaging design. *Journal of Marketing Management, 23*(9–10), 917–928.

Clynes, M. (1977). *Sentics: The touch of emotions.* Garden City: NY, Anchor Press

Cohen, E. (1998). Trussardi alla Scala. Interior Design, New York. *NY, 69*(5), 178–185.

Cohen, J. (1960). A coefficient of agreement for nominal scales. *Educational and Psychological Measurement, 20,* 37–46.

Crilly, N. (2011). Do users know what designers are up to? Product experience and the inference of persuasive intentions. *International Journal of Design, 5*(3), 1–15.

Csikszentmihalyi, M. (1990). *Flow: The psychology of optimal experience.* New York: Harper and Row.

Custers, P., de Kort, Y., IJsselsteijn, W., & de Kruiff, M. (2010). Lighting in retail environments: Atmosphere perception in the real world. *Lighting Research & Technology, 42*(3), 331–343.

Davis, T. C. (1991). Theatrical antecedents and the mall that ate downtown. *Journal Of Popular Culture, 24*(4), 1–15.

Donovan, R., & Rossiter, J. (1982). Store atmosphere: An environmental psychology approach. *Journal of Retailing, 58,* 34–57.

Dreze, X., & Hussherr, F. (2003). Internet advertising: Is anybody watching? *Journal of Interactive Marketing, 17*(4), 8–23.

Duchowski, A. (2007). *Eye tracking methodology: Theory and practice.* Berlin: Springer.

Eger, N., Ball, L. J., Stevens, R., & Dodd, J. (2007). Cueing retrospective verbal reports in usability testing through eye-movement replay. *People and Computers XXI—HCI ... but not as we know it.* Proceedings of HCI, British Computing Society

Folk, C. L., & Remington, R. (1998). Selectivity in distraction by irrelevant featural singletons: Evidence for two forms of attentional capture. *Journal of Experimental Psychology: Human Perception and Performance, 24*(3), 847–858.

Fox, R. J., Krugman, D. M., Fletcher, J. E., & Fischer, P. M. (1998). Adolescents attention to beer and cigarette print ads and associated product warnings. *Journal of Advertising, 27*(3), 57–68.

Gauri, D., Minakshi Trivedi, K., & Grewal, D. (2008). Understanding the determinants of retail strategy: An empirical analysis. *Journal of Retailing, 84*(3), 256–267.

Haney, W., Russell, M., Gulek, C., & Fierros, E. (Jan-Feb, 1998). Drawing on education: Using student drawings to promote middle school improvement. *Schools in the Middle, 7(3)*, 38- 43.

Hoch, S. J., & Deighton, J. (1989). Managing what consumers learn from experience. *Journal of Marketing*, *53*, (Apr), 1–20.

Holbrook, M. B., & Hirschman, E. C. (1982). The experiential aspects of consumption: Consumer fantasies, feelings and fun. *Journal of Consumer Research, 9*(2 Sept), 132–140.

Howard, J. A., & Sheth, J. (1969). *Theory of buyer behaviour.* New York: Wiley & Sons.

Hu, H., & Jasper, C. R. (2006). Social cues in the store environment and their impact on store image. *International Journal of Retail & Distribution Management, 34*(1), 25–48.
Hyde, K. (2006). Contemporary information search strategies of destination-naı¨ve international vacationers. *Journal of Travel and Tourism Marketing, 21*(2/3), 63–76.
Isaacowitz, D. M. (2005). The gaze of the optimist. *Personality and Society Psychology Bulletin, 31*(3), 407–415.
Israel, L. J. (1994). *Store planning/design: History, theory, process*. New York: John Wiley & Sons.
Istance, H., Bates, R., Hyrskykari, A., & Vickers, S. (2008). Snap Clutch, a moded approach to solving the midas touch problem. *ETRA*, Savannah, Georgia, March 26–28
Janakiraman, R., & Niraj, R. (2011). The impact of geographic proximity on what to buy, how to buy, and where to buy: Evidence from high-tech durable goods market. *Decision Sciences, 42*(4), 889–919.
Janiszewski, C. (1998). The influence of display characteristics on visual exploratory search behavior. *Journal of Consumer Research, 25,* 290–301.
Kent, R., & Allen, C. T. (1994). Competitive interference effects in consumer memory for advertising: The role of brand familiarity. *Journal of Marketing*, *58*(July), 97–105
Klonk, C. (2005). Patterns of attention: From shop windows to gallery rooms in early twentieth-century Berlin. *Art History, 28*(4), 468–496.
Kotler, P. (1974). Atmospherics as a marketing tool. *Journal of Retailing, 49*(4), 48–64.
Krippendorff, K. (2004). *Content analysis: An introduction to its methodology* (2nd ed). Thousand Oaks, CA: Sage.
Lee, A. Y., & Sternthal, B. (1999). The effects of positive mood on memory. *Journal of Consumer Research*, *26*(Sept), 115–127.
Lynch, J., & Srull, T. (1982). Memory and attentional factors in consumer choice: Concepts and research methods. *Journal of Consumer Research*, *9*(June), 18–37.
Markin, R., Lillis, C., & Narayana, C. (1976). Social-psychological significance of store space. *Journal of Retailing, 52,* 43–54.
Martineau, P. (1958). The Personality of the Retail Store. *Harvard Business Review, 36*(1), 47–55.
Maughan, L., Gutnikov, S., & Stevens, R. (2007). Like more, look more. Look more, like more: The evidence from eye-tracking. *Journal of Brand Management, 14*(4), 335–342.
Mehrabian, A., & Russell, J. A. (1974). *An approach to environmental psychology*. Cambridge, MA: MIT Press.
Miles, M. B., & Huberman, A. M. (1994). *Qualitative data analysis*. London: Sage Publications.
Nasar, J. L. (1987). Effect of sign complexity and coherence on the perceived quality of retail scenes. *Journal of the American Planning Association, 53*(4), 499–509.
Nayar, J. (2008). Light and leisure. *Contract (San Francisco, CA 2000)*, *50*(11), 52–56.
Paletta, L., & Rome, E. (2007). *Attention in cognitive systems: Theories and systems from an interdisciplinary viewpoint. 4th International Workshop on Attention in Cognitive Systems, WAPVC*. Berlin: Sprenger-Verlag.
Park, C. W., Easwar, S. I., & Smith, D. C. (1989). The effects of situational factors on in-store grocery shopping behavior: The role of store environment and time available for shopping. *Journal of Consumer Research*, *15*(Mar), 422–433.
Pieters, R., Wedel, M., & Zhang, J. (2007). Optimal feature advertising design under competitive clutter. *Management Science, 53*(11), 1815–1828.
Pieters, R., Wedel, M., & Batra, R. (2010). The stopping power of advertising: Measures and effects of visual complexity. *Journal of Marketing, 74*(5), 48–60.
Rook, D. W. (1987). The buying impulse. *The Journal of Consumer Research, 2*(14), 189–199.
Rosler, A., Ulrich, C., Billino, J., Sterzer, P., Weidauer, S., Bernhardt, T., Steinmetz, H., Frolich, L., & Kleinschmidt, A. (2005). Effects of arousing emotional scenes on the distribution of visuo-spatial attention: Changes with aging and early subcortical vascular dementia. *Journal of Neurological Science*, 229–230, 109–116.
Russo, J. E., & Leclerk, F. (1994). An eye-fixation analysis of choice processes for consumer non-durables. *Journal of Consumer Research, 21,* 274–290.

Shapiro, G., & Markoff, J. (1997). A matter of definition. In C. W. Roberts (Ed.), *Text analysis for the social sciences: Methods for drawing statistical inferences from texts and transcripts*. Mahwah, NJ: Lawrence Erlbaum Associates.

Spanenburg, E. R., Crowley, A. E., & Henderson, P. W. (1996). Improving the store environment: Do olfactory cues affect evaluations and behaviors? *Journal of Marketing*, *60*(Apr), 67–80.

Steenkamp, J.-B., & Baumgartner, H. (1992). The role of optimum stimulation level in exploratory consumer behaviour. *Journal of Consumer Research*, *19*(Dec), 434–448

Suhur, J., & Sorenson, H. (2010). The power of Atlas: Why in-store shopping behavior matters. *Journal of Advertising Research, 50*(1), 21–29.

Summers, A. T., & Hebert, P. R. (2001). Shedding some light on store atmospherics influence of illumination on consumer behaviour. *Journal of Business Research, 54,* 145–150.

Swaminathan, V. (2003). The impact of recommendation agents on consumer evaluation and choice: The moderating role of category risk, product complexity, and consumer knowledge. *Journal of Consumer Psychology*, *13*(1 and 2), 93–101.

Theeuwes, J. (1992). Perceptual selectivity for color and form. *Perception & Psychophysics, 51,* 599–606.

Turley, L. W., & Milliman, R. E. (2000). Atmospheric effects on shopping behaviour: A review of the experimental evidence. *Journal of Business Research*, *49*(2 August), 193–211.

Van der Lans, R., Pieters, R., & Wedel, M. (2008); salience. *Marketing Science, 27*(5), 922–931.

Vargo, S., & Lusch, R. (2004). Evolving to a new dominant logic for marketing. *Journal of Marketing*, *68*(Jan), 1–17.

Venkatesh, S. J., Inman, J., Mantrala, M., Kelley, E., & Rizley, R. (2011). Innovations in shopper marketing: Current insights and future research issues. *Journal of Retailing*, *87*, S29–S42.

Wang, L. C., & Hsiao, D. F. (2012). Antecedents of flow in retail store shopping. *Journal of Retailing and Consumer Services, 19*(4), 381–389.

Weber, R. P. (1990). *Basic content analysis* (2nd ed.). Newbury Park, CA: Sage.

Wedel, M., & Pieters, R. (2008a). Eye tracking for visual marketing. *Foundations and Trends in Marketing*, *1*(4), 231–320, 2006.

Wedel, M., & Pieters, R. (2008b). A review of eye-tracking research in marketing. In N. K. Malhotra (Ed.), *Review of marketing research*, *4* (pp. 123–147). Armonk, NY: M.E. Sharpe.

Xu, J., Chen, H., & Wyer, R. S. (2012). Does the distance between us matter? Influences of physical proximity to others on consumer choice. *Journal of Consumer Psychology, 22*(3), 418–423.

Yin, R. K. (1984). *Case study research: Design and methods*. Newbury Park: Sage Publications.

Zhang, J., Wedel, M., & Pieters, R. (2009). Sales effects of attention to feature advertisements: A Bayesian mediation analysis. *Journal of Marketing Research*, *46*(Oct), 669–681.

Private and Public: Eye Movement and Eye Tracking in Marketing

En Li, James Breeze, Mike Horsley and Donnel A. Briely

1 Growth in Eye Movement and Eye Tracking Research

There has been an exponential growth in eye research and concomitant use of eye-tracking methodologies in the last decade. Improvements in eye-tracking technologies, enhancements in the sophistication of eye-tracking data analysis software and the development of smaller, more portable and mobile eye trackers has immeasurably increased the range and volume of applications. Komorgortsev (2011), using Google Scholar, has demonstrated that eye-tracking-based research publications had reached over 2000 articles and papers in 2011.

2 Eye Tracking in Marketing Research

This exponential increase in applications of eye-tracking research methodologies and applications is understated in marketing and marketing-related disciplines. This is because as well as eye-tracking research and applications being undertaken and disseminated by the university and academic sector a whole eye-tracking marketing industry has arisen in the private sector. This private sector industry is both large and expanding, with hundreds of companies internationally conducting eye-tracking research and consulting for private and public sector clients.

E. Li (✉)
Central Queensland University, Rockhampton, QLD, Australia
e-mail: e.li@cqu.edu.au

J. Breeze
Objective Eye Tracking Pte Ltd, Sydney, NSW, Australia

M. Horsley
Central Queensland University, Noosaville, QLD, Australia

D. A. Briely
University of Sydney, Sydney, NSW, Australia

M. Horsley et al. (eds.), *Current Trends in Eye Tracking Research*,
DOI 10.1007/978-3-319-02868-2_15, © Springer International Publishing Switzerland 2014

A core development in this space has been the rise of what Duchowski (2007) has termed 'applied research organisations'. In many cases, such applied research organisations are wholly or partially owned by eye-tracking firms. Such organisations provide full service delivery for eye-tracking businesses: training and development, marketing research, consulting and the development of specialised software to solve industry problems.

These applied research organisations can also be eye-tracking technology service providers and also private marketing business specialising in neuroscience solutions for marketing problems. In addition, some large national and multinational firms have developed eye-tracking marketing research capabilities within their own marketing departments.

The reasons for these developments in eye-tracking marketing research were first outlined by Duchowski (2007), who noted that 'the motivation for utilising in market research stems from the desire to understand consumer actions. …eye tracking can provide insight into at least one aspect of the internal consumer model: how the consumer disperses visual attention over different forms of advertising' (Duchowski 2007, p. 262). Eye-tracking research can also be used to analyse and develop the consumer attention process and modify presentations and display to increase the capture of this attention. One part of marketing research using eye tracking is to develop usability research—to iteratively design more attention developing visual display through systematic collection of user response data.

3 The Rise of Applied Eye-Tracking Research Organisations

The importance of marketing in modern business and the significant and considerable investment in marketing and the evaluation of the costs and benefits of this investment have directly led to the rise of applied eye-tracking research organisations. These organisations and cognate private marketing businesses, however, rarely share their research. Usually the research is proprietary and confidential. Dissemination of results and methodologies is usually contractually restricted, commercial in confidence or has involved substantial intellectual property development—protected through contractual negotiations. As a result, much the actual marketing research work is not reported or released. Accordingly, there is a significant gap in understanding of the entire corpus of eye-tracking marketing research due to its proprietary nature and the restrictions around its dissemination and publication.

As well as providing an academic case study based on eye movement research, the chapter also provides a case study from a private sector organisation. This chapter presents a case study from an applied eye-tracking research organisation, which seeks to reflect the large volume of such eye-tracking marketing research and represent the eye-tracking marketing research activities conducted by such applied market research organisations. Central to the case study are the needs of the clients and the directed nature of the research to solve business problems in marketing.

4 Case Study 1: A Case Study from the Financial Services Industry on the Facilitation of Customer Design

4.1 Background to case study

Customer experience is a buzz word in the retail arena. The last decade has seen a tremendous change in the way customers interact with services. Ubiquitous technology has empowered the customer to stay connected and informed. This increased power and awareness has resulted in higher customer expectations of the retail experiences they are delivered. Businesses are, therefore, looking at user research for understanding their customers and the context of use of their services and, thus, design services that provide the enhanced experience the customer is expecting. Eye tracking is fast catching up as a preferred user research technique. Businesses are leveraging on eye tracking to understand the outreach of their marketing collaterals, the physical evidences in their retail outlets, the effectiveness of their service encounters, etc.

The Service Innovation practice of a bank has been propagating the value of service design and the need to design for customer experience. In this context, the bank partnered with Objective Eye Tracking Pte Ltd, an eye-tracking consulting company in conducting customer experience research for a major Financial Services Institution that offers retail banking services. The retail banking service provider was interested in understanding their customer, their context of use of the bank branch services and the performance of their products and merchandise in attracting the attention of customers and embedding their brand and products in their minds.

4.1.1 Eye Tracking

Tobii eye-tracking glasses were used in the research to record audio, video and eye gaze position of the person wearing the glasses. In the Tobii eye trackers, the image sensors collect visual information of the person and the reflection patterns generated on the cornea using the near-infrared diodes. Applying sophisticated image analysis and mathematics on this information, the exact gaze point is calculated every 30th of a second, thus revealing what the person is looking at.

The eye-tracking glasses provide data in the form of the audio/conversations, video/scene within the field of vision of the user and the gaze point. In addition to this information, it is necessary to capture the intention behind the attention given to an item of interest. This is done through a retrospective think-aloud interview session, where the user watches a playback of the video recorded and voices their intentions as they come across various points of interest. Using this technique, we review both their conscious and unconscious experiences to decode exactly why they behave in a particular way.

4.2 The Research

The bank has several branches situated island-wide typically serving customers who work/live in the neighbourhood. The bank branches, like any retail outlet, use their physical venue to display marketing merchandise that promotes the service products offered by the bank. The bank has modelled some of the branches to cater to certain demography of customers that it serves, thus displaying marketing merchandise relevant to that demography.

The objective of this research was to provide an understanding of how all the merchandising is working (or not working) within the branch. The bank had also identified some merchandise as their 'areas of interest', which need to be specifically analysed in this research. From this understanding, the bank planned to design its campaigns and services to benefit the customers, marketing teams and branch managers.

Based on the 'areas of interest' identified and the objective of the research, the metrics that need to be derived were identified. Participant percentage, fixation count and fixation length were some of the key metrics that were identified to provide insights into the 'area of interest' that captures and retains the attention of the bank's customers.

The research was conducted in four bank branches, two typical, one themed on a specific product and one remodelled with mobile gadgets to cater to digital natives and digital immigrants. The data collection was done over 2 weeks at these branches. The walk-in customers to the bank were intercepted and enrolled for the research. The enrolment included set-up with the Tobii eye-tracking glasses by fastening the glasses at the back of the head with a head strap and calibrating the glasses. Once enrolled, the participants proceeded to perform their transaction at the teller. The participants had to wait in queues before their turn to be served which allowed for rich data capture. While the participants are wearing the glasses and transacting, they are also passively observed for any specific action or intervention. On completing their transaction, the glasses were recovered from the participants and the recording was played back to them for their retrospective think-aloud interview. The participants were questioned on their thought process at various points in the video and these data were also recorded. The participants were compensated for their time with an incentive.

The video recording, the gaze-point information, the audio, the recording of the retrospective think aloud and notes made by the observer form the volume of data that were later analysed. The qualitative data collected were manually coded against the preset 'areas of interest' and statistical data were derived. On the other hand, the researchers also used affinity diagramming to synthesize and derive insights from the qualitative data gathered. The results from these derived data were put together and presented to the bank as a baseline for the performance of their various merchandise.

The results were presented to the bank in two stages. The first-stage presentation was in the form a workshop, with a purpose of providing the wider team with an overview of the methodology and headline findings. In the workshop, the audience was grouped into teams and provided with some of the issues uncovered in the research and guided through analysis and brainstorming sessions to identify possible solutions. The second and final presentation was given in the form of a detailed report which presented statistically significant evidences that confirm/refute hunches, convert early hypotheses to tenable theories and present new findings that establish a baseline for the marketing merchandise in the context of retail branch banking.

In this study, visual media (TV) emerged a clear leader as opposed to print media (brochures, posters, etc.) in capturing and retaining attention, with 85 % attention and 81.5 s dwell time, as opposed to their immediate competitor, menu boards having 75 % attention and 15.7 s dwell time.

This eye-tracking research has provided the bank with findings and evidences that are beneficial in establishing a baseline for their merchandise and strategizing their campaigns and services to better serve their purpose.

4.3 Customer Experience and Eye Tracking

It is seen that people choose products and services based on their experiences with them. Such experiences are formed over many touch points or interactions the customers have with an agent or artefact of the service or product. The experience the customers take away is unique and depends on them and their context. Thus, it is essential to understand the profile of the users, their expectations and the various contexts of use of the service/product. Eye tracking works as a favourable data collection mechanism to provide insights into the use of the services or products.

5 Case Study 2: A Case Study Exploring the Influence of Articulatory Suppression on the Reading Direction Effect

5.1 Background to Case Study

This case study presents a research project that examined the influence of articulatory suppression on an eye movement driven phenomenon, the "reading direction effect" (Li and Briley, 2011). This project is broadly representative of published eye movement marketing research emanating from university/academic marketing departments.

The past decade has seen noticeable research advances on eye tracking and eye movement processes in the academic marketing area. For example, Pieters and Wedel (2007) utilised an eye-tracking experiment to investigate the influences of consumer goals on their visual attention in viewing advertisements. They found that when a consumer aimed to memorize an advertisement, both its verbal and pictorial information was attended to a higher level; whereas when the consumer's goal was to learn the brand, only verbal component in the advertisement received increased attention (Pieters and Wedel, 2007). In another series of eye-tracking studies, Brasel and Gips (2008) examined the potential threat toward advertising effectiveness posed by consumer fast-forwarding behaviour (e.g., consumers utilising video recorders to fast-forward through the advertisement breaks) and they found that fast-forwarding viewers tended to constrain their attention primarily to the central screen position. Hence, they argue that marketers can alleviate the negative impacts of video recorders by placing brand information centrally in advertisements (Brasel and Gips, 2008). Atalay, Bodur, and Rasolofoarison (2012) also conducted eye-tracking studies to further explore the centrality bias where they demonstrated brands presented at horizontal centres were able to gather more attention and selected more frequently by consumers.

Despite the fruitful findings, the academic marketing research has so far mainly focused on gauging consumer visual attention under ideal laboratory environments. Little effort has been made to address the potential interference effects on eye movements or eye tracking by secondary tasks consumers engage in on daily basis. For instance, can consumers' eye movement processes be shaped by their concurrent oral muscle movements, including consumer daily activities such as eating or speaking? This case study aims to provide a preliminary answer to this question. Specifically, it looks into how eating activities can interfere with the eye movement processes underlie the reading direction effect, where consumers' attitudes toward a moving brand increase when the brand movement direction is consistent with their habitual reading direction (Li and Briley, 2011).

5.2 Previous Research on the Reading Direction Effect

Li and Briley (2011) proposed that two modes of eye movement could lead to the reading direction effect: a habitual and a situational mode. First, as the dominant eye movements in processing alphanumeric stimuli (e. g. brand names) are made in the habitual reading direction (Rayner 1998), eye movements in this habitual direction could act more fluently than in other directions. Secondly, when an alphanumeric stimulus moves horizontally across a consumer's field of vision, it tends to activate situational eye movements of identical direction (Palmer 1999). If the direction of activated eye movements coincides (vs. conflicts) with the consumer's habitual reading direction, s/he might experience heightened motor fluency. This process might lead to increased evaluations of the stimulus (i.e., the reading direction effect) when the consumer attributes the fluency experience to the quality of the stimulus itself (Bornstein and D'Agostino 1994).

Li and Briley (2011) presented compelling evidence for the reading direction effect. They demonstrated that this directional bias existed on both pure (i.e., moving brand name) and mixed alphanumeric stimulus (i.e., moving brand logo; Li and Briley, 2011). Furthmore, this effect should be driven by motor fluency in eye movements and procedural knowledge activated in reading (Li and Briley, 2011).

5.3 Current Research Project

Extent research suggests that motor fluency effects should be characterised by their effector dependency (e.g., Topolinski and Strack 2009). That is, such effects can only be blocked by engaging fluency-associated muscles. If that is the case, and if the reading direction effect is really driven by motor fluency, then a concurrent unrelated motor task, say, a manual motor task, should not cause any interference on the effect. Furthermore, as being exposed to words (e. g. moving brand name) automatically triggers reading activities (Stroop 1935), and the central effectors in reading are the oral muscles (McGuigan et al. 1964), the reading direction effect should be blocked by an articulatory suppression manipulation (Emerson and Miyake 2003), e. g. oral muscle movements such as gum chewing activities (Campbell et al. 1991).

The current research project examined conditions under which the reading direction effect was susceptible to a concurrent motor task. In this study, participants (all left-to-right readers) were assigned to one of four conditions, based on a two (brand logo movement: left-to-right vs. right-to-left) by two (concurrent motor task: manual vs. oral) between-participants design. Specifically, participants in the concurrent manual task condition were asked to watch and evaluate a moving brand logo while kneading a rubber ball, whereas participants in the concurrent oral task condition chewed a piece of gum while evaluating the moving brand. The results demonstrated a replicated reading direction effect under the concurrent manual task. That is, participants who kneaded the rubber ball indicated higher likings toward the moving brand when it moved from left-to-right rather than from right-to-left. However, this effect went away among participants engaging in gum chewing while evaluating the moving brand. Furthermore, no difference was found on the amount of attention paid to the brand evaluation task between participants in the concurrent manual and oral tasks, ruling out the alternative explanation of attention depletion. These results show that the reading direction effect is indeed effector dependent, adding evidence that motor fluency and procedural knowledge activated in reading are the underlying mechanisms behind this effect.

5.4 Implications for Eye Movement and Eye Tracking Research

The findings of the current research project also shed light on general practice and future directions in eye movement and eye tracking research. On the one hand, as visually presented language stimuli can only access phonological working memory through subvocalization (e.g., Eiter and Inhoff, 2010), any oral muscle movements that might block subvocalization could interfere with the language stimuli being processed and utilised. Therefore, oral activities that might cause articulatory suppression (e.g., eating or speaking) could bias eye movement effects or give rise to less than accurate eye tracking results. Future researchers should heed these potential biases especially when they run eye tracking studies in unobtrusive, natural settings which could cause a lack of control on participants' oral muscle movements. Moreover, literature also suggests that the powerfulness of acticulatory suppression varies depending on the type of language being read (Besner, 1987). Hence, eye movement and eye tracking researchers should pay particular attention on the validity of data collected cross-linguistically, or cross-culturally. Lastly, researchers should also explore bodily processes other than oral muscle movements and rule out any other potential interference factors in eye movement or eye tracking research.

6 Conclusion

This chapter has provided contrasting case studies that illustrate the types of research being conducted in the private sector and academic sectors. Although superficially similar—especially in regard to the use of similar research methodologies—the research illustrates the breadth of eye movement and eye research applications in marketing. The industry problem and focus determines the research methodologies and iterative nature of the first case study located in the private sector consulting and research industry. The hypotheses at the core of the second case study reflect an academic focus in seeking to support research problems expressed in academic discourse. The published eye movement and eye-tracking research applied to marketing seriously underrepresents the volume and range of eye-tracking-based marketing research conducted by the private sector.

References

Atalay, A. S., Bodur, H. O., and Rasolofoarison, D. (2012). Shining in the center: Central gaze cascade effect on product choice. Journal of Consumer Research, 39 (December), 848-866.

Besner, D. (1987). Phonology,lexical access in reading, and articulatory suppression: A critical review. The Quarterly Journal of Experimental Psychology Section A: Human Experimental Psychology, 39 (3), 467-478.

Bornstein, F. R., & D'Agostino, P. R. (1994). The attribution and discounting of perceptual fluency: Preliminary tests of a perceptual fluency/attributional model of the mere exposure effect. *Social Cognition*, *12*(Summer), 103–128.

Brasel, S. A. and Gips, J. (2008). Breaking through fast-forwarding: Brand information and visual attention. Journal of Marketing, 72 (November), 31-48.

Campbell, R., Rosen, S., Solis-Macias, V., & White, T. (1991). Stress in silent reading: Effects of concurrent articulation on the detection of syllabic stress patterns in written words in English speakers. *Language and Cognitive Processes, 6,* 29–47.

Duchowski, A. (2007). *Eye tracking methodology: Theory and Practice.* London: Springer.

Eiter, B. M. and Inhoff, A. W. (2010). Visual word recognition during reading is followed by subvocal articulation. Journal of Experimental Psychology: Learning, Memory, and Cognition, 36 (2), 457-470.

Emerson, M. J., & Miyake, A. (2003). The role of inner speech in task switching: A dual-task investigation. *Journal of Memory and Language, 48,* 148–168.

Komorgotsev, O. (2011). Importance and challenges of automated eye movement classification. Frankfurt: Tobii Conference, October, 2011.

Li, E. and Briley, D. (2011). Attitudes shaped by eye movements: The reading direction effect. Advances in Consumer Research, 39, 666-667.

McGuigan, F. J., Keller, B., & Stanton, E. (1964). Covert language responses during silent reading. *Journal of Educational Psychology, 55*(6), 339–343.

Palmer, S. E. (1999). *Vision science-photons to phenomenology*. Cambridge: The MIT Press.

Pieters, R. and Wedel, M. (2007). Goal control of attention to advertising: The Yarbus implication. Journal of Consumer Research, 34 (August), 224-233.

Rayner, K. (1998). Eye movements in reading and information processing: 20 years of research. *Psychological Bulletin, 124*(3), 372–422.

Stroop, J. R. (1935). Studies of interference in serial verbal reactions. *Journal of Experimental Psychology, 12,* 242–248.

Topolinski, S., & Strack, F. (2009). Motormouth: Mere exposure depends on stimulus-specific motor simulations. *Journal of Experimental Psychology: Learning, Memory, and Cognition, 35*(2), 423–33.

Eye Movement Evaluation of Signature Forgeries: Insights to Forensic Expert Evidence

Adrian G. Dyer, Bryan Found, Mara L. Merlino, Avni L. Pepe, Doug Rogers and Jodi C. Sita

1 Introduction

To find solutions to complex problems often requires the networking of experts from a variety of disciplines. This can be especially relevant in the application of eye movement studies to understand the complex ways in which humans make decisions for non-trivial tasks. For example, the use of eye movement investigations to understand how experienced forensic officers make decisions whilst evaluating case evidence has important implications for legal proceedings (Dyer et al. 2006, 2008; Merlino et al. 2008a). To conduct these types of studies firstly requires detailed knowledge of the legal context for which quality empirical evidence is required (Merlino et al. 2008b) and secondly how experiments need to be designed to model the real-world working environment of a forensic officer. By utilising the correct context setting information to design eye movement studies, it is possible to collect high-quality data to answer specific questions pertinent to certain disciplines. Below we discuss the value of multidisciplinary inputs for understanding FDE visual processing as a model for how eye movement studies can provide high-value empirical data on human visual expertise.

A. G. Dyer (✉)
RMIT University, Melbourne, VIC, Australia
e-mail: adrian.dyer@rmit.edu.au

B. Found
Victoria Police, Melbourne, VIC, Australia

M. L. Merlino
Kentucky State University, Frankfort, KY, USA

A. L. Pepe · D. Rogers
La Trobe University, Melbourne, VIC, Australia

J. C. Sita
Australian Catholic University, Melbourne, VIC, Australia

M. Horsley et al. (eds.), *Current Trends in Eye Tracking Research*,
DOI 10.1007/978-3-319-02868-2_16, © Springer International Publishing Switzerland 2014

2 Understanding the Legal Context of Performance Evaluations in Forensic Document Examiners

In vision sciences, the opportunity to make comparisons between experts and novice groups can be important for understanding visual processing mechanisms (Tanaka and Sengco 1997; Reingold et al. 2001; Nodine et al. 2002; Busey and Vanderkolk 2004), but having accurate information to classify groups may be difficult to obtain. Due to the context of how adversarial legal systems operate, we discuss the recent opportunities that have emerged for using eye tracking to collect high-quality empirical data. The use of opposing argumentation is a defining characteristic of the adversary system upon which justice in many countries, like the USA, UK, New Zealand and Australia, is based. For example, in the USA, it is the judge's responsibility to hear arguments from opposing counsel in legal disputes about the relevance and reliability of expert testimony and to determine whether or not a jury should be presented with evidence (Gottesman 1998). The responsibility for ruling on the admissibility of expert testimony is often referred to as a 'gatekeeping' role (Faigman 1995).

Whilst up to 1993 judges in the USA largely based admissibility decisions on a 'general acceptance' standard, in a US Supreme Court decision Daubert v. Merrell Dow Pharmaceuticals, Inc. (1993), the Justices held that the Federal Rules of Evidence superseded the general acceptance standard for scientific evidence in federal trials. In this landmark decision, guidelines were suggested for judges to determine the reliability of expert testimony, including knowledge of potential error rate in decisions, whether techniques have been exposed to peer review and publication and whether the claimed expertise had general acceptance in the relevant community. In subsequent cases, the US Supreme Court created a statutory obligation for judges to evaluate the merits of scientific, technical, or other specialized knowledge in rulings of expert testimony (Dixon and Gill 2002).

Since the Daubert ruling in 1993, research demonstrates that the factors used by judges when determining the admissibility of expert testimony are often related to whether the judge views the evidence as scientific, technical or other specialized knowledge (Gatowski et al. 2001; Dahir et al. 2005; Merlino et al. 2008a; Merlino et al. 2008b), and the precedence of new findings has an important role in determining the admissibility of expert testimony (Dixon and Gill 2002). Immediately following the Daubert decision, a number of US legal scholars and defence attorneys began challenging the admissibility of many areas of forensic science (Merlino et al. 2008a, b), including such long-accepted fields as latent fingerprint examination and forensic document examination. The earliest success of these initial challenges to forensic document examination came with the challenge to expert testimony in the US v. Starzecpyzel (880 F. Supp. 1027, S.D.N.Y. 1995). The judge of that case stated, 'if testing is possible, it must be conducted if forensic document examination is to carry the imprimatur of "science".' Consequently, forensic document examiners have increasingly become involved in research testing to access the reliability and validity of the discipline, for example, testing at the La Trobe University Handwriting Analysis and Research Laboratory (Merlino et al. 2008a).

3 Evidence of Expertise for Decision-Making in Forensic Document Examiners

The need to effectively communicate information about expert proficiency has led to the development of a new format for testimony to explain the methodology including documenting its reliability (Merlino et al. 2008b). This extended form of admissibility review has required experts in the forensic disciplines to provide information about the reliability of their methods and findings, the validity of the premises upon which their areas of expertise are based and the proficiency of the examiners (Merlino et al. 2008b). A consequence of this activity in understanding the validity of expert evidence has been increased research into the performance of FDEs, and subsequently there have been a number of studies that have quantified the expertise effect of FDEs in providing opinions based upon the visual inspection of questioned handwritten material. All studies reported to date show that FDE subjects are significantly more accurate at correctly indentifying questioned signatures than control subjects (Kam et al. 1994, 1997, 2001; Found et al. 1999; Sita et al. 2002; Dyer et al. 2006). For example, empirical studies on text-based writing identification have shown evidence of expertise in FDEs compared to control subjects (Kam et al. 1994, 1997) as the FDEs made significantly fewer errors and were more cautious than control subjects. This expertise effect in FDEs was also observed in studies using signatures as test stimuli (Found et al. 1999; Sita et al. 2002). Sita et al. (2002) used a multiple signature evaluation task to compare 17 FDEs from both Australian and New Zealand government bodies to 13 tertiary-qualified and well-motivated control subjects. The FDE subjects showed an expertise effect and were able to correctly call whether a questioned signature was genuine or a simulation (a simulation can include either a forged or disguised signature) with 94.2 % accuracy, which was significantly better than control subjects that scored 74.4 %. A third study by Kam and colleagues (Kam et al. 2001) compared 69 FDEs and 50 control subjects on questioned genuine and simulated signatures and also concluded that there was a significant difference in performance.

Taken together, these studies allow for the classification of FDEs as having an expertise effect in identifying signatures. As signatures can be classified as complex visual stimuli which are comprised of multiple elemental features like turning points, retraces and intersections (Found and Rogers 1998; Found et al. 1998), it is of high value to understand the basis of this expertise effect in visual processing. Previously, expertise in visual processing has been investigated for fields including face recognition (Tanaka and Sengco 1997), chess playing (Reingold et al. 2001), mammogram classification (Nodine et al. 2002) and forensic fingerprint analysis (Busey and Vanderkolk 2004). One measure of expertise common in these fields appears to be evidence that, with increasing exposure to complex stimuli, processing moves from an elemental or piecemeal form through to a more holistic or configural representation of how the elemental information is represented.

4 Understanding the FDE Work Environment

To understand the visual mechanisms that might allow FDEs to demonstrate an expertise effect for evaluating complex stimuli like signatures, and the potential value of this expertise to the legal system, it is very important to understand and model the real-world working environment in which FDEs make decisions about writing material. Many FDEs are associated with government laboratories and express opinions regarding the authorship of writings and also the process by which writings were produced. The skill and expertise of FDEs are drawn from a wide variety of sources including training programmes within laboratories or specialist society meetings, and the scientific literature relating to the discipline. Especially in the context of the Supreme Court rulings described above, FDEs regularly undergo independent testing programmes to provide feedback regarding the probative value of the skills that they utilize in casework.

The skills of FDEs rely on the appropriate identification and comparison of handwriting features, which are examined primarily by visual inspection of handwriting specimens. Handwriting is a complex motor behaviour that is learnt through experience and is carried out by most members of the population (Huber and Headrick 1999). Handwriting is a highly accessible and useful form of evidence in the forensic environment because it is typically a relatively stable behaviour for a given writer, but handwriting does vary quite markedly between writers, even for writers from the same educational background (Durina and Caligiuri 2009). As with most forms of forensic identification evidence, it is the 'inter-individual' variation that provides a useful tool to determine whether there is a nexus between a particular writing act and a person, or another writing act. Handwriting as a learnt behaviour is, however, subject to a number of variables. Examples of variables that FDEs must deal with include the intra-writer variation and the reality that people can change their handwriting purposefully (disguise) or copy the handwriting of others (forgery). One type of writing artefact that an FDE may inspect for is line fluency disturbances (Fig. 1). These disturbances occur when a writer trying to either forge a signature or disguise their own signature does not make the normal smooth writing movements with the pen, leading to these telltale signs of non-genuine writing movements. Line fluency disturbances may result from the writer not using a learnt and thus natural process of writing; thus, these indications are of high value in the determination of non-genuine signatures or other writings (Dyer et al. 2006; Found and Rogers 1999; Huber and Headrick 1999; Ellen 1989; Hilton 1982; Osborn 1929).

The approach taken by FDEs in order to form opinions regarding the authorship of a sample of writing relies on visual and cognitive processing of the evidence. Typically, examiners are provided with a handwriting sample (whether text based or signatures) whose author is considered unknown or disputed. This sample is termed 'questioned'. The questioned writing is then compared to specifically collected samples of writing which have been knowingly written by a particular person, termed the 'specimen' or 'exemplar' samples. Generally, the analysis is done as a

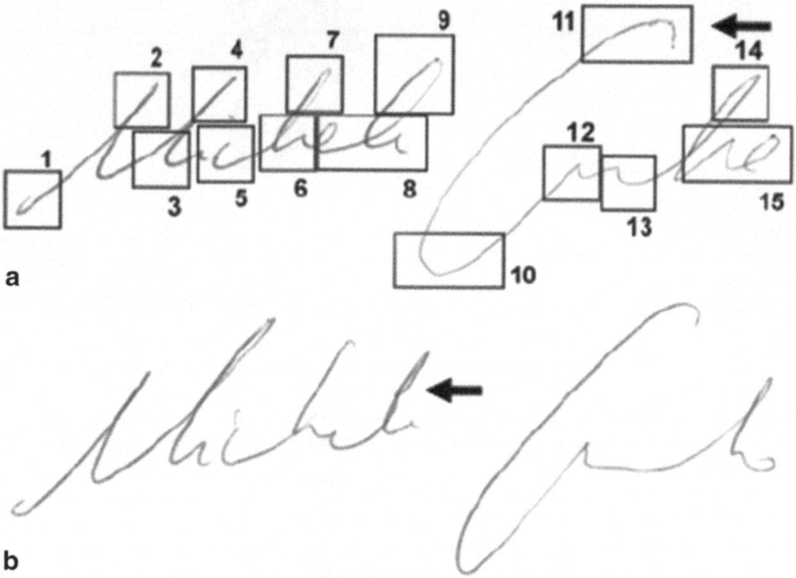

Fig. 1 An example of a questioned signature. (**a**) Also shown are 15 area of interest (AOI) boxes used to quantify fixations made to the different regions of the signatures (AOI not viewed by test subjects). Shown here is a forgery with evidence of disturbed line fluency (a predictor feature of the forgery process) in AOI 11. (**b**) Stimulus 17 which is a genuine signature of the specimen provider that also had some evidence of disturbed line fluency. The *bold arrows* highlight the line fluency disturbances

'side-by-side' comparison task (Fig. 2), and the criterion for decision-making by an FDE is in forming an accurate opinion, so there is no time limit imposed whilst they undertake their visual comparison.

Until recently, the process by which FDE subjects visually process questioned signatures (or other handwriting) has remained a 'black box' where questioned and specimen samples were provided to the examiners, and a formal report on a decision of genuine, not genuine or inconclusive evidence was returned. The use of eye movement recordings, discussed below, provides important insights into the process by which FDEs make decisions, and potentially the basis of the expertise effect discussed above.

5 Eye Tracking to Understand the Mechanisms of Expertise in FDEs

To conduct eye movement experiments to potentially understand mechanisms of expertise in FDEs in order to satisfy legal requirements, it is very important that the experiments closely match the normal operating environment of FDE practice.

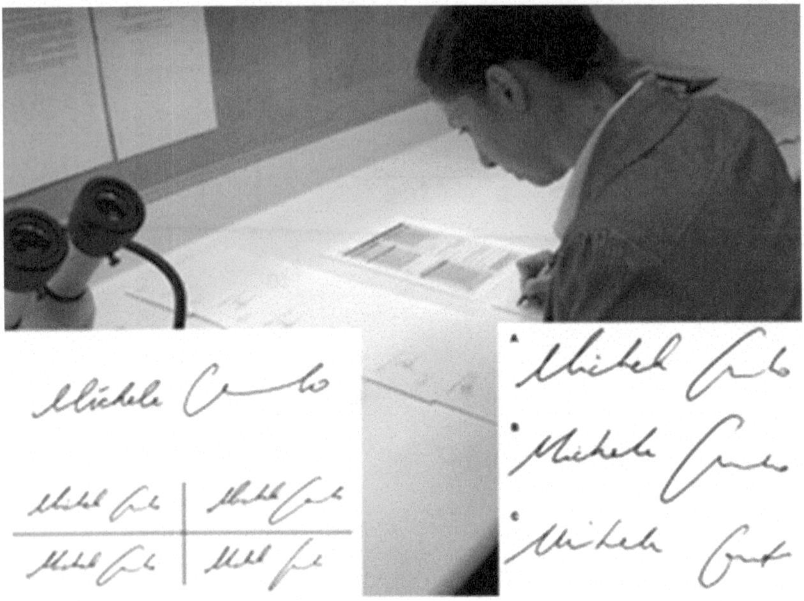

Fig. 2 A forensic document examiner simultaneously inspecting a questioned signature in comparison to known exemplars *Insert* (*left-hand side*) shows how this normal viewing condition environment was simulated on a display monitor to record eye movements whilst retaining sufficient spatial resolution to dissect attention to respective features within the signature. *Insert (right-hand side)* shows examples of a genuine (**a**), disguised (**b**) and forged (**c**) signature used to evaluate expertise during eye tracking

Four very important points can be drawn from the modelling of the FDE work environment described above:

1. Decisions are of high value to the legal system, and the criterion on accuracy for decision-making is paramount to how an experiment needs to proceed. Thus, instructions to test participants need to clearly state accuracy to avoid speed–accuracy trade-offs (Rival et al. 2003), and a performance scoring metric that reinforces the gravity of the decision-making process is important.
2. The type of visual tasks that document examiners typically have made decisions on is simultaneous comparisons, where both a questioned signature and specimens are available at the same time. The choice of stimulus display could have a significant effect on the type of data collected due to the likely occurrence of referral saccades (Fig. 3) between a questioned stimulus and specimens, compared to a single stimulus presentation.
3. In a real-world operating environment, since accuracy is paramount, there is no limit imposed on the time allocated by an FDE to complete a visual inspection.
4. The key features that are likely to be of interest for understanding expertise in FDEs could be turning points, and/or line fluency disruptions, which are thus likely to be the key AOIs for evaluating fixations in subsequent analyses.

In the first eye movement study (Dyer et al. 2006) that investigated visual attention and expertise for signature analysis in both FDEs and well-motivated control subjects, it was possible to meet these criteria by designing a stimulus set-up to closely match the way in which case work is typically conducted. The study tested nine FDEs (mean age 36.4+/−1.8 standard error of mean (SEM); years experience 11.7+/−2.3 SEM), eight of which were qualified by their organization to present expert evidence regarding signatures, whilst one FDE subject was a trainee with 1 year of experience. The control group consisted of 12 subjects (mean age, 22.0+/−3.5 SEM years) who were La Trobe University students who volunteered to participate in the study as part of a third-year student research subject. All subjects had better than 6/12 uncorrected visual acuity and no history of neurological disorders. The control subjects, as a group, were given a 30-min presentation on what a genuine, disguised and forged signature was and the role of an FDE in discriminating between these different classes of stimuli. The control subjects were given an opportunity within the talk to sign their own name a number of times so as to understand the degree of signature variability and also to attempt to simulate another person's signature. The control subjects were made aware of the potential consequences of an FDE making errors in a real-life scenario where the evidence could have important legal implications.

Since FDEs typically will only evaluate a questioned signature if it is possible to compare this to known exemplars (Fig. 2), the study presented stimuli where the questioned signature was at the top of the stimulus and four specimen signatures were simultaneously presented at the bottom of the stimulus (Fig. 2 insert). The subject instructions were that they may 'view a signature for as long as required in order to make an accurate decision about whether it is a genuine signature of the specimen provider or not'. This simple factor of time for data collection was vital to data analysis to understand the processing of visual information by FDEs. Data were collected with a Tobii 1750 binocular eye-tracking system (Tobii Technology, Stockholm, Sweden). Calibration for each subject to a 16-point reference grid was carried out providing for a resolution of subject gaze to better than 0.5° of visual angle. Stimuli were displayed using an integrated Tobii 1750 1280×1024 pixel thin-film-transistor (TFT) monitor. Subjects viewed the screen from a distance of 57 cm so that the visual angle of the screen was 33°×27° (W×H) and that the width of a typical questioned signature subtended a visual angle of approximately 28° which enabled sufficient spatial resolution to dissect the multiple AOIs for local features that a subject may view. Eye fixations were determined using criteria of eye position remaining within a 50-pixel area for a time of greater than 100 ms (values determined in preliminary pilot experiments). Data collection, fixation measurements and analysis of AOI data were determined with Tobii Clearview 2.1.2 software and analysed with custom-written software. To analyse subject attention to different features within the signature set, a range of AOIs were defined (Fig. 1). The study revealed several important points about how FDEs visually inspect signatures to make decisions. Consistent with previous studies (Kam et al. 2001; Found et al. 1999; Sita et al. 2002), there was a significant difference in the capacity of FDE subjects to make accurate decisions compared to the control subjects (Dyer et al. 2006).

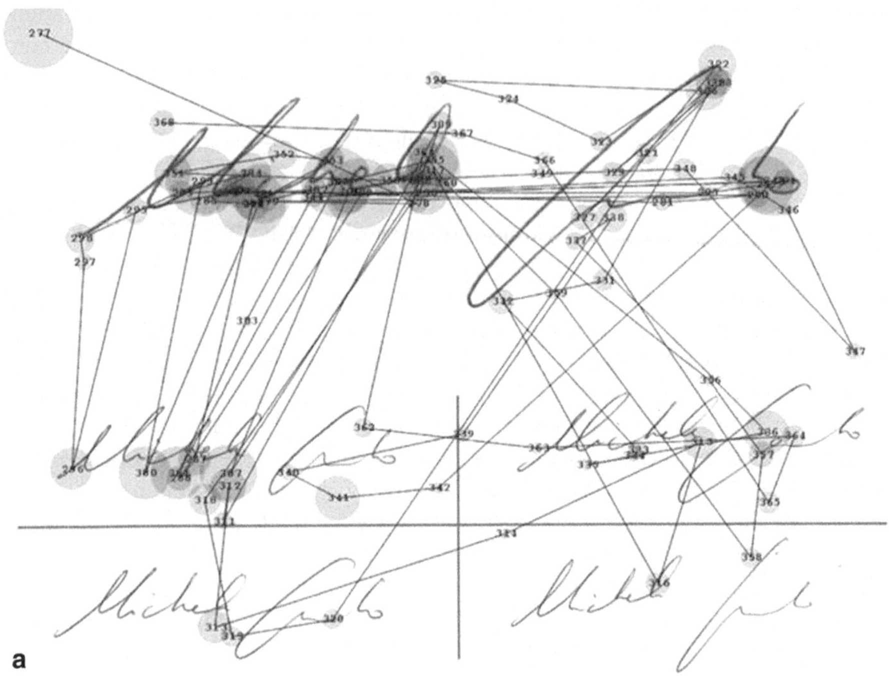

a

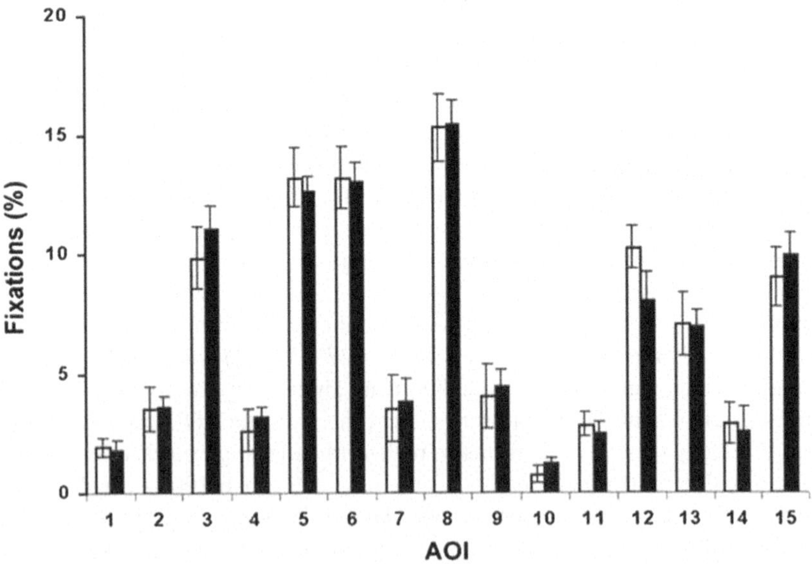

b

Fig. 3 (**a**) Typical eye movement recording of a forensic document examiner (FDE). This was done while they were making a decision about a questioned signature presented simultaneously with four (smaller) exemplar signatures; comparison referral saccades can be seen between questioned signature and exemplars. In this example, the signature was a genuine signature of the specimen provider, and the subject correctly called the signature. The response time of the subject to this stimulus was 25.4 s which is very similar to the mean response time of FDEs for genuine questioned signatures.

However, there was no significant difference on the time-respective subject groups allocated to particular comparison tasks, indicating a high level of motivation in control subjects. There was no significant correlation between response times and accuracy for either test group (Dyer et al. 2006). The study also showed that eye movements were of high value to understanding visual processing by FDEs, as if the stimuli were tachistoscopically presented there was a significant drop in subject performance, although performance for tachistoscopically presented stimuli was above chance levels suggesting that there was both global processing of the spatial properties of the entire signature that do not require eye movements and local processing of features fixated on during an evaluation which enables improved performance at correctly classifying signatures. Importantly, the study showed that the majority of errors made by the control subjects were regarding decisions made about genuine signatures, where the error rate was seven times higher than for the FDE subjects.

To more fully analyse the data to test for potential differences in performance due to what features were inspected, it was possible to allocate AOI over key stimuli elemental features like turning points and intersections which have been previously identified as important in writing evaluations (Found and Rogers 1998; Found et al. 1998). Somewhat surprisingly, the frequency with which FDE and control subject inspected specific features was remarkably similar (Fig. 3), and analyses of subject inspection suggested that both groups used a somewhat anarchic, reflexive-type search strategy (bottom-up search) rather than a top-down scanpath along the lines originally proposed by Norton and Stark (1971). The next question then was what might explain the significant difference in performance between the groups? It was possible to answer this question by comparing the AOI fixated just prior to a decision being made (this important factor would not have been possible if a set time limit had been used for stimulus presentation). By considering stimuli which contained line fluency evidence, it was possible to see that whilst both groups inspected the line fluency disturbance on the forged signature just prior to a decision, for the genuine signature the control group also used the line fluency disturbance to make a decision, whilst the FDEs did not fixate more on this information prior to a decision. The FDEs had observed fluency disturbance (as earlier eye movements revealed) but had weighted this information in relation to the rest of the available spatial information to render better decision-making. Thus, consistent with earlier studies on expertise and configural representations (Tanaka and Sengco 1997; Reingold et al. 2001; Nodine et al. 2002; Busey and Vanderkolk 2004), it appears a factor of expertise in FDEs is an enhanced capacity to process local feature information in the context of more global information. However, further work on the topic of FDE

Circles show fixations and *lines* indicate sequential saccadic eye movements. *Numbers* on the fixations indicate sequence order which begin at fixation 277 at the *top left-hand side* of the image, where a fixation cross had been displayed before the presentation of the questioned signature. (**b**) The mean percentage fixations (+/−SEM) to 15 different areas of interest (AOI; see Fig. 1a) by 12 control subjects (*white bars*) and nine FDEs (*black bars*) for a test including 32 questioned signatures. Data suggested that FDEs and controls use a very similar search strategy

expertise and eye movements will be of high value for determining both how experts in different countries inspect handwritten documents, and how different visual inspection methodologies may need to be considered depending upon the requirements of evidence evaluation by legal systems in respective countries.

Typically, FDEs in government laboratories (in Australasia at least) do not attempt to discriminate between signatures that are either forged or disguised. These types of signatures are referred to as 'simulations' or 'non-genuine' signatures. However, to investigate if it might be possible to make such discriminations, an extension of the study described above was conducted and showed that there was also an expertise effect for FDEs who could discriminate between the different classes of simulated signatures at a level significant from chance expectation (one sample *t*-test, $p<0.05$) whilst control subjects could not. There was also a significant difference between FDEs and control subjects (independent samples *t*-test, $p<0.05$); however, the measure of relative number of fixations to either forged or disguised signatures revealed very similar overall fixation frequencies between FDEs and controls, again supporting the conclusion that FDEs were predominately using a bottom-up search strategy and that expertise compared to the lay subjects resulted in the way that the local feature information was cognitively processed by the FDEs (Dyer et al. 2008).

Eye tracking has previously been used to understand the process of visual inspection and drawing (Coen-Cagli et al. 2009). To better understand the processing of information when a subject attempts to forge a specimen signature, one study (Pepe et al. 2012) has recently used a system of combining eye tracking (Tobii x50 eye tracker: 50 Hz) with a Panasonic NV-GS17 (25 Hz) Digital 'scene' camera, and a PTZ-1230 Wacom Intuos 3 digitising tablet (200 Hz; 0.25 mm resolution) to simultaneously record eye and hand movements (to separately quantify graphomotor output). Forging a signature is potentially a demanding cognitive task because it requires a subject to visually inspect a specimen, load this information into a form of working memory that can facilitate the planning of the sensory motor system required to execute the pen movement and, finally, produce the physical motor task of executing the forgery. This type of cognitive processing has been previously shown to be demanding for drawing and copying tasks (Kellogg 1996; Miall et al. 2009). It is possible that the requirement of cognitively processing such complex information may result in the errors like line fluency disturbances (Fig. 1) being produced that subsequently enable FDEs to detect the differences between genuine and simulated (non-genuine) writings (Dyer et al. 2006). As forging is a demanding cognitive task, it was hypothesised that task difficulty could stress the perceptual system in different ways depending upon the complexity of the signature to be forged. Pepe et al. (2012) thus considered how signatures were inspected during a forging process depending upon signature complexity which was defined as high or low using a previously described complexity classification model (Found and Rogers 1998). Subjects ($N=17$, mean age $=29$ y.o.) were required to produce multiple copies of 'high-complexity' and 'low-complexity' signatures. The eye-tracking data revealed a complex picture where a significantly greater number of fixations were made to the model signatures of both high and low complexity when compared to the forgery attempt ($p<0.01$), whilst the high-complexity model and forged signatures received significantly more fixations compared to the respective low-complexity signatures ($p<0.05$) (Pepe et al. 2012). These data are consistent with the complexity classification model proposed by Found et al. (1998); however, the opinions of test subjects on how difficult individual signatures were to forge did not fully support the hypotheses of the Found et al. (1998) model.

Prior to their forgery attempts, subjects opined that high-complexity signatures would be the most challenging to forge, but this opinion switched for most subjects following the forgery attempts (Pepe et al. 2012). By also analysing the handwriting measures of pen pressure, normalized jerk and writing velocity for both the high- and low-complexity signatures, it was found that significantly more pen pressure was applied during the high-complexity simulation attempts compared to the low-complexity attempts ($t(16)=3.93$, $p<0.01$), but the low-complexity simulation attempts were completed with a significantly greater writing velocity ($t(16)=3.05$, $p<0.01$). The mean pen pressure for the high-complexity model signature was 305 pressure units (pu) with a mean velocity of about 17.6 cm/s, whilst the low-complexity model signatures velocity was 32.7 cm/s at 236 pu. Importantly, the results for the normalized jerk suggest that subjects displayed significantly more dysfluent writing when producing the low-complexity compared to high-complexity signatures ($t(15)=2.13$, $p<0.05$). Thus, a comparison of subject opinions for high- and low-complexity signatures (based on visual inspection of the stimuli quantified with eye-tracking and also graphomotor data) suggests that whilst initial visual inspection and ranking by subjects reveal one type of result, the quantification of pen movements is actually more consistent with opinions of subjects following an attempt to forge signatures. This study thus shows that whilst signature simulation difficulty may be accessed by using eye-tracking analyses for looking at forging behaviour, the data need to be very carefully interpreted by also considering well-designed questionnaires and other factors to fully evaluate subject perceptions both prior to and following an experiment. This work shows that using eye movement recording technologies to fully understand learning and human performance may require input from several fields of expertise in order to produce focussed questions on the most important factors to enable meaningful quantitative analyses.

In conclusion, eye movement research has recently entered an exciting phase enabled by innovative technologies that can accurately track the eyes of subjects performing a wide range of real-world tasks (Vassallo et al. 2009; Tatler et al. 2011; Sasson et al. 2012; Chetwood et al. 2012; Tomizawa et al. 2012). These new technologies potentially open the door to understanding human expertise, learning processes and visual attention in a wide range of fields, with potential flow on benefits for teaching, automating procedures or validating current practices. However, a risk in the application of some of these new techniques is that it may be difficult to dissect the mechanisms of visual behaviour by simple gaze plot analyses. Understanding the characteristics of expert performance is crucial in the forensic sciences. The experiments described here utilised the specialist input from a variety of disciplines. In terms of experimental design, it is important that attention be directed toward gaining a detailed knowledge of the context in which expertise is important, how experts actually conduct their normal work practice and how the visual and other sensory processes operate. We hope that this model of collaboration discussed here may be useful for using eye tracking in other disciplines to better understand the nature of how humans carry out complex visual tasks.

Acknowledgments We are grateful to the ARCDP0878968 for funding of A. G. Dyer.

References

Busey, A., & Vanderkolk, J. R. (2004). Behavioral and electrophysiological evidence for configural processing in fingerprint experts. *Vision Research, 45,* 431–448.

Chetwood, A. S. A., Kwok, K. W., Sun, L. W., Mylonas, G. P., Clark, J., Darzi, A., & Yang, G. Z. (2012). Collaborative eye tracking: a potential training tool in laparoscopic surgery. *Surgical Endoscopy, 26*. DOI: 10.1007/s00464-011-2143-

Coen-Cagli, R., Coraggio, P., Napoletano, P., Ferraro, M., & Boccignone, G. (2009). Visuomotor characteristics of eye movements in a drawing task. *Vision Research, 49*, 810–818.

Dahir, V. B., Richardson, J. T., Ginsburg, G. P., Gatowski, S. I., Dobbin, S. A., & Merlino, M. L. (2005). Judicial application of Daubert to psychological syndrome and profile evidence: A research note. *Psychology, Public Policy, and Law, 11*, 62–82.

Daubert v. Merrell Dow Pharmaceuticals, Inc., 509 U.S. 579, 113 S.Ct. 2786 (1993).

Dixon, L., & Gill, B. (2002). Changes in the standards for admitting expert evidence in federal civil cases since the Daubert decision. *Psychology, Public Policy, and Law, 8,* 251–303.

Durina, M. E., & Caligiuri, M. P. (2009). The determination of authorship from a homogenous group of writers. *Journal of the American Society of Questioned Document Examiners, 12*(2)

Dyer, A. G., Found, B., & Rogers, D. (2006). Visual attention and expertise for forensic signature analysis. *Journal of Forensic Sciences, 51*, 1397–1404.

Dyer, A. G., Found, B., & Rogers, D. (2008). An insight into forensic document examiner expertise for discriminating between forged and disguised signatures. *Journal of Forensic Sciences, 53,* 1154–1159.

Ellen, D. (1989). *The scientific examination of documents: methods and techniques*. West Sussex: Ellis Horwood Limited,

Faigman, D. I. (1995). The evidentiary status of social science under Daubert: Is it 'scientific,' 'technical,'or 'other' knowledge? *Psychology, Public Policy, and Law, 1*, 960–979.

Found, B., & Rogers, D. (1998). A consideration of the theoretical basis of forensic handwriting examination: The application of "Complexity Theory" to understanding the basis of handwriting identification. *International Journal of Forensic Document Examiners, 4*, 109–118.

Found, B., & Rogers, D. (1999). Documentation of forensic handwriting comparison and identification method: A modular approach. *Journal of Forensic Document Examination, 12*, 1–68.

Found, B., Rogers, D., Rowe, V., & Dick, D. (1998). Statistical modelling of experts' perceptions of the ease of signature simulation. *Journal of Forensic Document Examination, 11*, 73–99.

Found, B., Sita, J., & Rogers, D. (1999). The development of a program for characterising forensic handwriting examiners' expertise: Signature examination pilot study. *Journal of Forensic Document Examination, 12*, 69–80.

Gatowski, S. I., Dobbin, S. A., Richardson, J. T., Ginsburg, G. P., Merlino, M. L., & Dahir, V. (2001). Asking the gatekeepers: Results of a national survey of judges on judging expert evidence in a post-Daubert world. *Law and Human Behavior, 25*, 433–458.

Gottesman, M. H. (1998). From barefoot to daubert to joiner: Triple play or double error? *Arizona Law Review, 40*, 753–780.

Hilton, O. (1982). *Scientific examination of questioned documents.* Elsevier New York: Science Publishing Co., Inc.

Huber, R. A., & Headrick, A. M. (1999). *Handwriting identification: Facts and fundamentals*. Boca Raton: CRC Press.
Kam, M., Wetstein, J., & Conn, R. (1994). Proficiency of professional document examiners in writer identification. *Journal of Forensic Sciences, 39*, 5–14.
Kam, M., Fielding, G., & Conn, R. (1997). Writer identification by professional document examiners. *Journal of Forensic Sciences, 42*, 778–786.
Kam, M., Gummadidala, K., Fielding, G., & Conn, R. (2001). Signature authentification by forensic document examiners. *Journal of Forensic Sciences, 46*, 884–888.
Kellogg, R. T. (1996). A model of working memory in writing. In C. M. Levy & S. Ransdell (Eds.), *The science of writing: Theories, methods, individual differences and applications* (pp. 57–72). Mahwah, NJ: Erlbaum.
Merlino, M. L., Springer, V., Kelly, J. S., Hammond, D., Sahota, E., & Haines, L. (2008a). Meeting the challenges of the daubert trilogy: Refining and redefining the reliability of forensic evidence. *Tulsa Law Review, 43*, 417–445.
Merlino, M. L., Murray, C. I., & Richardson, J. T. (2008b). Judicial gatekeeping and the social construction of the admissibility of expert testimony. *Behavioral Sciences and the Law, 26*, 187–206.
Miall, R. C., Gowen, E., & Tchalenko, J. (2009). Drawing cartoon faces—a functional imaging study of the cognitive neuroscience of drawing. *Cortex, 45*, 394–406.
Nodine, C. F., Mello-Thoms, C., Kundel, H. L., & Weinstein, S. P. (2002). Time course of perception and decision making during mammographic interpretation. *American Journal of Roentgenology, 179*, 917–923.
Norton, D., & Stark, L. (1971). Scanpaths in eye movements during pattern perception. Science, 171, 308–311.
Osborn, A. S. (1929). Questioned documents (2nd ed.). Chicago: Nelson-Hall Co.
Pepe, A., Rogers, D., & Sita, J. C. (2012). A consideration of signature complexity using simulators' gaze behaviour. Journal of Forensic Document Examination. In press.
Reingold, E. M., Charness, N., Pomplun, M., & Stampe, D. M. (2001). Visual span in expert chess players. *Psychological Science, 12*, 48–55.
Rival, C., Oliver, I., & Ceyte, H. (2003). Effects of temporal and/or spatial instructions on the speed accuracy tradeoff of pointing movements in children. *Neuroscience Letters, 336*, 65–69.
Sasson, N. J., & Elison, J. T. (2012). Eye tracking young children with autism. *Journal of Visualized Experiments,* (61), e3675 10.3791/3675, DOI: 10.3791/3675.
Sita, J., Found, B., & Rogers, D. (2002). Forensic handwriting examiners' expertise for signature comparison. *Journal of Forensic Sciences, 47*, 1117–1124.
Tanaka, J., & Sengco, J. A. (1997). Features and their configuration in face recognition. *Memory & Cognition, 25*, 583–592
Tatler, B. W., Hayhoe, M. M., Land, M. F., & Ballard, D. H. (2011). Eye guidance in natural vision: Reinterpreting salience. *Journal of Vision, 11*(5), 1–23. http://www.journalofvision.org/content/11/5/5,doi:10.1167/11.5.5.
Tomizawa, Y., Aoki, H., Suzuki, S., Matayoshi, T., & Yozu, R. (2012). Eye-tracking analysis of skilled performance in clinical extracorporeal circulation. *Journal of Artificial Organs, 15*(2), 146–157. DOI: 10.1007/s10047-012-0630-z.
US v. Starzecpyzel, 880 F. Supp. 1027 - Dist. Court, SD New York 1995
Vassallo, S., Cooper, S. L. C., & Douglas, J. M. (2009). Visual scanning in the recognition of facial affect: Is there an observer sex difference? *Journal of Vision, 9*, 1–10.

A Role for Eye-Tracking Research in Accounting and Financial Reporting?

Lyn Grigg and Amy L. Griffin

1 Introduction

Eye tracking is well used in usability studies (Goldberg and Wichansky 2003), for marketing and ecommerce (Wedel and Pieters 2006), as well as for fundamental research in neuroscience and visual processing (Blair et al. 2009) and developmental psychology (Johnson et al. 2004). However, it can also be applied to disciplines such as accounting. In this chapter, we describe why we believe behavioural studies supported by eye tracking are essential for improving financial decision making, both in Australia and globally. We provide examples from a pilot study to support our contention.

Today there is a plethora of legislation, standards and guidelines on how to report financial information, with oversight by numerous government and professional accounting bodies at both national and international levels. For example, Australian public companies are required to generate financial reports under the Corporation Act 2001 (Cwlth), in compliance with the International Financial Reporting Standards (IFRS), the Australian Accounting Standards and regulatory guides that require disclosure of non-IFRS financial information (e.g. RG230), issued by the Financial Reporting Council (FRC), which was established under the Australian Securities and Investments Commission (ASIC) Act 2001.

Regardless of *how* information is required to be reported, there has also been an increase in the complexity and amount of *what* information is reported. Furthermore, there has been a shift towards developing a vast array of business, performance and key indicator reports to support traditional financial reporting (Guthrie and Boedker 2006). For example, see the latest initiatives under the International Integrated Reporting Committee (IIRC), which come out of the United Nations Global Reporting Initiative (GRI) and the Prince Charles Accounting for Sustainability project.

The users of this plethora of financial information are at a tremendous disadvantage. The basis of financial reporting is a series of technical documents coming

A. L. Griffin (✉) · L. Grigg
University of New South Wales, Canberra, NSW, Australia
e-mail: a.griffin@adfa.edu.au

M. Horsley et al. (eds.), *Current Trends in Eye Tracking Research*,
DOI 10.1007/978-3-319-02868-2_17, © Springer International Publishing Switzerland 2014

out of a process that records a business's financial transactions. These documents, by their nature, are not necessarily user-centred reporting documents designed for non-experts. Hence, the accounting and financial industries rely on and demand that users (e.g. company directors and audit committee members) have the skills to analyse and interpret these documents to make decisions (McDaniel et al. 2002). Given the complexity of these technical documents, their use requires a high degree of financial literacy. This view that users must 'fit the system' pervades much of the thinking concerning the development of accounting and financial reporting, and related information for decision making.

This chapter reports on our initial attempts to build a research programme dedicated to conducting user studies of financial reports in order to derive guidelines for producing truly user-centred financial and business reports.

2 Methodology

We are not interested in changing or confronting the legal and professional requirements imposed in financial reporting. Rather, we seek to initiate greater understanding of how users perceive these and other forms of financial and related non-financial information that guide decision making by stakeholders. To this end, eye tracking provides the opportunity to do behavioural research that provides valuable information in an unobtrusive way. To date, we have found no published literature that uses eye tracking for studying financial reports or related business reports. Notwithstanding this, Locarna (2012) does report on their website some work in progress at Simon Fraser University in which an eye tracker was used to document visual attention patterns when users viewed several new proposed accounting reports. They noted that eye tracking helped to develop a financial report that was easier to read, showed key information faster and with greater accuracy.

We have conducted a small pilot study to investigate the extent to which eye tracking can provide information that is of value in understanding what participants see in financial reports. In the pilot study, we showed participants a standard public company Statement of Financial Position (Balance Sheet). The company's name was removed and the financial years were changed to prevent users from guessing the identity of the company whose financial information was reported, avoiding users' reliance on memory of the company's history and requiring the use of information in the financial document to complete each task. Four participants, who had varying levels of professional and formal accounting education and/or experience in accounting, ranging from 0 to 5 years, looked at the financial statement.

We gave each participant the same scenario; he/she was a director of a company reviewing the company's financial statement. We asked them to answer five questions on key issues in financial analysis. The questions related to the company's ability to pay debts when they fall due, the efficiency of management in a key operational area, whether the business was debt or equity funded, if the business was profitable and whether the business appeared sound for the future.

While participants answered these open-ended questions, we recorded their reading behaviours using a Tobii x120 eye tracker, recording at 60 Hz. We displayed the financial report on a 21″ LCD monitor using Adobe Acrobat. We used the Tobii Clearview fixation filter to identify fixations. Our behavioural measures included participants' answers to the questions, the time it took them to answer questions and a variety of metrics describing their fixation patterns and scanpaths, including fixation density maps, time to first fixation and visit durations (Duchowski 2007; Holmqvist et al. 2011; Pernice and Nielsen 2009).

3 Results

Our pilot study data suggest that eye tracking will be very helpful for measuring and describing the readability of financial reports, and identified several possible issues for users of these documents. These included the ease of finding information, and problems of accuracy and relevance. Furthermore, as would be expected, experience and education matter, leaving important questions to ask about the level of financial skills that can reasonably be expected of, for example, board members and what role education plays in skill development.

Figure 1 shows that accuracy and relevance are an issue for financial report users. Here, we asked participants to assess the level of working capital the company had. All participants answered the question correctly. However, it is also clear that all the participants missed the full story. To understand where the working capital was coming from, they would need to examine the receivables' notes, shown in the figure in the large green area of interest (AOI). The two accounting educated participants (red, blue; Table 1) did not examine the area at all. The non-accounting educated participants (orange and grey) reviewed the additional data, but their answers to later questions indicated that they did not have the background knowledge to understand the implications of this information.

Another question asked participants to identify if the management of the business was efficient as seen through accounts receivable (debtors) management. Figure 2 compares the participant with the most accounting experience (red), with each of the other participants. Here, we aimed to examine the effect of skill and knowledge on the ability to analyse and interpret a financial statement. The green AOIs indicate areas of the statement that are relevant to answering this question.

The panel at the upper left (red/blue; Fig. 2) shows that both participants saw all of the relevant information with a reasonable level of efficiency. This is perhaps not surprising given that both have an accounting degree and experience in the industry. The novice (upper right panel; grey) does find all of the relevant information, but the experienced accountant (red) was much more efficient than the novice. The participant with no accounting degree, but some experience in the field (right panel), is the least effective at finding the relevant information. This participant spends most of her time reading the labels rather than looking at the actual financial data and does not find all of the needed data.

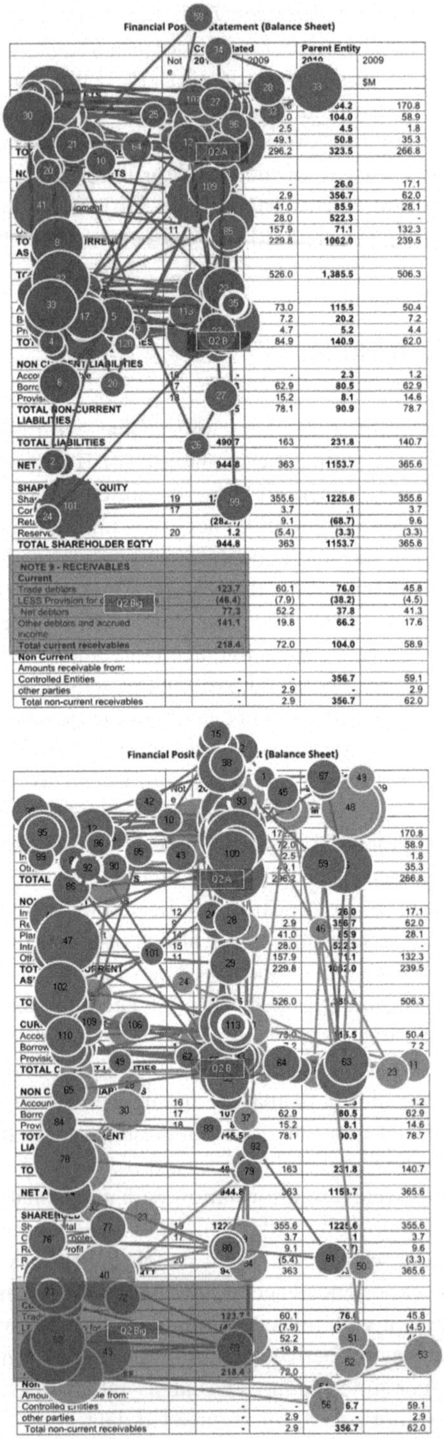

Fig. 1 Identifying all relevant information

Table 1 Pilot study participant characteristics

Colour	Years worked in accounting	Has an accounting degree
Red	More than 5	Yes and professional accounting body accreditation
Blue	1–2	Yes
Orange	More than 5	No
Grey	None	No

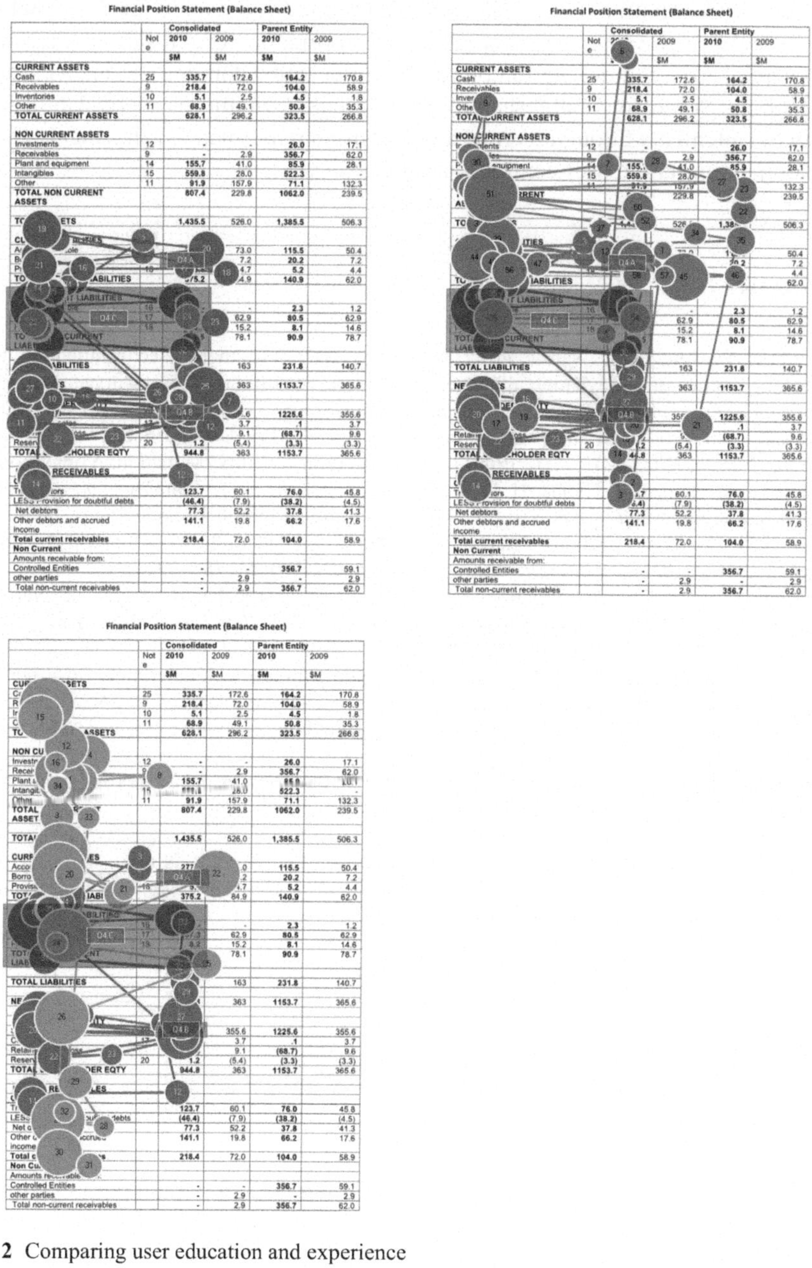

Fig. 2 Comparing user education and experience

4 Conclusion

We undertook this pilot study to determine whether eye tracking was likely to provide information that would be useful for improving the usability of financial reports. Our data indicated that financial statements that come out of a technical process are not particularly user centred and the statement was not easy to read. In terms of decision-making usefulness, even trained and experienced accountants had difficulty in finding the most relevant information. Our results demonstrate the potential for eye tracking to contribute to our understanding of financial decision making. Some potential research opportunities lie in improving readability and clarity of statements through identifying how users read financial statements and in developing alternative user-centred formats.

References

Blair, M. R., Watson, M. R., Walshe, R. C., & Maj, F. (2009). Extremely selective attention: Eye-tracking studies of the dynamic allocation of attention to stimulus features in categorization. *Journal of Experimental Psychology: Learning, Memory, and Cognition, 35*(5), 1196–1206.

Duchowski, A. T. (2007). *Eye tracking methodology* (2nd ed.). London: Springer.

Goldberg, J. H., & Wichansky, A. C. (2003). Eye tracking in usability evaluation: A practitioner's guide, In R. Radach, J. Hyönä, H. Duebel (Eds.), *The mind's eye: Cognitive and applied aspects of eye movement research* (pp. 493–516). Amsterdam: North-Holland.

Guthrie, J., & Boedker, C. (2006). Perspectives on "new" models of business reporting: A reflective note. *Accounting, Auditing & Accountability Journal, 19*(6), 785–792.

Holmqvist, K., Nyström, M., Andersson, R., Dewhurst, R., Jarodzka, H., & van de Weijer, J. (2011). *Eye tracking: A comprehensive guide to methods and measures*. New York: Oxford University Press.

Johnson, S. P., Slemmer, J. A., & Amso, D. (2004). Where infants look determines how they see: Eye movements and object perception performance in 3-month-olds. *Infancy, 6*(2), 185–201.

Locarna (2012). *More transparent accounting*, accessed at http://www.Locarna.com/applications.html 1 July 2012.

McDaniel, L., Martin, R. D., & Maines, L. A. (2002). Evaluating financial reporting quality: The effects of financial expertise vs. financial literacy, *The Accounting Review, 77*(Suppl. 2002), 139–167.

Pernice, K., & Nielson, J. (2009). *Eyetracking methodology,* Fremont, CA. Nielson Norman Group.

Wedel, M., & Pieters, R. (2008). *Eye tracking for visual marketing, Foundations and trends in marketing*, Hanover, MA: nowPublishers Inc.

Eye Tracking During a Psychosocial Stress Simulation: Insights into Social Anxiety Disorder

Nigel T. M. Chen and Adam J. Guastella

1 Background

Social anxiety disorder (SAD) is a debilitating mental illness primarily characterized by an excessive fear of negative social evaluation. Cognitive models of SAD (e.g. Rapee and Heimberg 1997) suggest that this condition is maintained and exacerbated by various information processing biases, which may occur from early automatic stages of processing, through to later interpretive stages of processing.

With regard to the former, a strong base of literature suggests that SAD is associated with an attentional bias to threat (Bar-Haim et al. 2007). That is, the preferential attentional processing of stimuli which may indicate feared negative social evaluation. Such stimuli may include disapproving or angry facial expressions exhibited by other individuals in a given social situation. This threat bias may maintain social anxiety by increasing arousal, falsely confirming various negative beliefs, and may provoke safety behaviours and avoidance, thus, exacerbating this condition (Hofmann 2007; Rapee and Heimberg 1997).

Although the relationship between social anxiety and attentional bias to threat is well established, the attentional components which underpin such a bias necessitate empirical inquiry. For instance, theoretical distinctions have been made between the attentional engagement with a stimulus, and the subsequent disengagement from the stimulus. Therefore, an attentional bias to threat may either pertain to either a facilitated engagement with the threat stimulus, or some form of disruption to the disengagement from the threat. Studies employing reaction time (RT)-based measures of attentional engagement and disengagement, commonly the modified Posner cuing task (Fox et al. 2001), have consistently observed RT data interpreted as disrupted disengagement from threat in high-anxious individuals. However, the validity of this measure has been questioned. It has alternatively been suggested that the same RT data may reflect an additive effect of facilitated engagement and a general slowing in

N. T. M. Chen (✉) · A. J. Guastella
University of Sydney, Sydney, Australia
e-mail: nigel.chen@sydney.edu.au

M. Horsley et al. (eds.), *Current Trends in Eye Tracking Research*,
DOI 10.1007/978-3-319-02868-2_18, © Springer International Publishing Switzerland 2014

the presence of threat for high-anxious individuals (Mogg et al. 2008). The modified Posner cuing task cannot differentiate between these two alternate conjectures.

Recently, there has been a growing interest in the use of eye tracking to assess attentional bias, conceivably due, in part, to the limitations of RT-based measures. For instance, when a stimulus is presented for a duration of 500 ms, it is entirely possible that multiple shifts of attention may occur during this interval. However, RT is not sensitive to such shifts, only providing a snapshot of attention (Bradley et al. 2000). In contrast, eye movement (EM) may provide a continuous and relatively direct measure of visual attention. Previous studies using EM assessments have found facilitated engagement with threat (Mogg et al. 2000) or both threat and positive stimuli (Calvo and Avero 2005; Garner et al. 2006). In addition, facilitated disengagement from positive stimuli, relative to threat, has been observed in clinically socially anxious individuals (Chen et al. 2012).

In addition to engagement and disengagement biases, avoidant attentional styles in high-anxious individuals have been observed from EM studies. That is, when social stimuli are presented for a long duration, such as 3 s, anxious individuals have shown reductions in the sustainment of attention over time to either threat stimuli (Calvo et al. 2005; Rohner 2002) or both threat and positive stimuli (Chen et al. 2012; Garner et al. 2006). It has been suggested that while anxiety-linked attentional bias to threat may reflect an automatic process, attentional avoidance may represent a strategic effort, in which a socially anxious individual may avoid attending to emotional social stimuli in an attempt to regulate their emotional state (Cisler and Koster 2010).

While recent research has incorporated eye tracking to assess attentional selectivity, a further advantage of eye tracking is its capacity to unobtrusively assess visual attention during realistic simulations. For instance, in the field of human computer interaction, eye tracking has been utilized in aviation and driving simulations to assess information selection and management, and situational awareness (Duchowski 2002). Such simulations possess high ecological validity. In contrast, the majority of anxiety and attentional bias research has been conducted in contrived laboratory settings. While this setting allows for rigorous experimental control, the artificial nature of these laboratory settings begs the question of whether their findings generalize to real-world situations. However, in clinical research, the use of eye-tracking realistic simulations remains a novel methodology.

2 Proposed Paradigm: Eye-Tracking Speech Simulation

Public speaking is a common fear for socially anxious individuals, as it requires them to engage in social performance and exposes them to potential negative social evaluation from the audience. Given this, a public speaking simulation, combined with eye tracking, may provide critical insights into the attentional processes that occur during conditions of psychosocial stress for individuals with SAD, in comparison to low-anxious control participants.

Fig. 1 Layout of audience display

2.1 Design

Participants are required to give a brief, 5-minute speech on a topic of their own choice, in front of a large display playing a pre-recorded video of an audience, while EM at the display is continuously recorded. The audience consists of eight confederates, assigned to either express socially positive or negative gestures, or remain neutral throughout the speech (see Fig. 1). Socially positive gestures may include a smile, or an agreeing nod, whereas socially negative gestures may include a disagreeing shake of the head, or a sigh of boredom. Participants receive one of two counterbalanced audience videos, in which the emotional confederates are switched for valence.

An initial 40-s neutral period is presented, in which all confederates are neutral. Following this a number of trials are presented, one trial every 10 s. Each trial begins with a flashing cross cue presented for 1 s on either an emotional face, or a neutral face. Immediately following this, a gesture pair consisting of one positive and one negative confederate make a dynamic social gesture for 6 s, and then return to neutral. Participants are asked to look at the cue whenever it appears. This method of presentation derives from the recent development of an engagement disengagement cuing (EDC) task which uses a cue to secure attention in order to assess attentional engagement and disengagement (Chen et al. 2012).

For trials assessing engagement (see Fig. 2a), the cross cue is presented on the neutral face. The participant's attention is, therefore, secured in between and equi-

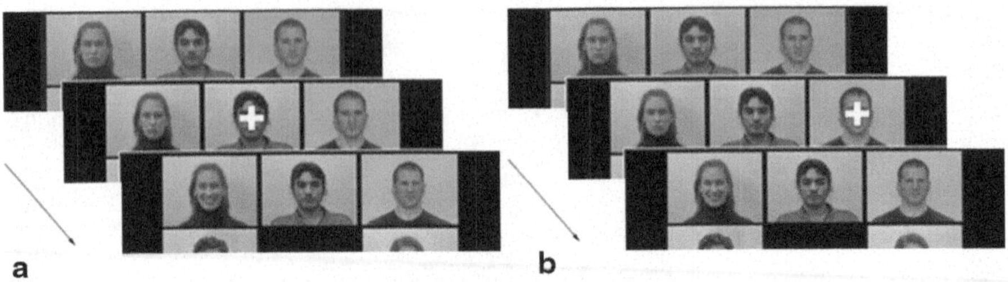

Fig. 2 Example of trials assessing engagement (a, *left*) and disengagement from threat (b, *right*)

distant from the positive and negative gesture. Following gesture onset, the propensity and speed of initial EM orientation may be assessed to provide measures of engagement propensity and engagement speed respectively, to positive and negative stimuli.

For trials assessing disengagement (see Fig. 2b), the cross cue appears on either the positive or the negative face. The participant's attention is, therefore, secured at the location of the emotional face. Following gesture onset, the latency to saccade away from the face may be assessed to provide a measure of disengagement speed from positive and negative stimuli.

2.2 Data preparation

Given that speaking may cause momentary disruptions to the EM samples. A two-sample noise reduction filter is first applied to the EM data (Stampe 1993), followed by an interpolation filter to smooth out brief signal loss for gaps smaller than 100 ms where EM is held within 1°visual angle (VA) immediately before and after the gap. Subsequently, fixations are defined as EM samples held within 1°VA for a minimum duration of 100 ms. Due to the noise introduced by concurrent speaking, fixation detection algorithms are preferable to velocity and acceleration-based saccade detection algorithms.

2.3 Engagement Propensity and Engagement Speed

Trials assessing engagement are included for analysis if (a) fixation is present at the location of the cross cue immediately prior to gesture onset, (b) fixation occurs on at least one of the emotional faces before offset, and (c) this critical fixation occurs at least 100 ms following gesture onset (to remove anticipatory saccades). From the included trials, engagement propensity may be calculated as the relative likelihood to initially orient to positive and negative stimuli, and engagement speed as the mean latency of initial orientation to positive and negative stimuli.

2.4 *Disengagement speed*

Trials assessing disengagement are included for analysis if (a) fixation is present at the location of the cross cue immediately prior to gesture onset, (b) a saccade away from the emotional face is made before offset, and (c) this critical saccade occurs at least 100 ms following gesture onset. From the included trials, disengagement speed may be calculated as the mean latency to saccade away from positive and negative stimuli.

2.5 *Sustained attentional processing*

To assess for the manner in which attention is sustained throughout the speech, total fixation time to positive, negative, neutral and non-face regions of the display are summed for the duration of the speech. The non-face region refers to the display area in between and around the confederate faces.

3 Expected Outcomes and Implications

If SAD is associated with an engagement bias to threat, it is anticipated that socially anxious individuals, relative to controls, will exhibit a greater propensity to initially orient to threat, possibly at a faster speed. If SAD is associated with a disengagement bias from threat, then it is expected that socially anxious individuals, relative to controls, will be slower to disengage from threat gestures. In addition, if socially anxious individuals employ avoidance strategies in an attempt to self-regulate, it is likely that this will be reflected in reduced total fixation time to emotional social stimuli, and a relative increase in the total fixation time towards non-face regions in between and around the confederate faces. Such anxiety-linked findings would provide validation of previous empirical and theoretical literature using a novel simulation-based methodology. Thus, it will be possible to elucidate whether such expected findings generalize to realistic settings. Moreover, given the consistent and evidently causal relationship between social anxiety and attention (e.g. Amir et al. 2009; MacLeod et al. 2002), the online assessment of attentional selectivity under conditions of psychosocial stress may provide utility as a potential marker for SAD, and as a quantitative heuristic for symptom reduction in the clinical treatment of SAD.

References

Amir, N., Beard, C., Taylor, C. T., Klumpp, H., Elias, J., Bums, M., & Chen, X. (2009). Attention training in individuals with generalized social phobia: A randomized controlled trial. *Journal of Consulting and Clinical Psychology, 77*(5), 961–973. doi:10.1037/a0016685.

Bar-Haim, Y., Lamy, D., Pergamin, L., Bakermans-Kranenburg, M. J., & van Ijzendoorn, M. H. (2007). Threat-related attentional bias in anxious and nonanxious individuals: A meta-analytic study. *Psychological Bulletin, 133*(1), 1–24. doi:10.1037/0033-2909.133.1.1.

Bradley, B. P., Mogg, K., & Millar, N. H. (2000). Covert and overt orienting of attention to emotional faces in anxiety. *Cognition & Emotion, 14*(6), 789–808. doi:10.1080/026999300501566 36.

Calvo, M. G., & Avero, P. (2005). Time course of attentional bias to emotional scenes in anxiety: Gaze direction and duration. *Cognition & Emotion, 19*(3), 433–451. doi:10.1080/02699930441000157.

Chen, N. T. M., Clarke, P. J. F., MacLeod, C., & Guastella, A. J. (2012). Biased attentional processing of positive stimuli in social anxiety disorder: An eye movement study. *Cognitive Behaviour Therapy, 41*(2, Sp. Iss. SI), 96–107. doi:10.1080/16506073.2012.6665620.

Cisler, J. M., & Koster, E. H. W. (2010). Mechanisms of attentional biases towards threat in anxiety disorders: An integrative review. *Clinical Psychology Review, 30*(2), 203–216. doi:10.1016/j.cpr.2009.11.0030.

Duchowski, A. T. (2002). A breadth-first survey of eye-tracking applications. *Behavior Research Methods Instruments & Computers, 34*(4), 455–470. doi:10.3758/bf03195475.

Fox, E., Russo, R., Bowles, R., & Dutton, K. (2001). Do threatening stimuli draw or hold visual attention in subclinical anxiety? *Journal of Experimental Psychology-General, 130*(4), 681–700. doi:10.1037//0096-3445.130.4.681.

Garner, M., Mogg, K., & Bradley, B. P. (2006). Orienting and maintenance of gaze to facial expressions in social anxiety. *Journal of Abnormal Psychology, 115*(4), 760–770. doi:10.1037/0021-843x.115.4.760.

Hofmann, S. G. (2007). Cognitive factors that maintain social anxiety disorder: A comprehensive model and its treatment implications. *Cognitive Behaviour Therapy, 36*(4), 193–209. doi:10.1080/16506070701421313.

MacLeod, C., Rutherford, E., Campbell, L., Ebsworthy, G., & Holker, L. (2002). Selective attention and emotional vulnerability: Assessing the causal basis of their association through the experimental manipulation of attentional bias. *Journal of Abnormal Psychology, 111*(1), 107–123. doi:10.1037//0021-843x.111.1.107.

Mogg, K., Holmes, A., Garner, M., & Bradley, B. P. (2008). Effects of threat cues on attentional shifting, disengagement and response slowing in anxious individuals. *Behaviour Research and Therapy, 46*(5), 656–667. doi: 10.1016/j.brat.2008.02.011.

Mogg, K., Millar, N., & Bradley, B. P. (2000). Biases in eye movements to threatening facial expressions in generalized anxiety disorder and depressive disorder. *Journal of Abnormal Psychology, 109*(4), 695–704. doi: 10.1037//0021-843x.109.4.695.

Rapee, R. M., & Heimberg, R. G. (1997). A cognitive-behavioral model of anxiety in social phobia. Behaviour Research and Therapy, 35(8), 741–756. doi: 10.1016/S0005-7967(97)00022-3.

Rohner, J. C. (2002). The time-course of visual threat processing: High trait anxious individuals eventually avert their gaze from angry faces. *Cognition & Emotion, 16*(6), 837–844. doi: 10.1080/02699930143000572.

Stampe, D. M. (1993). Heuristic filtering and reliable calibration methods for video-based pupil-tracking systems. *Behavior Research Methods, 25*(2), 137–142. doi: 10.3758/bf03204486.

Using Saccadic Eye Movements to Assess Cognitive Decline with Ageing

Alison Bowling and Anja Draper

1 Introduction

In viewing our environment, our eyes make rapid eye movements (*saccades*) from one location to another and then rest briefly (fixations) while our eyes and brain derive a "snapshot" of the world at the point of gaze. About three successive saccades and fixations take place every second. We are rarely aware of these saccadic eye movements, but they are influenced by a wide range of cognitive processes, and over the last 30 years saccadic eye movements have been widely used by researchers to enhance our understanding of neural processes in the brain (Hutton 2008).

The neural basis of saccadic eye movements is well understood (McDowell et al. 2008). The motor neurons which signal the muscles which control the eye are located in the brainstem (Sparks 2002). These neurons receive signals from the superior colliculus, in which the location of saccade destinations is systematically mapped (Everling et al. 1999). Neural input from higher areas of the brain such as the basal ganglia, the dorsolateral prefrontal cortex and frontal eye fields further provides for the control and modification of eye movements. Consequently, any influence that affects the function of the brain regions involved in the execution of saccades, such as pharmacological interventions, brain lesions, or neurological conditions, may affect some aspect of saccadic control. For example, patients suffering from Parkinson's disease, which results from neural deterioration in the basal ganglia, have difficulty in voluntary saccade execution (Amador et al. 2006).

Although saccadic eye movements normally occur without conscious awareness and are usually considered reflexive, they can be controlled intentionally and saccades may be made voluntarily or inhibited when necessary. For example, objects appearing in the periphery of vision capture attention, and a saccade is frequently reflexively generated towards the novel object. However, it is possible to inhibit this saccade, such as occurs when a driver attending to the road ahead avoids looking at advertising billboards and other roadside objects which may distract attention.

A. Bowling (✉) · A. Draper
Southern Cross University, Coffs Harbour, NSW, Australia
e-mail: Alison.Bowling@scu.edu.au

M. Horsley et al. (eds.), *Current Trends in Eye Tracking Research*,
DOI 10.1007/978-3-319-02868-2_19, © Springer International Publishing Switzerland 2014

Several paradigms have been devised to test how well participants are able to inhibit reflexive saccades. One of these is the antisaccade task, which is a very useful tool for investigating cognitive deficits in psychopathology (Hutton and Ettinger 2006). The antisaccade task requires a voluntary saccade to be made in the opposite direction to a stimulus which appears on one side of a computer screen. Participants are instructed to look away from this peripheral stimulus towards its mirror position (Hallett 1978; Munoz and Everling 2004). To successfully perform an antisaccade, participants have to suppress the reflexive saccade to the stimulus, before generating the intentional saccade to the alternative position (Everling and Fischer 1998). When a participant is unable to inhibit the reflexive saccade, a prosaccade error is produced. The percentage of errors made by participants is one measure of inhibitory control. An increase in overall prosaccade error rate with increasing age has frequently been reported (Butler et al. 1999; Klein et al. 2000; Nieuwenhuis et al. 2000; Olincy et al. 1997), suggesting that older people have greater difficulty in preventing responding to distracting influences in their environment than younger people.

Most prosaccade errors are immediately corrected. However, sometimes the prosaccade errors are not corrected. Bowling, Hindman and Donnelley (2012) found that uncorrected errors in particular correlated with neuropsychological tests such as spatial working memory.

Another saccade task requiring the inhibition of an unwanted saccade is the oculomotor capture task. This requires a saccade to be made to a target in the presence of an onset distractor which appears abruptly during the presentation of the target. In this paradigm, participants frequently direct an eye movement to the onset distractor even when they are not aware of its presence (Kramer et al. 1999, 2000). These erroneous saccades are referred to as "captures" and reflect a failure to automatically inhibit overt attentional capture by the distractor. Kramer et al. (2000) found no differences between older and younger age groups when the onset distractor was the same luminance as other elements in the display, and participants may not have been aware of its presence. However, when the onset distractor was brighter than other distractors in the display, older participants produced a greater proportion of captures than younger participants. Kramer et al. (2000) hypothesised that when participants are unaware of the presence of a distractor, inhibition of a saccade towards it is relatively automatic. However, when participants are aware of the distractor (its salience is high), they must intentionally inhibit an eye movement towards it, which involves higher-level cognitive processes (Kramer et al. 2005).

Participants are necessarily aware of the stimulus in the antisaccade task, since they must pay attention to it to compute the location and direction of the antisaccade. If both the antisaccade task and the oculomotor capture task with a highly salient distractor require intentional inhibitory control, we would expect those people who have a high error rate on one task to similarly have a high error rate on the other. Participants who perform poorly on both tasks may consequently have difficulty inhibiting responses to distractors in their environment, indicative of a reduction in cognitive capacity. In this chapter, we present evidence to suggest that a combination of saccadic eye movement tests may effectively detect individuals with a reduced capacity to inhibit distracting stimuli. We compared the performance of a group of older adults with a group of younger adults on the antisaccade and oculomotor capture tasks and found that a subset of the older group performed more poorly than the younger group and the remainder of the older group. The participants in this subset were all more than 70 years old, and their poor performance may be indicative of a reduction in cognitive capacity in this subgroup.

2 Method

2.1 *Participants*

Twenty-seven younger adults (age range18–34) and 23 older adults (age range 60–79) completed the two eye movement tasks. Younger adults were recruited by means of an invitation letter or email, and included 11 community members and 17 students from the university. Older adults all lived independently and were recruited by means of an invitation letter or communication placed on the University of the Third Age bulletin board. All participants self-reported as having normal or corrected-to-normal vision, including normal colour vision. Participants with multifocal lenses, nystagmus, cataracts, or recent cataract surgery were excluded from the study due to difficulty in calibrating the eye-tracking system. No incentives were offered and all participants provided written, informed consent. Approval for this study was granted by Southern Cross University's Human Research Ethics Committee.

2.2 *Apparatus*

We used an Eyelink 1000 eye-tracking system (SR Research) to collect antisaccade and oculomotor capture data. Specifically, eye movements were recorded by monitoring pupil position and corneal reflectance with a sampling rate of 1,000 Hz and spatial resolution of 0.05°. Eye-movement stimuli were created using SR Research Experiment Builder 1.1.2. All tasks were presented on a 19-in. CTX computer monitor with a resolution of 1,024 × 768 pixels at 60 Hz. A nine-point calibration of gaze position on the eye tracker was performed prior to each block of trials. Saccades were detected when eye velocity was 30°/s, acceleration exceeded 8,000°/s^2 and position changed by >0.15°. All variables were extracted from the recordings using SR Research Dataview 1.7.5.

2.3 *Stimuli*

2.3.1 Antisaccade Task

Six 0.5°-diameter grey dots were positioned on the periphery of an imaginary peripheral circle 8° from the centre of the monitor. All stimuli were presented on a

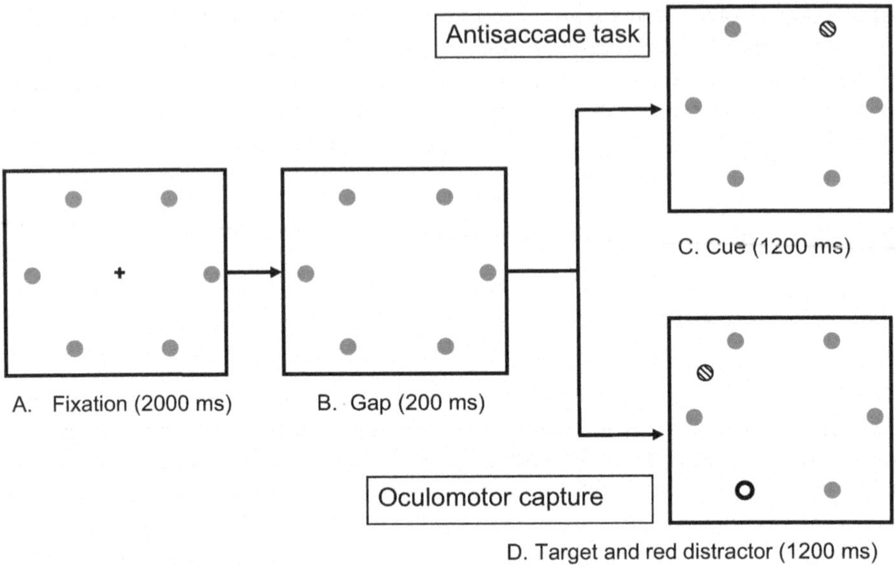

Fig. 1 Stimulus display and sequence of events in the antisaccade and oculomotor capture tasks. In the antisaccade task, participants were required to look at the dot immediately opposite the red dot (shown patterned here). In the oculomotor capture task, participants looked at the blue dot (shown unfilled here). The onset distractor was an additional red or grey dot which appeared between two of the static dots.

white background in clock positions of 1, 3, 5, 7, 9 and 11 as shown at the left of Fig. 1. Each trial was preceded by a single cross in the centre of the screen and a trial was initiated once this cross was fixated. A 2,000-ms presentation of the fixation display was followed by a 200-ms gap during which the fixation cross was absent. One of the six grey dots was then replaced by a red one. At the appearance of the red dot, participants were required to move their eyes to the grey dot diagonally opposite the red one. After 1,200 ms, the stimulus display was replaced by the central cross, in preparation for the next trial. There were 96 antisaccade trials presented in two blocks.

2.3.2 Oculomotor Capture Task

The distractor in this task was either a grey (low-salience condition) or a red (high-salience condition) dot, which could appear at any one of the six clock positions 2, 4, 8, 10, on the periphery of the same imaginary 8°-radius circle as the six original dots. The target was a blue dot with the same luminance as the grey ones.

As for the antisaccade task, each trial was initiated by fixation on a central fixation cross. A trial commenced with a display consisting of the six grey dots for 2,000 ms. After the subsequent 200-ms gap, a dot at any one of the six target locations was replaced by the blue target dot, and participants were instructed to look at the blue target as quickly as possible, ignoring the other dots on the screen. A red or grey dot appeared simultaneously with the target in one of the distractor locations. The display returned to the single fixation cross after 1,200 ms. There were 48 trials for each of the salience conditions.

3 Data Analysis

For both eye movement tasks, the primary saccade after target onset was recorded as the saccadic response. Trials were excluded if the primary saccade was initiated 80 ms before the onset of the target (anticipatory eye movements) or did not meet the amplitude criterion of 2° of visual angle. For the antisaccade task, a trial was scored as correct if the direction of the primary saccade deviated less than 15° from either side of the target location. The trial was scored as a corrected error if the primary saccade was directed to within 15° of the red cue location followed by a subsequent saccade towards the target location. An oculomotor capture trial was scored as a capture if the direction of the primary saccade was directed to within 15° of the distractor position. The total number of antisaccade errors and oculomotor captures were computed and converted to a percentage of the total number of valid trials in each condition.

Antisaccade errors and oculomotor captures were analysed using a general linear mixed-effect analysis since these data do not fulfil the assumptions for an analysis of variance (Dixon 2008). Age group (younger vs. older) was entered as a fixed factor into the model and participant as a random factor. For the oculomotor capture task, distractor salience was also included as a fixed factor in the model.

4 Results

For the antisaccade task, older participants produced a significantly higher percentage of corrected prosaccade errors ($M=36.15$, standard deviation (SD)=28.40) than younger participants, ($M=14.95$, SD=14.26), $F(1,45)=8.73$, $p=0.005$. For the older age group, there was a significant correlation between age and corrected error rate, $r(22)=0.62$, $p=0.002$. The oldest participants made the highest number of corrected errors, and all participants younger than 70 years old scored in the same range as the younger group.

Figure 2 shows the percentage of captures in the oculomotor capture task for older and younger participants for the grey and red onset distractor conditions. There was a significant effect of age group, with older participants producing more captures than younger participants, $F(1, 46)=8.85$, $p=0.005$. There was also a significant main effect of distractor, there being a greater number of captures for the red

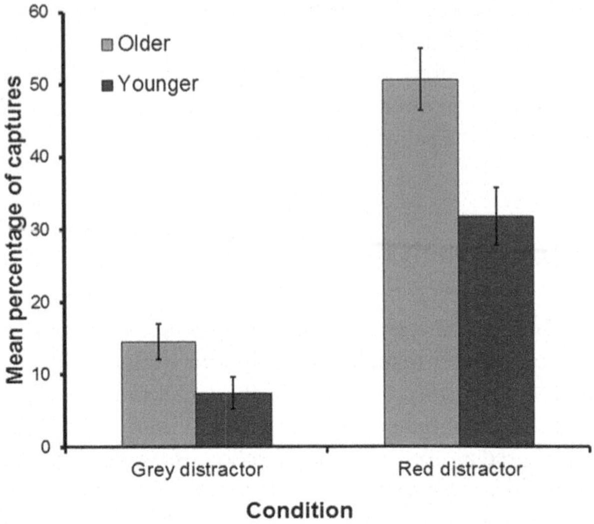

Fig. 2 Mean percentage captures for older and younger adults in the grey and red onset distractor conditions of the oculomotor capture task. Error bars represent ± 1 standard error of the mean

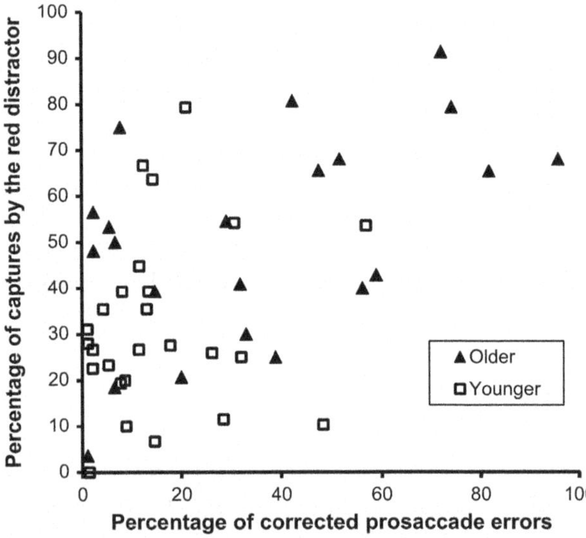

Fig. 3 Scatter plots showing percentage captures by a high-salience red distractor plotted against corrected prosaccade error rate in the antisaccade task

than for the grey distractor, $F(1,31)=244.52$, $p<0.001$). The interaction between age group and distractor was not significant, $F(1, 31)=0.15$, $p=0.70$.

Figure 3 shows the percentage of captures for the red onset distractor on the oculomotor capture task plotted against percentage of corrected prosaccade errors in the antisaccade task. For the older group, the percentage of corrected errors on the antisaccade task correlated with captures by the red distractor on the oculomotor capture task, $r(20)=0.51$, $p=0.016$, and in general, participants who performed poorly on one task also tended to perform poorly on the other. This suggests that both tasks provide a measure of inhibitory control (Butler and Zacks 2006).

5 Discussion

On average, older participants made more than twice the number of corrected pro-saccade errors on the antisaccade task than younger adults. Similarly, with the oculomotor capture task, there were approximately twice as many captures for older adults than for younger adults for both low-and high-salience onset distractors. The overall increase in the mean number of reflexive saccades for the older group appears to be mainly due to a small number of individuals whose error rate was substantially higher than that of the remainder of the older participants. As is evident from Fig. 3, not all older adults showed evidence for a decline in inhibitory control, with high error rates on both tasks being confined to adults older than 70 years of age. All older participants were functioning well in the community, and many were highly educated, participating in teaching and study through the University of the Third Age. Consequently, the observed differences between the age groups and older individuals are unlikely to be attributed to differences in intellectual ability or educational level.

Our finding that older adults looked towards the grey onset distractor more frequently than younger adults conflicts with previous research which obtained no differences between age groups in capture rate with a low-salience distractor (Kramer et al 2000). This difference in the outcomes of the two studies may be due to methodological differences. Unlike the stimuli used by Kramer et al. (1999, 2000), in which all distractors in the display changed colour, in our display only the target changed colour, in a similar manner to the to setup used by Godijn and Theeuwes (2002). We also introduced a 200-ms gap between the fixation offset and the target/distractor presentation. This has the effect of releasing saccade neurons in the superior colliculus from inhibition, increasing the possibility of triggering a reflexive saccade to a task-irrelevant stimulus (Dorris and Munoz 1995).

The finding that a subset of the older participants in this study exhibited comparably poor performance on two saccadic eye movement tasks designed to test the ability to inhibit a reflexive saccade suggests that these participants may be showing early signs of a reduction in inhibitory control, which is one aspect of cognitive functioning. Further studies using saccadic eye movement paradigms to test performance of older adults on these tasks and to compare the outcome with neuropsychological test results are necessary to determine whether these eye movement tasks may be used as a tool to detect early signs of cognitive decline. Follow-up studies of individuals showing poor performance on the tasks are also necessary to establish whether saccadic eye movement tests can predict dementia.

References

Amador, S. C., Hood, A. J., Schiess, M. C., Izor, R., & Sereno, A. B. (2006). Dissociating cognitive deficits involved in voluntary eye movement dysfunctions in Parkinson's disease patients. *Neuropsychologia, 44,* 1475–1482.

Bowling, A. C., Hindman, E. A., & Donnelly, J. (2012). Prosaccade errors in the antisaccade task: Differences between corrected and uncorrected errors and links to neuropsychological tests. *Experimental Brain Research, 216*(2), 169–179. doi:10.1007/s00221-011-2921-7.

Butler, K. M., & Zacks, R. T. (2006). Age deficits in the control of prepotent responses: Evidence for an inhibitory decline. *Psychology and Aging, 21*(3), 638–643.

Butler, K. M., Zacks, R. T., & Henderson, J. M. (1999). Suppression of reflexive saccades in younger and older adults: Age comparisons on an antisaccade task. *Memory & Cognition, 27*(4), 584–591.

Dixon, P. (2008). Models of accuracy in repeated measures designs. *Journal of Memory and Language, 59,* 447–456. doi:10.1016/j.jml.2007.11.004.

Everling, S., Dorris, M. C., Klein, R. M., & Munoz, D. P. (1999). Role of primate superior colliculus in preparation and execution of anti-saccades and pro-saccades. *The Journal of Neuroscience, 19*(7), 2740–2754.

Everling, S., & Fischer, B. (1998). The antisaccade: a review of basic research and clinical studies. *Neuropsychologia, 36*(9), 885–899.

Hallett, P. E. (1978). Primary and secondary saccades to goals defined by instructions. *Vision Research, 18,* 1279–1296.

Hutton, S. B. (2008). Cognitive control of saccadic eye movements. *Brain and Cognition, 68*(3), 327–340. doi:10.1016/j.bandc.2008.08.021.

Hutton, S. B., & Ettinger, U. (2006). The antisaccade task as a research tool in psychopathology: A critical review. *Psychophysiology, 43*(3), 302–313.

Klein, C., Fischer, B., Hartnegg, K., Heiss, W., & Roth, M. (2000). Optomotor and neuropsychological performance in old age. *Experimental Brain Research, 135*(2), 141–154.

Kramer, A. F., Gonzalez de Sather, J. C. M., & Cassavaugh, N. D. (2005). Development of attentional and oculomotor control. *Developmental Psychology, 41*(5), 760–772. doi:10.1037/0012-1649.41.5.760.

Kramer, A. F., Hahn, S., Irwin, D. E., & Theeuwes, J. (1999). Attentional capture and aging: Implications for visual search performance and oculomotor control. *Psychology and Aging, 14*(1), 135–154. doi:10.1037/0882-7974.14.1.135.

Kramer, A. F., Hahn, S., Irwin, D. E., & Theeuwes, J. (2000). Age differences in the control of looking behaviour: Do you know where your eyes have been? *Psychological Science, 11*(3), 210–217.

McDowell, J. E., Dyckman, K. A., Austin, B. P., & Clementz, B. A. (2008). Neurophysiology and neuroanatomy of reflexive and volitional saccades: Evidence from studies of humans. *Brain and Cognition, 68*(3), 255–270.

Munoz, D. P., & Everling, S. (2004). Look away: The antisaccade task and the voluntary control of eye movement. *Nature Reviews: Neuroscience, 5,* 218–228.

Nieuwenhuis, S., Ridderinkhof, K. R., de Jong, R., Kok, A., & van der Molen, M. W. (2000). Inhibitory inefficiency and failures of intention activation: Age-related decline in the control of saccadic eye movements. *Psychology and Aging, 15*(4), 635–647.

Olincy, A., Ross, R. G., Young, D. A., & Freedman, R. (1997). Age diminishes performance on an antisaccade eye movement task. *Neurobiology of Aging, 18*(5), 483–489.

Sparks, D. L. (2002). The brainstem control of saccadic eye movements. *Nature Reviews. Neuroscience, 3*(12), 952–964.

Comparing Personally Tailored Video- and Text-Delivered Web-Based Physical Activity Interventions—The Medium and the Message: An Eye-Tracking Study

Corneel Vandelanotte, Stephanie Alley, Nayadin Persaud and Mike Horsley

No between-group differences were recorded in terms of message recall and higher-order cognition immediately after receiving the intervention. The main message was recalled in both groups, but the details were largely forgotten and higher-order cognition was low. These results suggest that video-tailored messages are preferred, though more research is needed to optimise their impact so that recall and higher-order cognition are increased.

1 Introduction

Insufficient levels of physical activity double the risk of cardiovascular disease, type 2 diabetes and obesity, and increase the risk of breast and bowel cancer, depression and anxiety (Blair and Morris 2009). It is estimated that lack of physical activity causes more than 16,000 people to die prematurely annually in Australia (Medibank Private 2008). As a result, lack of physical activity remains the second largest disease burden in Australia (6.6%) (Begg et al. 2008), costing the health care system Australian $ 1.6 billion and the economy Australian $ 13.8 billion each year. Less than half of the adult Australian population is sufficiently active to obtain health benefits from physical activity (Bauman et al. 2001; Vandelanotte et al. 2010).

Hence, new and innovative physical activity interventions, such as web-based physical activity interventions, which can effectively reach large populations at low cost, are required to address the lack of physical activity in the Australian population. However, current research on web-based physical activity interventions has

C. Vandelanotte (✉) · S. Alley
Institute for Health and Social Science Research, Centre for Physical Activity Studies, Central Queensland University, Rockhampton, QLD, Australia
e-mail: c.vandelanotte@cqu.edu.au

N. Persaud · M. Horsley
Learning and Teaching Education Research Centre, Central Queensland University, Noosaville, QLD, Australia

M. Horsley et al. (eds.), *Current Trends in Eye Tracking Research*,
DOI 10.1007/978-3-319-02868-2_20, © Springer International Publishing Switzerland 2014

predominantly demonstrated short-term effectiveness (Vandelanotte et al. 2007, 2010; Spittaels et al. 2007a; ; Steele et al. 2007). Many studies have documented strong declines in website usage as the web-based physical activity interventions progressed. This usage decline limited their long-term effectiveness (Marshall et al. 2003; Eysenbach 2005). Low levels of website interactivity have been suggested to explain modest participant engagement and web intervention retention rates in these studies (Vandelanotte et al. 2007; Leslie et al. 2005; Marcus et al. 2009; Van Den Berg et al. 2007). A study associated with one of the largest physical activity interventions in Australia, the 10,000 Steps program, evaluated web usage patterns prospectively over a 2-year period (www.10,000steps.org.au). This study demonstrated that participants, who used interactive features of physical activity web interventions (such as the i-challenge), were more engaged with the website over a longer period of time when compared to those who did not (Davies et al. 2012).

Physical activity intervention websites that provide 'tailored', or individually adapted, feedback about physical activity are more interactive and show good effectiveness (Vandelanotte et al. 2005; Spittaels et al. 2007a; Kroeze et al. 2006; Noar et al. 2007). Users find tailored information more interesting and engaging because it is personally relevant (Vandelanotte and De Bourdeaudhuij 2003). Computer-tailored interventions can provide large numbers of people with individualised behaviour change information at low cost (de Vries and Brug 1999). Compared with non-tailored messages, tailored messages are more likely to be read and remembered, saved and discussed with others (Spittaels et al. 2007b; Skinner et al. 1999). To date, computer-tailored interventions have typically provided the personalised feedback in written text, which can accumulate to be several pages long.

However, a different stream of research has shown that text-based information is not effectively transmitted via the Internet, because people do not fully read text on the Internet but rather 'scan' and 'skim' the content (Liu 2005). Internet-based reading behaviour is characterised by more time spent on browsing and scanning, keyword spotting, non-linear reading and more selective reading (grazing, flick and click), while less time is spent on in-depth and concentrated reading (Liu 2005; Sutherland 2002). One eye-tracking-based study, including 232 participants, concluded that people do not 'read' websites as they would a book or newspaper. Instead, participants quickly scanned websites in an F-shaped pattern, in which they primarily focussed on the top-left quadrant of the page and this pattern was consistent across many different types of websites (Nielsen and Pernice 2010).

With the rise of broadband Internet connections, web-based video has become an increasingly popular part of users' online experiences (Anderson et al. 2011). For example, 71 % of adults use video-sharing websites such as *YouTube* and 28 % of adults access these sites daily (Moore 2011). Popular websites can not only be highly interactive but they also provide most of their content in a visually attractive and engaging format (images or video). Health promotion physical activity intervention websites are competing with more popular entertainment and social networking websites for the attention of Australians. As a result, physical activity web interventions will need to be equally interactive and engaging. To fully harness the potential of computer-tailored physical intervention websites, they will need to be made interactive. As such, providing personally relevant physical activity information using engaging videos might be more effective in producing behaviour change, when compared to providing the same content in traditional text-based formats, which computer-tailored interventions have predominantly applied to date. Video messages have been used in other health behaviour change interventions; however, the videos applied were generic and not tailored to individual recipients (Frenn et al. 2005; Swartz et al. 2006; Alexander et al. 2010; Campbell et al. 2004; Vandelanotte and Mummery 2008; Vandelanotte et al. 2007).

As such, the current study aims to evaluate, by means of an eye-tracking methodologies and interviews, whether a video-tailored physical activity intervention results in higher level attention, recall and understanding in participants when compared to a traditionally text-tailored intervention. Currently, there is no published research applying cutting-edge eye-tracking technology to assess optimal delivery modes of information in health behaviour intervention websites. It is hypothesised that the video-tailored intervention will produce superior attention, recall and understanding in participants when compared to a text-tailored intervention.

2 Methods

2.1 Study Design and Procedures

The research used a two-group descriptive, exploratory design in which participants were asked to view either a text-based or a video-based website that provides personally relevant physical activity information after completing a brief online survey. Participants were English-speaking adult (more than 18 years of age) volunteers, with adequate computer skills, who were recruited locally (via e-mail) from staff and students of a regional university in Queensland. They were invited to sit behind the computer fitted with eye-tracking equipment in a research laboratory where they would not be disturbed by anyone entering the room. Once seated, participants' eyes were calibrated with the eye-tracking equipment. Following the calibration, participants completed a series of basic demographic questions that were provided as an electronic questionnaire on the computer. Participants were then instructed to follow the link to the test website, being the 'My Physical Activity Advice' website (www.mypaa.com). Each participant was given a login name and password code that randomised them automatically to receive the personalised physical activity feedback either in text-based format or video-based format. The structure and content of the activity advice was identical in both groups, with the exception of how the personalised feedback was delivered (text or video). The researcher supervising the procedure was not aware of the group allocation that was randomly assigned to the passwords and logins provided to participants. Next, the participants were exposed to the online physical activity intervention (see 'Intervention' section for a detailed description of the interventions administered to both groups), during which all eye movements were tracked, and website user statistics where gathered. Participants were also video-taped during this time. Finally, when exposure to the intervention was complete, participants in both groups were asked a number of questions to recall the contents of the intervention they had just been exposed to. The questions were administered by the researcher supervising the procedure and the answers were audio taped, so that they could be transcribed and analysed afterwards. The entire process was designed to take approximately 20–25 min of the participants' time. Low-risk ethical clearance was obtained from the Central Queensland University ethics committee.

The sequence of the eye-tracking test was:

a. Introduction to the nature of the test
b. Calibration of the Tobii 120 to the participant's gaze
c. Completion of demographic questions built into the gaze test using Tobii Studio software
d. Visit the MyPaa website, complete the questions to receive the personalised feedback
e. Read (text-tailoring) or watch (video-tailoring) the personalised advice provided
f. Answer recall questions pertaining to the physical activity feedback that was received

2.2 *Intervention*

The intervention was based on previous internet-delivered and computer-tailored studies that successfully increased physical activity (Vandelanotte and Bourdeaudhuij 2003; Spittaels et al. 2007a; Spittaels et al. 2007b; Vandelanotte et al. 2007, 2008); however, in addition to the previously developed text mode, a video mode was developed for the purpose of this study. To inform the development of the video-tailored intervention, focus groups and a state-wide survey were conducted to explore perceived appropriateness of the new delivery mode and volume of information presented (Vandelanotte and Mummery 2011). The computer-tailored content of the two intervention modes was identical, only the intervention delivery mode was different. A series of screenshots provide an impression of the intervention and show the home page (Fig. 1), an example of survey questions (Fig. 2), an example of text mode (Fig. 3), an example of video mode (Fig. 4).

The intervention was largely based on the Theory of Planned Behavior (Aijzen 1991) and the Stage of Change Concept (Prochaska and Velicer 1997). Constructs of the Theory of Planned Behavior were presented through provision of personally relevant feedback on attitudes, self-efficacy, intentions, benefits and barriers in relation to their physical activity level. The intervention content was also adapted based on participants' stage of change, and 'normative feedback' (whether or not participants meet the physical activity recommendation) (DHAC 1999) was provided in a graph. Other non-theoretical tailored variables were age, body mass index (BMI), work environment and the distance to much visited places. In order to

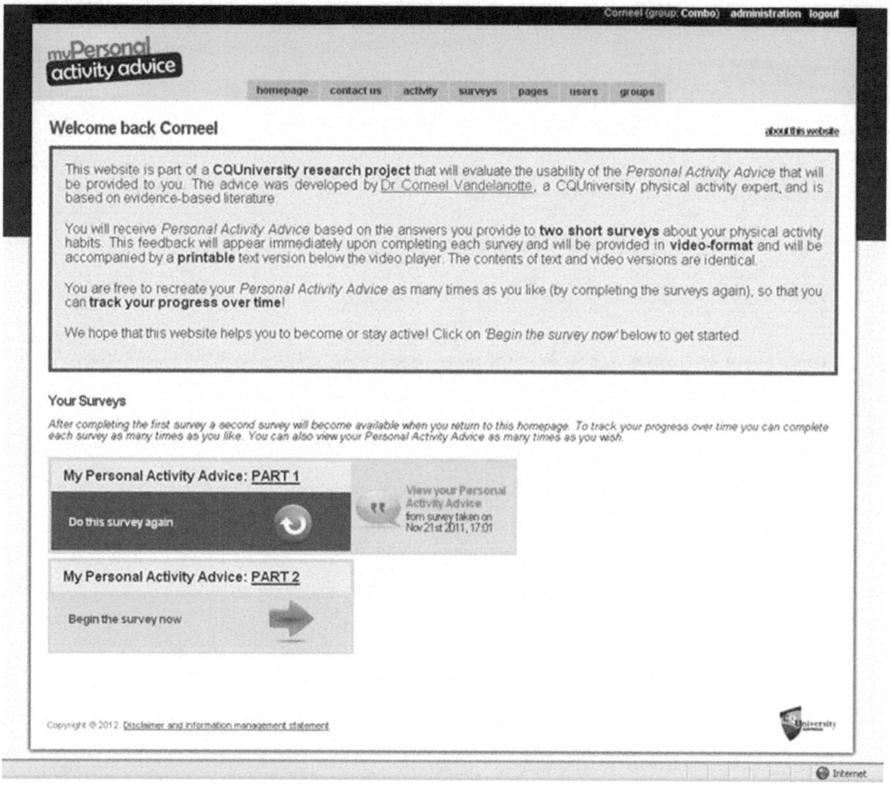

Fig. 1 MyPaa website home page

receive personalised physical activity advice, participants first had to complete a short questionnaire about their physical activity levels, after which the personal advice immediately appeared on screen (Part 1). Participants who did not meet the physical activity recommendation were encouraged to receive more feedback by completing additional questions related to the psychosocial correlates of physical activity (Part 2). A more detailed overview of the information participants received is provided below.

First part of the feedback: The first part of the feedback started with an introduction, where the aim of the personalised feedback was explained. Next, the individual average amount of physical activity a day was compared to the guideline of 30 min a day. This was also shown in a bar chart. If someone did less than five sessions of physical activity a week, extra feedback about this was provided. The feedback was accompanied by information about the health benefits of meeting the guidelines for physical activity. Participants with a BMI over 25 kg/m^2 received tailored feedback about this, in which the health risks of being overweight or obese were explained. Participants who were older than 45 years and not meeting the

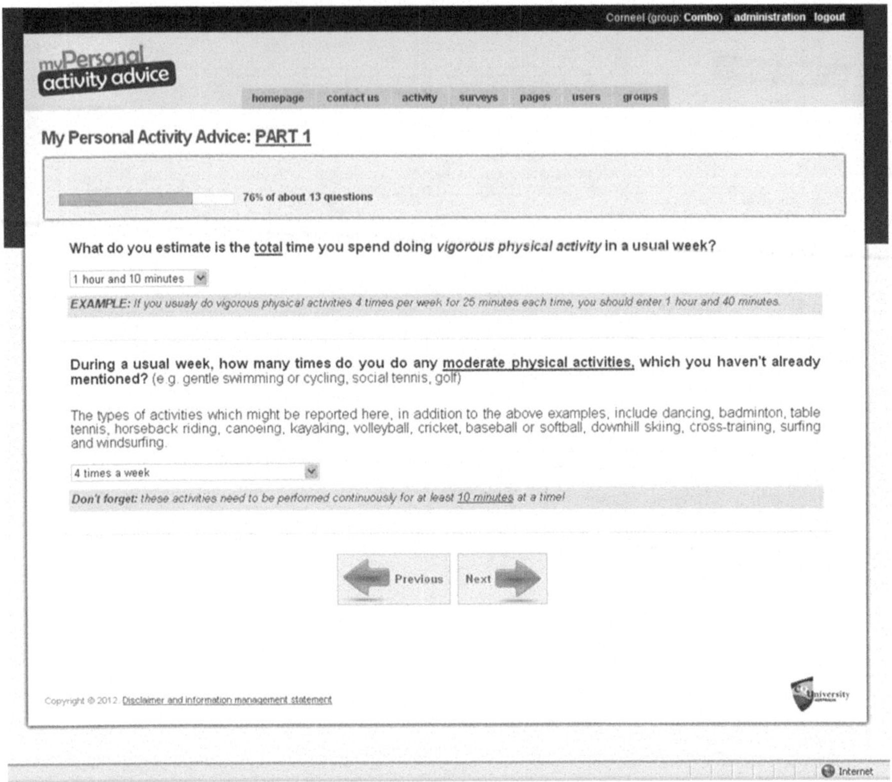

Fig. 2 Example of survey question

physical activity recommendations received extra feedback about the importance of physical activity even at an older age. After this, the participants who did not meet the guidelines were provided with information how to increase their physical activity (e.g. it is possible to do blocks of 10 min of physical activity). In the last section of the first part of the feedback, participants were directed to the next survey if they did not meet the guidelines for physical activity. Participants who were already active enough were told that it is not necessary to go to the next survey, but that they are still allowed to continue.

Second part of the feedback: The second part of the feedback also started with an introduction, which was tailored to the current level of physical activity and the intention to become more active. For participants who were already active enough, the feedback in this part was shortened. The participants were provided with feedback on how to achieve an active lifestyle (e.g. to incorporate physical activity in all parts of your daily life). Information was given about how to be physically active around the house and garden, at work, in transport, and during leisure time. The feedback was tailored to: the possession of a garden, having a job with short or long lunch breaks,

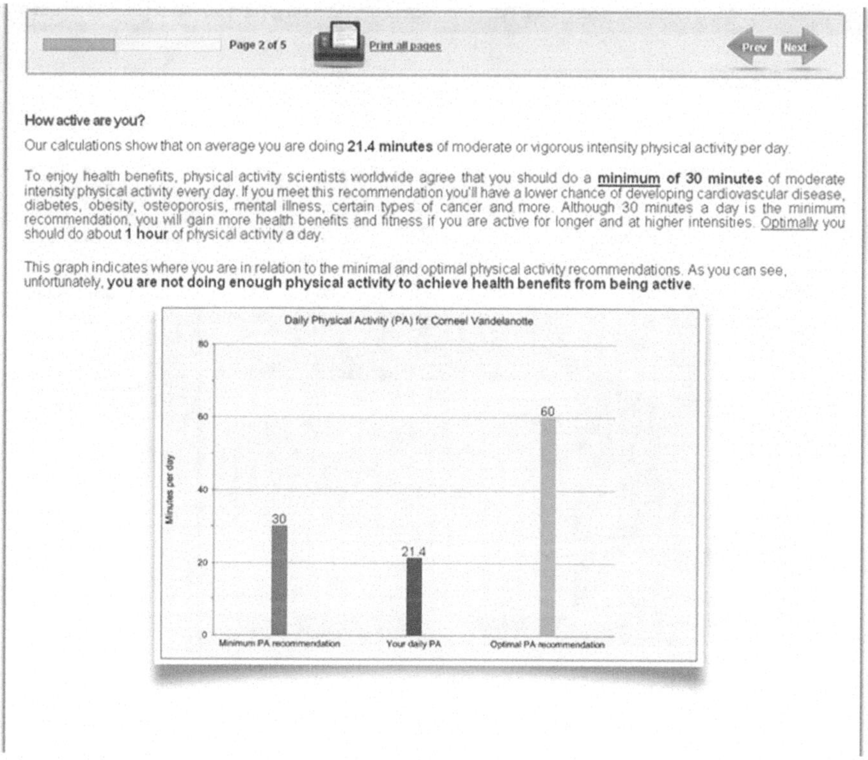

Fig. 3 Example of text-tailored physical activity advice

facilities of the workplace (e.g. showers and gym), the living distance to regular visited places, and the (intensity of the) current physical activity levels. Next, the individual benefits of being physically active (e.g. having fun) were discussed and confirmed. Information was provided on how to overcome relevant barriers of being physically active (e.g. too hot, cold or rainy weather). Further, tailored feedback on the participants' self-efficacy was provided. In the last section of the second part of the feedback, participants were wished good luck staying or becoming more active.

2.3 *Measures*

2.3.1 Demographics

Participants were asked to complete a short demographic survey at the start of the test. The following demographic information was collected: gender, age (18–30, 31–50, >50 years), height and weight (to calculate BMI: normal or overweight), motivation to increase physical activity (motivated, not motivated), education (sec-

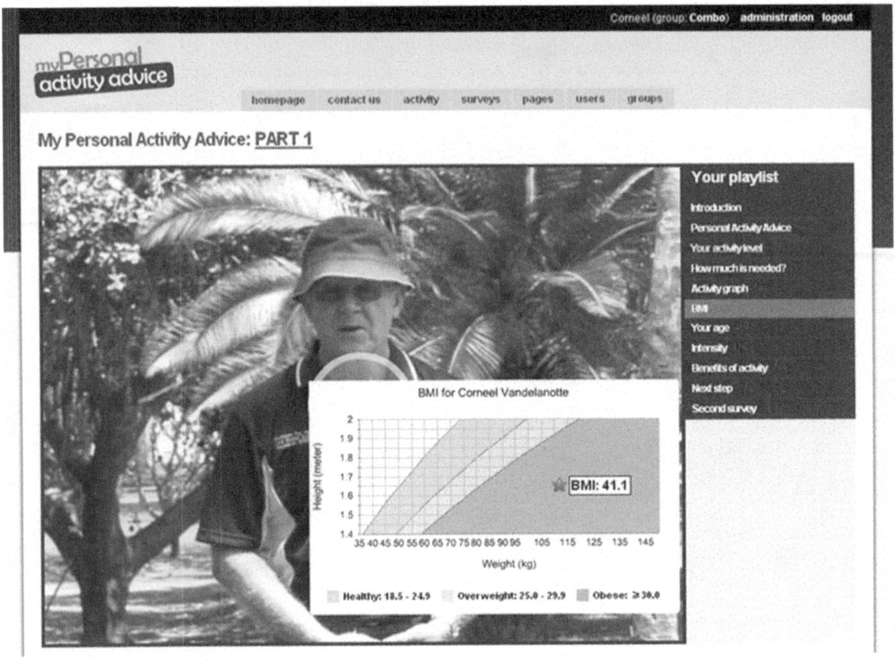

Fig. 4 Example of video-tailored physical activity advice

ondary school, TAFE, university), employment status (full time, part time, unemployed, student) and income (< Australian $ 30,000, Australian $ 30,001–70,000, >Australian $ 70,000).

2.3.2 Gaze Concentration and Click Stream

Participants' eye movements and gaze patterns were recorded on a Tobii 120 eye-tracking device, and the data collected and analysed by Tobii Studio software. The Tobii 120 tracks eye movements at a resolution of 1,280 pixels at a controller refresh rate of 60–75 Hz. The Tobii 120 eye tracker allows 15 degrees of head movement 60 cm from the screen, and tracks corneal movements to identify gaze focus, fixation and saccade at 15 ms. Total number of clicks made whilst receiving the feedback was also recorded.

2.3.3 Interview

A brief interview (nine questions) was conducted after each participant received the physical activity advice for each of the conditions (video based or text based). The items in the interview were constructed to explore recall of information provided to participants in the personalised feedback and to explore lower- to higher-order thinking in relation to the physical activity advice (evaluative response). Each interview was approximately 1–4 min in duration. Participant responses were recorded using an audio digital recorder (LiveScribe Pen) and transcribed in Microsoft Word.

2.4 *Data Analysis*

All analyses were conducted using Statistical Package for Social Sciences (SPSS) version 19. Significance level was set at $p<0.05$. A chi-square analysis was conducted to compare group baseline participant characteristics. Gaze duration (as an indicator of level of attention) was measured by the duration of all eye gazes whilst receiving the physical activity feedback within: (a) the total computer screen and (b) the area on the computer screen where the feedback was presented. To examine differences between the groups (text and video), analyses of variance (ANOVAs) were conducted for duration of eye gazes within the total screen (minutes), duration of eye gazes within the feedback area only (minutes) and number of clicks made whilst receiving the feedback. ANOVAs were also conducted to examine gender (males vs. female), age (<39 vs. ≥39), BMI (normal vs. overweight) and motivation to increase physical activity levels' (motivated vs. not motivated) differences for duration of eye gazes within the entire computer screen, duration of eye gazes within the feedback area and number of clicks. The participant responses for each question of the interview conducted after receiving the personalised feedback were assigned to a category of can't say/incorrect, recall response and, where appropriate, high-order response. Participant responses were coded as recall when they correctly recalled the physical activity advice. The responses were coded as high-order when they applied the physical activity advice to themselves. The high-order response category was not included for all questions if it was not possible for participants to respond in this way. A chi-square analysis was conducted to determine whether responses to each question differed between the video and text groups. A total recall score was also calculated for each participant by summing the total number of correct recall or high-order responses the participant gave. A Mann–Whitney *U*-test was conducted in SPSS to compare the recall scores in video and text participants.

3 Results

3.1 *Participants*

The demographic details of the participants are documented in Table 1. Data were collected from 41 participants. Overall, 66 % of participants were women, 61 % had a university education and 76 % said they would like to increase their physical activity. Participants were randomly assigned to the video ($n=21$) or text group ($n=20$).

Table 1 Participant characteristics by intervention group

	Total *n* (%) (*n*=41)	Video *n* (%) (*n*=21)	Text *n* (%) (*n*=20)
Gender			
Males	14 (34)	7 (33)	7 (35)
Females	27 (66)	14 (67)	13 (65)
Age			
18–30	8 (40)	3 (15)	11 (28)
31–50	6 (30)	12 (60)	18 (45)
>50	6 (30)	5 (25)	11 (27)
BMI			
Normal (<25)	23 (57)	11 (55)	12 (60)
Overweight (≥25)	17 (43)	9 (45)	8 (40)
Motivation			
Motivated	31 (76)	16 (76)	15 (75)
Not motivated	10 (24)	5 (24)	5 (25)
Education			
Secondary School	9 (22)	5 (24)	4 (20)
TAFE	7 (17)	3 (14)	4 (20)
University	25 (61)	13 (62)	12 (60)
Employment			
Full time	13 (32)	7 (33)	6 (30)
Part time	17 (42)	8 (38)	9 (45)
Unemployed	3 (7)	1 (5)	2 (10)
Student	8 (20)	5 (24)	3 (15)
Income			
<30,000	16 (39)	9 (43)	7 (35)
30,001–70,000	11 (26)	5 (24)	6 (30)
>70,001	14 (35)	7 (33)	7 (35)

There were no baseline differences between the two intervention groups in participant characteristics, including gender (x^2 (1)=0.01, p=0.91), age (x^2 (1)=4.36, p=0.11), BMI (x^2 (1)=0.10, p=0.75), motivation to increase physical activity (x^2 (1)=0.01, p=0.93), education (x^2 (1)=1.66, p=0.65), employment (x^2 (1)=0.95, p=0.81) and income (x^2 (1)=0.39, p=0.57).

3.2 Gaze Concentration and Click Stream

Table 2 shows descriptive statistics of gaze duration (minutes) within the total screen, gaze duration within the feedback area only and clicks by intervention group, gender, age, BMI and motivation. As shown in Fig. 5, both gaze duration within the total screen and gaze duration within the feedback area only were significantly higher in the video group than in the text group ($F(1,37)=61.38$ $p<0.001$, $\eta^2=0.630$; $F(1,37)=32.7$, $p<0.001$, $\eta^2=0.476$, respectively). Although the proportion of gaze duration within the feedback area compared to gaze duration within the total screen was lower in video participants (n=21, M=82 %, standard deviation

Table 2 Descriptive statistics for gaze duration and number of clicks by group and participant characteristics

	Gaze duration total screen (min)			Gaze duration feedback only (min)			Number of clicks		
	n	M	SD	*n*	M	SD	*n*	M	SD
Group									
Video	21	9.32	1.70	21	7.69	1.96	21	3.62	6.98
Text	17	4.07	2.42	17	3.60	2.45	19	10.16	1.86
Gender									
Males	13	6.03	3.24	13	4.85	2.60	21	6.00	6.92
Females	25	7.46	3.34	25	6.38	3.08	19	7.07	5.81
Age									
Younger	20	6.98	3.19	20	5.89	2.63	21	5.90	5.49
Older	17	6.83	3.64	17	5.79	3.51	18	7.50	6.97
BMI									
Normal	21	6.34	3.43	21	5.30	2.85	23	6.30	5.43
Overweight	16	7.66	3.21	16	6.56	3.18	16	7.12	7.28
Motivation									
Motivated	28	7.44	3.24	28	6.41	2.89	30	6.73	6.25
Not motivated	10	5.67	3.39	10	4.31	2.82	10	6.70	6.04

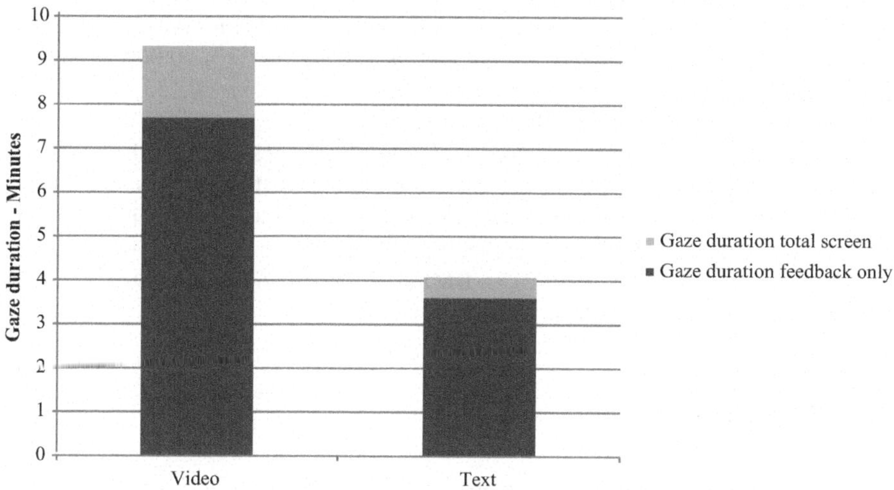

Fig. 5 Gaze duration in minutes during feedback for intervention groups

(SD) = 1 %) than in text participants (n = 17, M = 89 %, SD = 2 %), this was not statistically significant (F(1,36) = 0.884 p = 0.353, η^2 = 0.024).

Women had a significantly longer gaze duration within the total screen (F(1,37) = 3.83 p = 0.049, η^2 = 0.101) and within the feedback area (F(1,37) = 4.17 p = 0.049, η^2 = 0.109) than men (Fig. 6). Women had an average gaze duration within the feedback area 1.2 min longer than men in the video group and 1.8 min longer in the text group; however, no significant interaction between group and

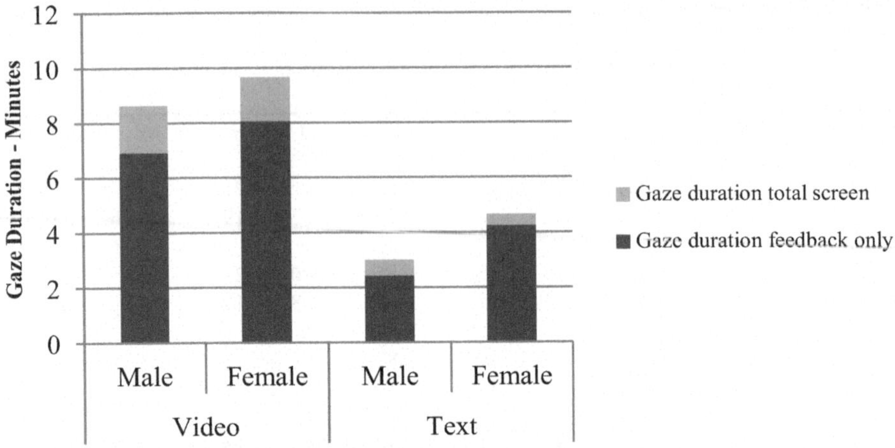

Fig. 6 Gaze duration in minutes during feedback by intervention group and gender

gender was found (F(1,37)=0.196, p=0.661, η^2=0.006). No differences were seen for either gaze duration within the total screen or within the feedback area by BMI (F(1,37)=1.407, p=0.244, η^2=0.039; F(1,37)=1.61, p=0.213, η^2=0.044, respectively), age (F(1,37)=0.018, p=0.894, η^2=0.001; F(1,37)=0.009, p=0.925, η^2=0.000, respectively) and motivation (F(1,37)=2.13, p=0.153, η^2=0.056; F(1,37)=3.95, p=0.055, η^2=0.099, respectively).

Participants in the text group used their mouse to click through the feedback more frequently than the video group (see Fig. 7). This difference was statistically significant (F(1,38)=15.652, p<0.001, η^2=0.292). There were no statistically significant differences in clicks by gender (F(1,38)=2.65, p=0.610, η^2=0.007), age (F(1,38)=0.640, p=0.429, η^2=0.017), BMI (F(1,38)=0.163, p=0.689, η^2=0.004) or motivation (F(1,38)=0.000, p=0.988, η^2=0.000).

3.3 *Interview*

The interview after receiving the personalised physical activity advice was designed to reveal recall of the information presented in the personal advice and provide data on the type of learning. Descriptive statistics of responses in video and text participants are presented in Table 3.

- **Question 1: What is the goal of the advice?** The majority of participants (63 %) answered this question correctly by recalling what the advice was about, but did not apply the advice to themselves (see Table 3). There were no significant differences in responses between the video and text groups (x^2 (2)=0.05, p=0.97). An example of a recall response from a participant in the video group is 'to become more active and ways to become more active', and an example of

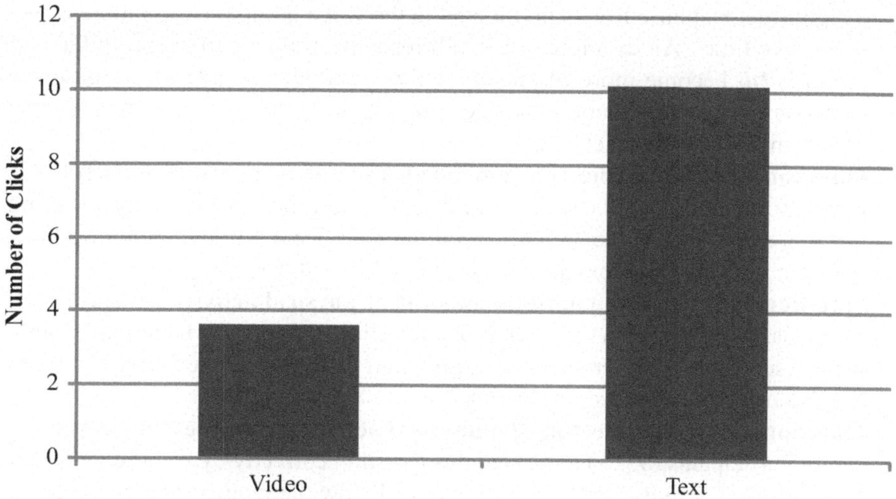

Fig. 7 Number of clicks during the feedback by intervention group

Table 3 Responses to the post-intervention interview by intervention group

		Total *n* (%)	Video *n* (%)	Text *n* (%)
Q1. What is the goal of the advice?	Can't say/incorrect	2 (5)	1 (5)	1 (5)
	Recall	26 (63)	13 (62)	13 (65)
	High order	13 (32)	7 (33)	6 (30)
Q2. What is the recommended amount of physical activity?	Can't say/incorrect	16 (39)	9 (43)	7 (35)
	Recall	25 (61)	12 (57)	13 (65)
Q3. What is the optimal amount of physical activity per day?	Can't say/incorrect	21 (51)	11 (52)	10 (50)
	Recall	20 (49)	10 (48)	10 (50)
Q4. Are you meeting the physical activity guidelines?	Can't say/incorrect	3 (7)	2 (10)	1 (5)
	Recall	38 (93)	19 (96)	19 (95)
Q5. Exactly how many minutes of physical activity do you do on a weekly basis?	Can't say/incorrect	20 (49)	11 (52)	9 (45)
	Recall	21 (51)	10 (48)	11 (55)
Q6 and Q7. What was presented in the graph? What was each of the bars showing?	Can't say/incorrect	21 (51)	9 (43)	12 (60)
	Recall	20 (49)	12 (57)	8 (40)
Q8. How will meeting the physical activity recommendations benefit you?	Can't say/incorrect	4 (10)	2 (10)	2 (10)
	Recall	24 (59)	14 (67)	10 (50)
	High order	13 (32)	5 (24)	8 (40)
Q9. What chronic diseases can be prevented?	Can't say/incorrect	15 (37)	9 (43)	6 (30)
	Recall	26 (63)	12 (57)	14 (70)

a high-order response from a participant in the video group is 'to get more active in *my* free time'. An example of a recall response from a participant in the text group is 'to become more physically active', and an example of a high-order response from a participant in the text group is 'to try to become more active by increasing *my* physical activity'.

- **Question 2: What is the recommended amount of physical activity**? The majority of participants (61 %) answered this question correctly by recalling '30 min a day' (see Table 3). There were no significant differences in responses between the video and text groups (x^2 (1)=0.27, p=0.61).
- **Question 3: What is the optimal amount of physical activity per day?** Half of the participants (49 %) answered this question correctly by recalling '60 min a day' (see Table 3). There were no significant differences in responses between the video and text groups (x^2 (1)=0.02, p=0.88.)
- **Question 4: Are you meeting the physical activity guidelines?** A clear majority of participants (93 %) answered this question correctly by recalling whether their physical activity levels were above or below the guidelines (see Table 3). There were no significant differences in responses between video and text groups (x^2 (1)=0.31, p=0.58). Examples of correct recall responses from participants in the video group are 'I'm doing more than I should be', and 'meeting minimum guidelines'. Examples of correct recall responses from participants in the text group are 'I am beating it', and 'I was meeting them'.
- **Question 5: Exactly how many minutes of physical activity do you do on a weekly basis?** About half of the participants (55 %) answered this correctly by recalling their estimated amount of physical activity presented in the graph (see Table 3). There were no significant differences in responses between the video and text groups (x^2 (1)=0.22, p=0.64).
- **Questions 6 and 7: What was presented in the graph? What was each of the bars showing?** A higher percentage of participants in the video group (57 %) gave a correct recall response to this question in comparison to the text group (40 %; see Table 3). However, the difference in responses between the video and text groups for this question was not statistically significant (x^2 (1)=1.21, p=0.27). Examples of correct recall responses from participants in the video group are 'minimum, then mine, then optimum', and 'my level of activity compared to optimal and minimum'. Examples of correct recall responses from participants in the text group are 'what the minimum is, what I do, and what the recommended physical activity is', and 'minimum, then mine, then the optimum'.
- **Question 8: How will meeting the physical activity recommendations benefit you?** A higher percentage of participants in the text group (40 %) answered this question with a high-order response when compared to the video group (24 %; see Table 3). However, the difference in responses between the video and text groups for this question was not statistically significant (χ^2 (2)=1.34, p=0.51). An example of a recall response from a participant in the video group is 'better health', and an example of a high-order response from a participant in the video group is 'to stay fit and I won't be susceptible to chronic diseases'. An example of a recall response from a participant in the text group is 'healthier, less chance

of getting sick, look and feel better', and an example of a high-order response from a participant in the text group is 'feel healthier and fitter, and feel better about myself'.

- **Question 9: What chronic diseases can be prevented?** A higher percentage of participants in the text group (70 %) were able to correctly recall at least one of the chronic diseases that physical activity helps to prevent than the video group (57 %; see Table 3). However, the difference in responses between the video and text groups for this question was not statistically significant (χ^2 (1) = 0.73, $p = 0.39$).
- **Total recall score:** A total recall score was calculated for each participant by summing the total number of correct recall or high-order responses the participant gave on all questions on the post-intervention interview. Possible scores ranged from 0 to 8. The mean response was 5.4 (SD = 1.57) in the video group and 5.6 (SD = 1.39) in the text group. This difference was not statistically significant according to a Mann–Whitney *U*-test ($p = 0.67$).

4 Discussion

Gaze duration (total screen and feedback area only) was significantly higher in the video group when compared to the text group. A way of measuring attention is to examine the amount of time that participants typically spend on different tasks and then to correlate this time with learning outputs. As such, Carol et al. (1965) have demonstrated that time on task is a strong predictor of learning from task. Gaze duration in the current study can be considered as a proxy measure for the level of attention that participants pay to the personalised physical activity message. This is important as attention is a critical component of remembering and learning. When people are concentrated, focussed and their senses attuned, then they are likely to learn more. Moreover, in a world where people are bombarded by hundreds of messages, advertisements and stimuli each hour and day, the struggle for attention is the key to having messages accepted processed and internalised. The messages delivered in video format are therefore more likely to be remembered and result in health behaviour change (increased physical activity) when compared to the messages delivered in text format. This is also in line with outcomes from previous website-delivered physical activity interventions that show that higher exposure to intervention contents leads to higher intervention effectiveness (Davies et al. 2012; Van Den Berg et al. 2007).

It was remarkable to observe that women had significantly longer gaze duration and, thus, paid more attention, when compared to men, irrespective of the group they were allocated (text or video). This is in line with other research that shows that women are keener to participate in web-based physical activity interventions and are also less likely to drop out of intervention trials (Vandelanotte and Spathonis et al. 2007). This research thus indicates that even when men do agree to participate in a web-based trial, their engagement is lower than that of women. It is crucial to

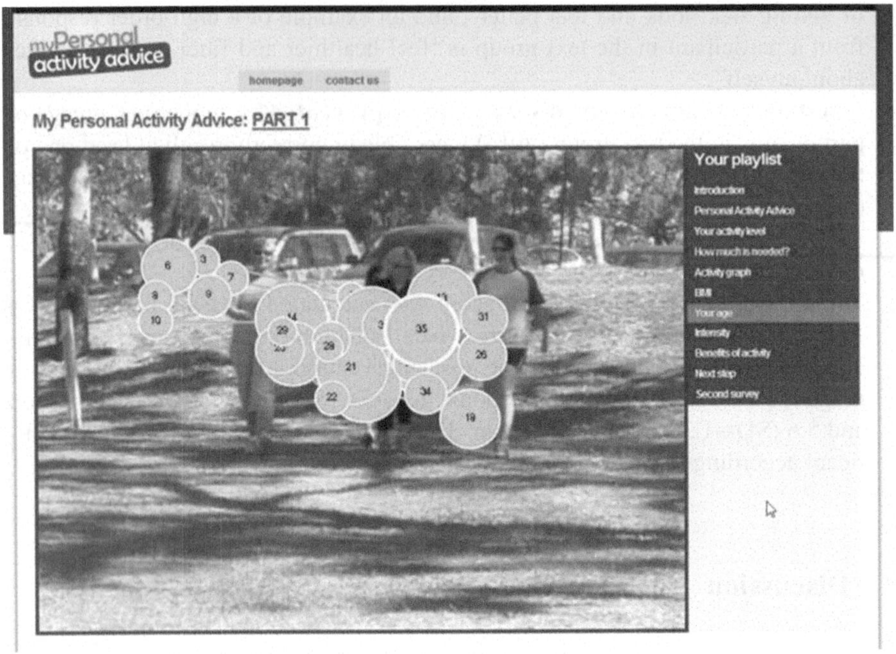

Fig. 8 Gaze plot of participant interacting with the video-based personal activity advice

find physical activity interventions that are more appealing to men in order to produce behaviour change (Vandelanotte et al. 2012). Whilst women outperform men in both the video and text groups, it was encouraging to observe that, the gender gap was a lot smaller in the video group (as illustrated in Fig. 6).

The observation that video-group participants pay more attention to the message is important in terms of achieving behaviour change, though that does not explain why the group differences in gaze duration are so large. A first explanation might stem from the differences in 'mental effort' expended in interacting with the personal physical activity advice between both groups (Kahneman 2011). In this context, mental effort refers to the concentration and focus that participants assigned to: (1) the message they were receiving; (2) their interaction with this message and (3) the way the message was presented to them. A key difference in the provision of physical activity feedback between the two groups was that the video participants did not have to read and process information. Rather, they listened and attended to the video and this reduced the mental effort on participants' short-term memory. In contrast, in the text-based group participants had to read and comprehend the text before they could move to the next page, requiring a higher mental load and a faster reduction of attention to the message. This is illustrated in the gaze plots (Figs. 8 and 9). More central and less dispersed video gaze plots indicate that the attention of the viewer is focussed on the content (Fig. 8), whereas less central and more dispersed text gaze plots conform to skim, scan and click processes (Fig. 9).

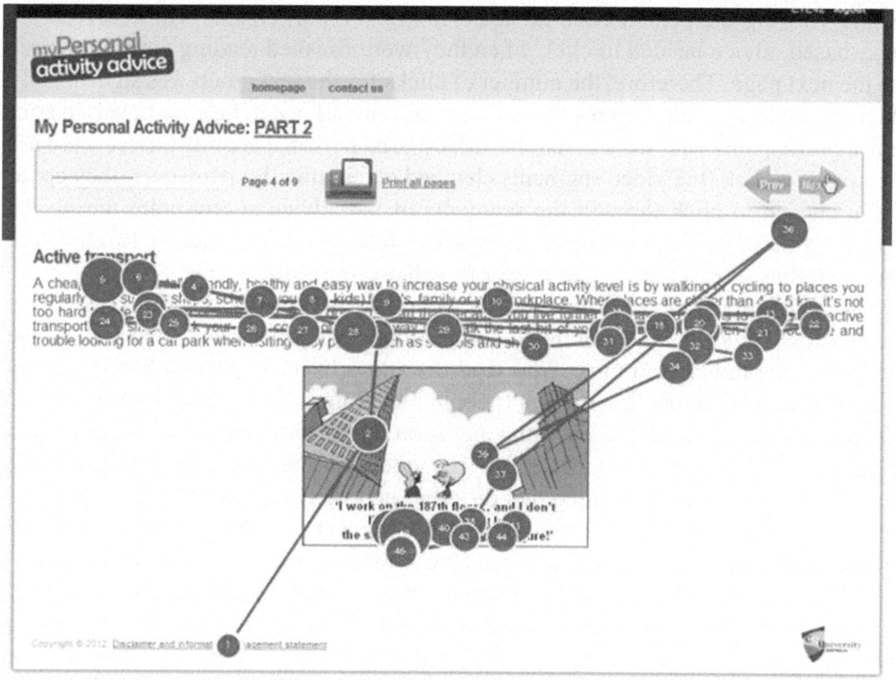

Fig. 9 Gaze plot of participant interacting with the text-based personal activity advice

A second explanation for gaze duration differences might stem from learned differences in the 'control' people perceive when reading text or watching a video. Participants in both groups could control their interaction with the message through: (1) setting the pace of which new information was presented to them (participants could control the speed at which new pages of text or videos would appear); (2) setting the duration of exposure to the new information (participants could shorten or extend the message on screen at any time) and (3) the review of information already received (participants could return to any part of the personalised advice at will). In short, participants in both groups could control what they saw, how much they saw, and whether they wished to see any part of the personal activity advice again. However, people have learned to control what appears on websites through previous 'flick and click' experiences that predominantly relate to text-based websites. This 'skimming and scanning' behaviour is best described as unfocussed gazing, attention to pop-ups and search for items that provoke interest or an emotional response, all at a rapid pace. In contrast, whilst participants in the video-based condition had similar control over content pace, duration and review, they did not take the opportunity to exercise this control but simply viewed the video to its conclusion. This reduced gaze duration in text-based websites compared to video-based websites is evidenced by the number of clicks performed in each condition. Participants in text group clicked more than twice as much when compared to those in the video group while viewing the personalised advice (see Fig. 7). Participants who received the text-based advice needed to click when they were finished reading a page to move to the next page. Therefore, the number of clicks for the text group was significantly higher as this was required for the personal activity advice to be completed. In contrast, participants who were given the video-based personal activity advice were not required to click (the video segments changed automatically). However, they could have chosen to click through the segments of the advice at any point during the video, as they had the same level of control as text-group participants. But they did not, and thus gaze duration was higher in video-group participants.

A final explanation of differences in gaze duration relates to the provision of a higher level of social and emotional connection to the participants who viewed the video advice rather than those who read the text advice. The video-based activity advice presented appealing images (people bicycling, walking and gardening) and created a sense of social presence and inclusion for the participant by featuring: (1) a personalised greeting; (2) human audio (the video was narrated with a human voice) and (3) human video (the video had different people speaking and participating in activities). The creation of this social context was entirely absent in the text-based advice and might have urged participants to 'move on' rather than 'hang around'.

Previous studies have developed taxonomies of cognition to explore the type of learning which has occurred (Bloom et al. 1956). These taxonomies provide a guide to analyse the types of leanings that occurred in the different media, such as text- or video-based information (Anderson et al. 2001). Generally, these taxonomies argue that lower order cognition occurs when learners are able to recall and comprehend facts and details. Higher-order cognition occurs when learners apply concepts to new situations, synthesise information and use their judgement in assessing its effectiveness for them and others. The interview questions were designed to reflect both higher and lower order cognition to gauge the thinking level of the participants. Some interview questions asked for specific recall of facts and numbers while others required analyses and judgement on the application of key concepts. When considering recall of specific bits of information, the level of abstraction is very low (Krathwhol et al. 1964). The interview questions were designed to compare the relative balance of higher and lower order thinking between participants who received the text- and video-based physical activity feedback. However, despite the greater level of attention provided by those in the video group, the interview questions did not indicate any significant differences between recall and higher-order cognition between the two groups. It is difficult to interpret these surprising findings. It might be that the physical activity messages were so simple and basic that a lower level of attention in the text group was still sufficient to get the message across. Alternatively, it might be that the interview questions were not sensitive enough to measure a difference between the groups or that participants in both groups had high knowledge about physical activity prior to taking part in the experiment. It is worth mentioning that video-group participants were better (though not significantly so) at interpreting the information displayed in the graph when compared to those in the text group. It is possible that the explanation provided by the human audio over the graph is more helpful in interpreting this information when compared to a text transcript.

Irrespective of group allocation, the interview was able to identify points of weakness and strength in the message communicated to the participants. Most participants understood the goal of the advice, whether they were meeting the physical activity guidelines or not and how meeting the physical activity guideline might benefit them. However, only about half of participants were able to recall the guideline itself, exactly how active they are, what information was presented in the graph and what chronic diseases can be prevented by being active. This seems to be in line with Posner's 'Theory of Attention' (1971) which poses that events that are linked to our emotions are better remembered and get our attention. Emotion and cognition are linked so that people often get the main message rather than the small details such as facts and numbers. This is directly in line with the outcomes of our study (e.g. participants remember whether or not they meet the physical activity guideline but find it harder to recall the actual number of minutes of activity they reported; people understand that meeting the guidelines will benefit them, but find it harder to recall which chronic diseases it will prevent) and provides us with a clear direction how to improve the feedback so participants might be able to remember more of what was provided to them.

In terms of higher-order cognition and internalisation, the feedback (across groups) did not score strong, with only about 30 % of participants showing any sign of higher-order cognition or internalisation. Internalisation is defined as the process through which a phenomenon or value successively and pervasively becomes a part of the individual (Klopfer 1976); a good example of internalisation was the video-group participant who responded 'to have fun' when asked 'What is the goal of the advice?' This might be due to our questions not being sensitive enough to register higher-order cognition or the fact that the process of internalisation takes time. When a person is first confronted with new information (the personalised physical activity advice), it might actually take time before this person accepts the message that was communicated and apply it to themselves; they might also need confirmation of the same message through different channels before they will start the process of internalisation.

5 Conclusions

Several conclusions can be drawn from this study. Firstly, the video-based physical activity advice provoked greater gaze duration and attention. Secondly, participants in both groups had equal control over the message but only those in the text-based group choose to exercise it. Thirdly, despite the greater gaze duration and attention in the video group, there was no difference in message recall or higher-order cognition between the groups. The main message was communicated, but the details were largely forgotten in both groups and higher-order cognition was low in both groups. These results suggest that video-tailored messages are preferred, though more research is needed to optimise their impact so that recall and higher-order cognition are increased. The results from this research provide valuable information to improve this new and innovative physical activity intervention.

References

Ajzen, I. (1991). The theory of planned behavior. *Organizational Behavior and Human Decision Processes, 50,* 179–211.

Alexander, G., et al (2010). Online F & V interventions. *American Journal of Public Health, 100,* 319–316.

Anderson, J., et al. (2011). *The changing needs of the online consumer*. Forrester Research.

Anderson, L.W., Krathwohl, D.R., Airasian, P.W., Cruikshank, K.A., Mayer, R.E., Pintrich, P.R., Raths, J., & Wittrock, M.C. (2001). A taxonomy for learning, teaching, and assessing: A revision of Bloom's taxonomy of educational objectives. New York: Longman.

Bauman, A., Ford, I., & Armstrong, T. (2001). Levels of physical activity in Australia, 1997, 1999 and 2000. In *A. Bauman, I. Ford, T. Armstrong*. Canberra: Australian Sports Commission.

Begg, S. J., et al. (2008). Burden of disease and injury in Australia. *Medical Journal of Australia, 188,* 36–40.

Blair, S., & Morris, J. (2009). Healthy hearts-and the universal benefits of being physically active: Physical activity and health. *Annals of Epidemiology, 19,* 253–256.

Bloom, B S. Taxonomy of Educational Objectives (1956). Published by Allyn and Bacon, Boston, MA.

Campbell, M., Carbone, E., & Farrell, D. (2004). A tailored nutrition education CD-ROM program for women receiving food assistance. *Journal of Nutrition Education and Behavior, 36,* 58–66.

Carol et al. 1965= Carroll JB, Burke ML. Parameters of paired-associate verbal learning: length of list, meaningfulness, rate of presentation and ability. Journal of Experimental Psychology, 1965, 69(6): 543–553.

Davies, C., Vandelanotte, C., & Mummery, K. (2012). Associations between intervention components and engagement in a physical activity website. *Journal of Medical Internet Research, 14*(1), e4.

de Vries, H., & Brug, J. (1999). Computer-tailored interventions motivating people to adopt health promoting behaviours. *Patient Education and Counseling, 36,* 99–105.

Department of Health and Aged Care (DHAC). (1999). National physical activity guidelines for Australians. *Commonwealth department of health and ageing*. Canberra: Department of Health and Ageing.

Eysenbach, G. (2005). The law of attrition. *Journal of Medical Internet Research, 7,* e11.

Frenn, M., Malin, S., Brown, R. L., et al. (2005). Changing the tide: An Internet/video exercise and low-fat diet intervention with middle-school students. *Applied Nursing Research, 18,* 13–21.

Kahneman D. Thinking, Fast and Slow (2011). New York: Farrar, Straus, and Giroux.

Klopfer, L. E. (1976). A structure for the affective domain in relation to science education. *Science Education, 60,* 299–312.

Krathwohl, D. R., Bloom, B. S., & Masia, B. S. (1964). *Taxonomy of Educational Objectives, Handbook ZZ: Affective Domain.* New York: David McKay Co.

Kroeze, W., Werkman, A., & Brug, J. (2006). A systematic review of randomized trials of computer-tailored physical activity and dietary interventions. *Annals of Behavioral Medicine, 31,* 205–223.

Leslie, E., et al. (2005). Engagement and retention in a PA website. *Preventive Medicine, 40*(1), 54–59.

Liu, Z. (2005). Reading behavior in the digital environment. *Journal of Documentation, 61,* 700–712.

Marcus, B., Ciccolo, J., & Sciamanna, C. (2009). Using electronic/computer interventions to promote physical activity. *British Journal of Sports Medicine, 43*(2), 102–105.

Marshall, A., et al. (2003). Print vs website PA programs. *American Journal of Preventive Medicine, 25*(2), 88–94.

Medibank, Private. (2008). *The cost of physical inactivity*. Medibank Health Foundation, Australia.

Moore, K. (2011). *The use of video-sharing sites*. Washington: PEW Research Center.

Nielsen, J., & Pernice, K. (2010). Eyetracking web usability. New Riders Press, ISBN 0-321-49836-4. http://www.useit.com/alertbox/reading_pattern.html.

Noar, S., Benac, C., & Harris, M. (2007). Does tailoring matter? Meta-analytic review of tailored print health behaviour change interventions. *Psychological Bulletin, 133,* 673–693.
Posner, M. I., & Boies, S. J. (1971). Components of attention. *Psychological review, 78*(5), 391.
Prochaska, J. O., & Velicer, W. F. (1997). The transtheoretical model of health behavior change. *American Journal of Health Promotion, 12*, 38–48. [PMID: 10170434].
Skinner, C., et al. (1999). How effective is tailored communication? *Annals of Behavioral Medicine, 21,* 290–298.
Spittaels, H., De Bourdeaudhuij, I., & Vandelanotte, C. (2007a). Evaluation of a website-delivered computer-tailored physical activity intervention. *Preventive Medicine, 44,* 209–217.
Spittaels, H., De Bourdeaudhuij, I., Brug, J., & Vandelanotte, C. (2007b). Effectiveness of an online computer-tailored physical activity intervention. *Health Education Research, 22,* 385–396.
Steele, R., Mummery, K., & Dwyer, T. (2007). Using the Internet to promote physical activity: A randomized trial of intervention delivery modes. *Journal of Physical Activity and Health, 4,* 245–260.
Sutherland-S, W. (2002). Changes in reading from page to screen. *The Reading Teacher, 55,* 662–669.
Swartz, L., Noell, J., Schroeder, S., & Ary, D. (2006). A randomised control study of a fully automated internet based smoking cessation programme. *Tobacco Control, 33,* 7–12.
Van Den Berg, M., Schoones, J., & Vlieland, T. (2007). Internet-based physical activity interventions: A systematic review of the literature. *Journal of Medical Internet Research, 9*(3), e26.
Vandelanotte, C., & De Bourdeaudhuij, I. (2003). Acceptability and feasibility of a computer-tailored physical activity intervention. *Health Education Research, 18,* 304–317.
Vandelanotte, C., & Mummery, W. K. (2011). Qualitative and quantitative research into the development and feasibility of a video-tailored physical activity intervention. *International Journal of Behavioral Nutrition and Physical Activity, 8*, 70. [PMID: 21718546].
Vandelanotte, C., et al. (2005). Efficacy of sequential or simultaneous interactive computer-tailored interventions for increasing physical activity. *Annals of Behavioral Medicine, 29,* 138–146.
Vandelanotte, C., De Bourdeaudhuij, I., & Brug, J. (2007). Two-year follow-up of sequential and simultaneous interactive computer-tailored interventions for increasing physical activity and decreasing fat intake. *Annals of Behavioral Medicine, 33*, 213–219. [PMID: 17447874].
Vandelanotte, C., Spathonis, K., Eakin, E., & Owen, N. (2007). Website-delivered physical activity interventions. A review of the literature. *American Journal of Preventive Medicine, 33*(1), 54–64.
Vandelanotte, C., Reeves, M. M., Brug, J., & De Bourdeaudhuij, I. (2008). A randomized trial of sequential and simultaneous multiple behavior change interventions for physical activity and fat intake. *Preventive Medicine, 46*, 232–237. [PMID: 17707079].
Vandelanotte, C., Duncan, M., & Mummery, K. (2010). Physical activity trends in Queensland (2002–2008). *Australian and New Zealand Journal of Public Health, 34,* 248–254.
Vandelanotte, C., Van Itallie, A., & Mummery, K. (2010). The development of a web-based cardiac rehabilitation intervention: A Delphi study. *BMC Cardiovas Disord, 10,* 27.
Vandelanotte, C., Caperchione, C., Ellison, M., George, E., Maeder, A., Kolt, G., Duncan, M., Karunanithi, M., Noakes, M., Hooker, C., Viljoen, P., & Mummery, K. (2012). (accepted 17-01-2012). What kind of website and mobile phone delivered physical activity and nutrition interventions do middle-aged men want? *Journal of Health Communication.*

Benefits of Complementing Eye-Tracking Analysis with Think-Aloud Protocol in a Multilingual Country with High Power Distance

Ashok Sivaji and Wan Fatimah Wan Ahmad

1 Introduction

Sherman (2009) reported the various user experience (UX) techniques practised across 34 countries. The 1,786 respondents comprise of UX practitioners, usability professionals, user researchers, user-centred design (UCD) practitioners, interface designers and others. Among the techniques reported in Fig. 1, this study includes eye tracking and lab-based usability testing (LBUT), as was reported by Sivaji et al. (2013).

2 Problem Statement

During the LBUT performed by Sivaji et al. (2011, 2013), Goh et al. (2013) and Abdollah et al. (2013), an eye tracker was used to capture the user's feedback such as audio, video and eye gaze data (fixation and saccades). The eye tracker keeps tracks of user's eye movement while they are performing the task. The LBUT also involves encouraging users to think aloud (TA) while they perform some task. However, from this, it has been found that the users have difficulty with the TA method. There are few reasons for this.

2.1 Hofstede's Power Distance

Hofstede (2005) and Yammiyavar et al. (2008) have shown that in countries like Malaysia and India, there exists high power distance as per the Hofstede's model. The power index score for Malaysia was recorded as 104 by Hofstede, being the highest

A. Sivaji (✉)
MIMOS Berhad, Kuala Lumpur, Malaysia
e-mail: ashok.sivaji@mimos.my

W. F. Wan Ahmad
Universiti Teknologi PETRONAS, Perak, Malaysia

M. Horsley et al. (eds.), *Current Trends in Eye Tracking Research,*
DOI 10.1007/978-3-319-02868-2_21, © Springer International Publishing Switzerland 2014

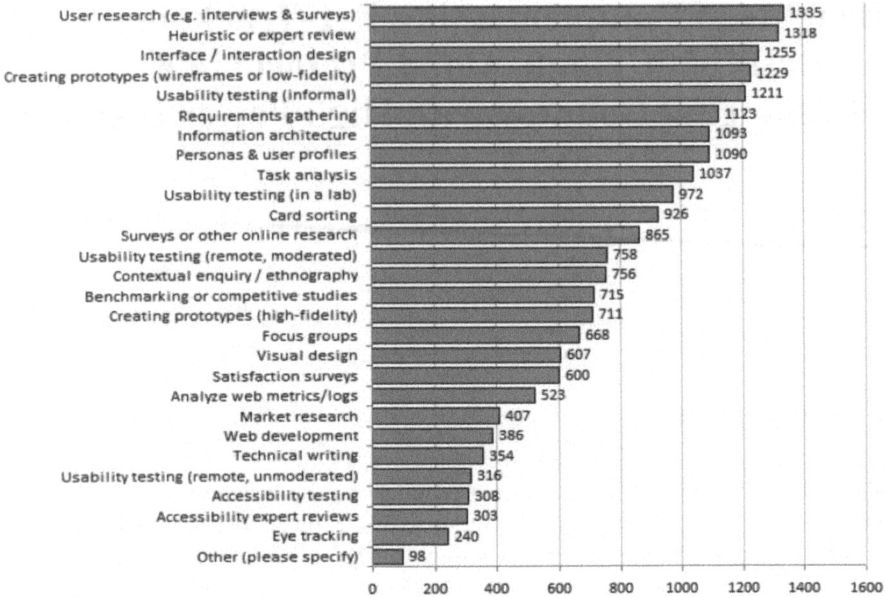

Fig. 1 User experience (UX) techniques employed by usability practitioners worldwide

around the world. In the context of usability testing, power distance could be defined as the relationship between the user who is being tested and the moderator who is facilitating the testing. In most cases, the moderator will recruit the users, after assessing his/her suitability to perform the testing based on a user screening questionnaire. During the testing, the user is expected to provide feedback based on the usability test session conducted by the moderator. Due to the power distance that is already present in the Malaysian culture, the user, during the TA process, sees the moderator as a supervisor and hence has a tendency to be afraid in disagreeing in the effectiveness, efficiency and satisfaction of degree of usability of a website under test. This is one reason why TA technique alone may not be suitable and reliable in usability studies in Malaysia. Yammiyavar et al. (2008) and Goh et al. (2013) also found that there is a rich amount of non-verbal behavioural data such as eye, hand and head gestures that are collected during the LBUT. It is almost impossible for the moderator to capture all these information during the LBUT. Sivaji et al. (2011) found that although the subjects have been encouraged to TA, some subjects are reluctant to do so, as they are afraid that a failure of completion of a given task would reflect poorly on their performance. This is despite the moderator briefing the user at the beginning of the task that the purpose of the LBUT is to assess the web interface and not the users themselves.

2.2 *Persuasive Power*

During the LBUT, when the moderator transcribes the TA feedback and raises the problem faced by the users as a defect, it is common for the design and development team to be sceptical and defensive. They even go to the extent of requesting for further evidence. This shows that LBUT and TA are not sufficient to convince the design and development team of the validity of the defects. In this case, it is important to complement the findings from LBUT and TA with a more visual method. Eye-tracking analysis has high persuasive power. Blandford et al. (2008) has identified persuasive power as one of the important criteria of assessment from multiple UX methods. This study aims to show how eye tracking can increase persuasive power of highlighting usability defects to the stakeholders.

2.3 Multilingual Society

Malaysia is a multicultural, multiracial and multilingual country. According to the 2010 Population and Housing Census of Malaysia by the Department of Statistics Malaysia, the citizens comprise of 67.4 % Bumiputera, 24.6 %Chinese, 7.3 % Indians and 0.7 % others. The official language of Malaysia is Malay or also known as Bahasa Malaysia. English remains as a second language and is taught in school. Malaysian English sees wide use in business, along with Manglish, which is a colloquial form of English with heavy Malay, Chinese and Tamil influences. It is common that during the usability moderation, some participants may tend to highlight some words in Malay, Chinese or Tamil. The moderator should translate this into English for the benefit of the international audience. However, there are chances for misinterpretation or miscoding of the TA feedback between the moderator and the users.

3 Objectives

The objectives of this study are as follows:

- Assess the level of the power distance index (PDI) in Malaysia
- Propose a new LBUT methodology that will incorporate both TA and eye tracking
- Perform an LBUT case study on a website based on the Malaysian demography
- Analyse and asses impact of the newly derived LBUT methodology

4 Current Study

4.1 Power Distance Assessment

Previous studies (Hofstede 2005 and Oshlyansky 2007) have performed surveys among Malaysians to gauge PDI. In order to validate these findings in the current context, Values Survey Module (VSM) 1994 questionnaire was distributed among Malaysians. The study revealed a PDI score of 36.8 from the 44 respondents. This shows that our findings (PDI of 36.8) is significantly lower than Hofstede (2005) having a PDI of 104. Our result, however, is closer to that reported by Oshlyansky (2007) being 23.47. One significant difference between the subjects for this study is that the sample was more targeted to working adults (working experience ranging from 1 to 35 years) as compared to postgraduates and undergraduates Oshlyansky (2007).

Although we found a lower PDI score of 36.8, we still believe there existed a significant power distance, as a closer look at Question 14 of the VSM revealed that significant power distance still existed as the average rating is "Sometimes"; for the response to "How frequently, in your experience, are subordinates afraid to express disagreement with their superiors?" This has a significant impact when users perform usability testing as there could be some reluctance in providing an honest opinion or feedback, especially negative feedback. Additionally, the moderators from MIMOS UX Lab who were involved in previous UX studies, such as Sivaji et al. (2011, 2013), Goh et al. (2013) and Abdollah et al. (2013), have also observed this high power distance response whereby subjects feel that their responses are being recorded and could be used against them when an issue is reported although anonymity in recruitment and reporting is practised.

4.2 LBUT with Eye Tracker

The process flow for the LBUT methodology employed in this study is shown in Fig. 2.The users are recruited as per the Malaysian demography who have participated in the VSM survey. Out of the six users, four were Malay, one was Chinese and another Indian. The seventh user was a native English speaker who was chosen as a control sample for this study. The websites chosen for this study was the Nielsen Norman Group (http://www.nngroup.com/) whereby the users were tasked to determine the registration details for a particular event.

The steps that are shaded in grey (Fig. 2) involve using an eye tracker. The eye tracker that was used in this study is the Tobii T60 with Tobii Studio 2. This version enables retrospective think aloud (RTA) where the moderator and users can play back the video to view the session. The test environment setup involves arranging a logical sequence of task, selecting a website URL to be tested, designing of the subjective ratings questionnaire. These information are then configured into URANUS, that would automatically link to the Tobii Studio. URANUS is an open source software that is developed by Sivaji et al. (2012) to facilitate usability testing of websites and any user interface. It is developed in such a way that practitioners are able to integrate with Tobii Studio. After the briefing session, subject will start to perform the first task. This will involve the subjects concurrently thinking aloud (CTA) as they perform the task. These will be recorded by the eye tracker. The subjects gaze patterns and mouse clicks are recorded for further analysis. The moderators would rate the effectiveness and efficiency of the task performed by the subjects. The subjects would also be able to provide some feedback based on a questionnaire. After all tasks are completed, the moderator will play back the recordings to confirm with the users on why they have reacted in a particular manner. For instance, if they really liked an interface, more details could be asked on what elements of the interface that attracted them. Conversely, if they had problems

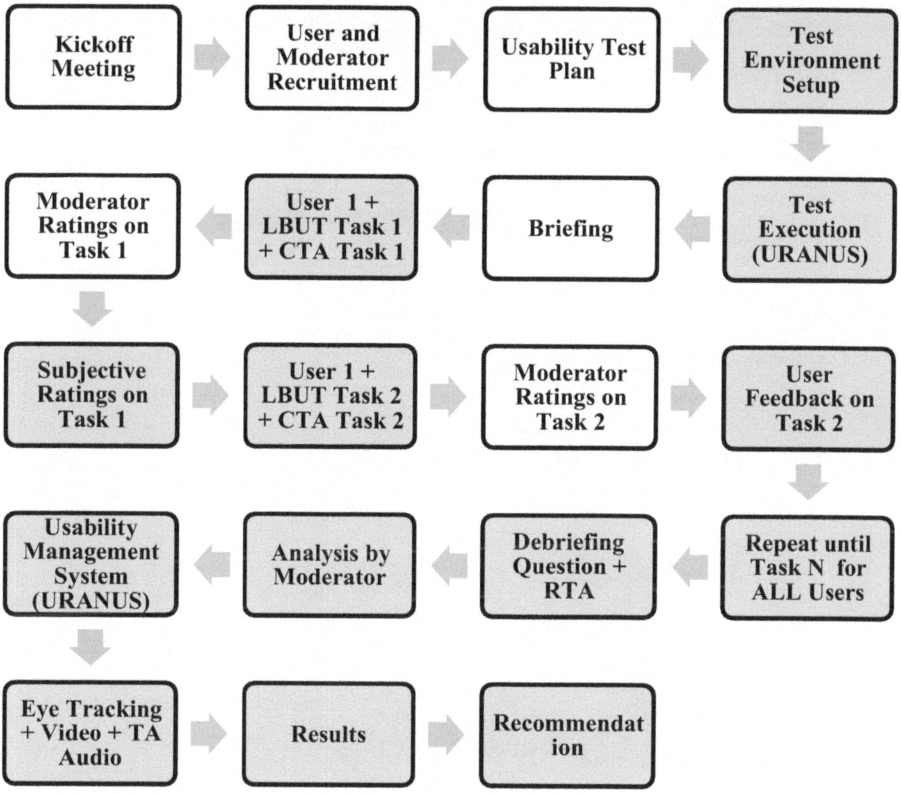

Fig. 2 Process flow for lab-based usability testing (LBUT) with eye tracker

with the interface, it will be interesting to determine the reason for that. In addition to that, the moderators would also gather the feedback from the observers to validate certain segments of the task with the users to gain a thorough understanding of the subject reaction towards the interface. This is called RTA. The moderators would also use the eye tracker to perform analysis, and obtain recommendation for fixes.

5 Results and Discussion

During the LBUT, an eye tracker was used to capture the user's feedback such as audio, video and eye gaze data (fixation and saccades). The eye tracker keeps track of the user's eye movement while they are performing the task. Although it is not mandatory to use an eye tracker in an LBUT study, there are some benefits to using one. If a study is carried out without an eye tracker, the moderator only has to rely on the feedback obtained from TA. With the eye tracker, the moderator can now support the TA feedback with one of the human biometric feature, which in this case is the user's eye. This increases the data integrity obtained from all users.

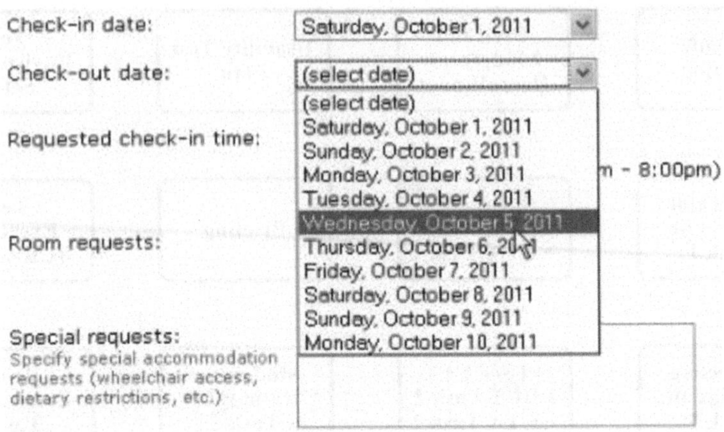

Fig. 3 Original image to be analysed

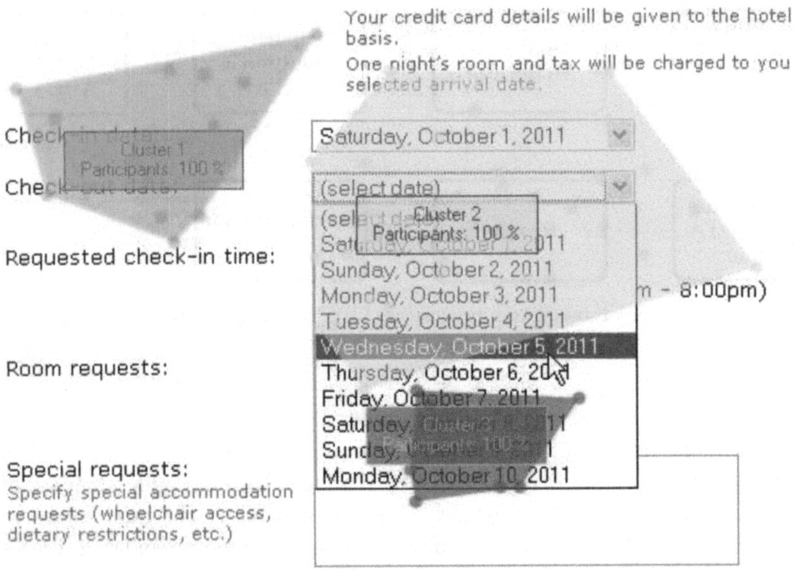

Fig. 4 Cluster plot

Figures 3–7 illustrate the various eye-tracking analysis features that are used in this study. Figure 3 has been chosen to be analysed, as it has some insightful TA feedback. Using the Tobii Studio's visualisation tools, the following observations could be made:

- The cluster plot (Fig. 4) shows the areas of high concentration of gaze data points when this task was performed. Based on this, the moderator could mark certain areas of interest (AOI) for further analysis.

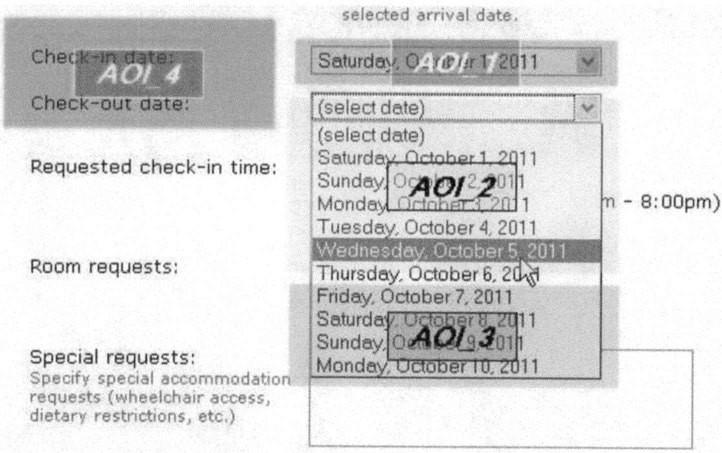

Fig. 5 Areas of interest

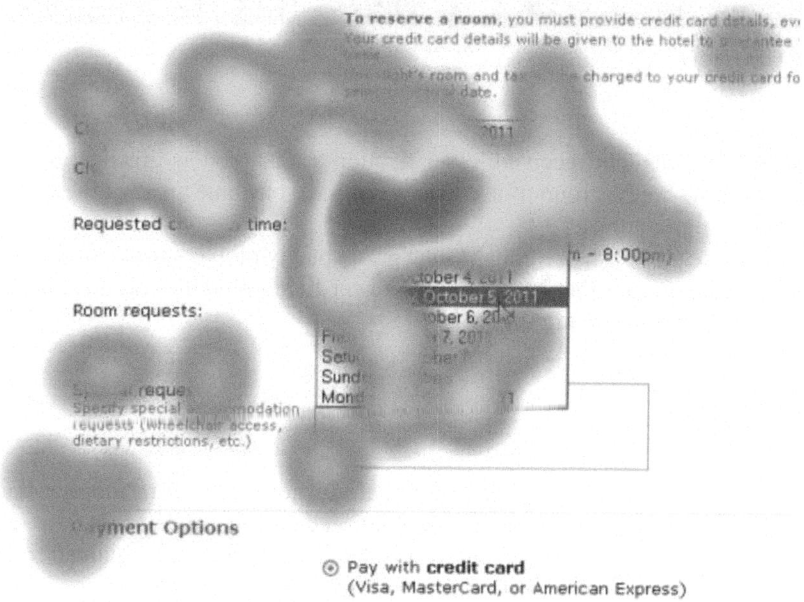

Fig. 6 Heat map

- Figure 5 shows the four AOIs that have been marked with AOI 1–4. AOI 4 corresponds to cluster 1, while AOI 3 corresponds to cluster 3. Since cluster 2 spans a larger space, it is divided down to 2 AOIs, namely AOI 1 and AOI 2.
- The heat map as shown in Fig. 6 highlights areas based on the fixation duration and fixation count. Areas that receive more fixation concentration is shown as

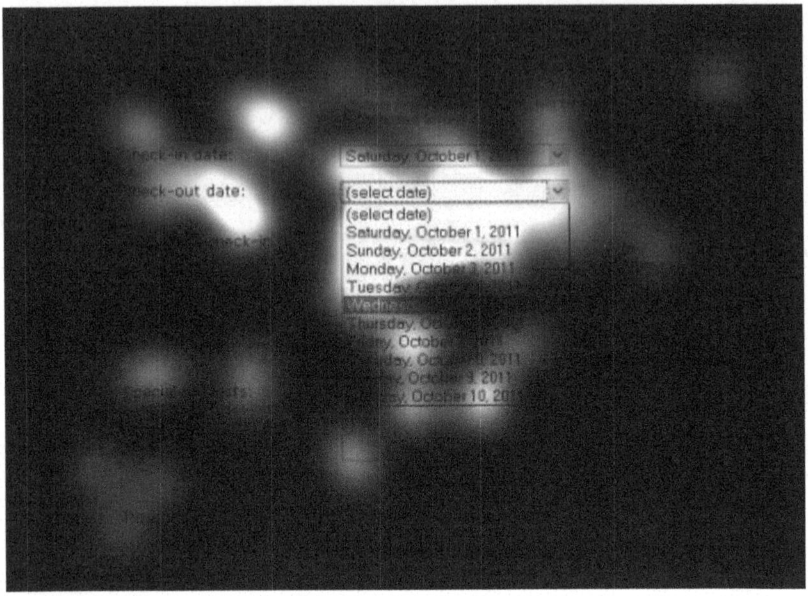

Fig. 7 Gaze opacity

red, while areas that received lesser fixation are marked as green. It is a quick way to show how a group of users have focussed on some areas of the web page that is of high interest.

- The inverse plot of the heat map is the gaze opacity plot as shown in Fig. 7. This plot hides out areas that have received fewer fixations. This way the areas with the most fixations are highlighted.

6 Analysis

Some of the other descriptive statistics that are generated from the eye tracker, based on these AOIs, are fixation count, fixation duration, time to first fixation and mouse click-related data. It is useful to correlate the TA feedback obtained from the users with the descriptive statistics obtained from the eye tracker.

6.1 Correlation of TA with Eye Tracker

When subjects (user 1–7) were performing the registration task on one of the eight websites, the TA feedback as shown in Table 1 was recorded from the task being carried out.

Table 1 Concurrent think aloud (CTA) feedback for the interface used to display the date to register for the conference

User	Concurrent think aloud
1	"Wow, I like this feature, it is different from other sites, others sites would have used a calendar, and expected me to scroll up and down to the landing page to look for the conference dates to register, …I could have even selected the wrong conference date"
2	"Date is Sunday October 2011"
3	"Found the date"
4	"Sunday October 2011"
5	""
6	"tak dapat jumpa tarikh"
7	""

6.2 *TA and Eye Tracker's Visualisation*

Figure 7 clearly shows that users were attracted to the drop-down menu, which actually displayed the relevant content that is the "Check Out Date" data of the conference. Even the heat map (Fig. 6) highlighted the drop-down menu with red, indicating areas of high fixation. This shows that the AOI 2 was the areas of high interest.

6.3 *TA and Eye Tracker's Descriptive Statistics*

The eye tracker can also support descriptive statistics as shown in the following tables. From Table 2, it could be seen that AOI 2 had 25 fixation counts. As pointed out by Ehmke and Wilson (2007), longer fixation indicates that the elements are more noticeable and/or more important. It also had the longest fixation duration of 9.22 s as compared to the remaining AOIs. Table 2 also shows that the visit duration for AOI 2 is 10.45 s, which is much higher than the remaining AOIs. This is consistent with the TA feedback from the users " Wow, I like this feature…".

The correlation and consistency between the TA and eye tracking enable the moderator to conclude that this site has a better navigation strategy than other sites. Just like the above segment of TA, many other parts of the user's recording could be further analysed with an eye tracker to gain insightful data of the web page design. From the results shown in Table 2, the moderator can highlight this feature as desirable and present it to the project team with visual evidence. Since the project team has the visual evidence to support their design, they could reuse this interface and incorporate it as a design best practice for their organization. Eventually, other designers will be able to adopt this practice, since it has been validated. Without the eye tracker, the moderator will not be able to recall visually what was happening, just based on the moderator notes which were gathered from the audio recordings from the TA. Now they have video and audio evidence with user's eye as a biometric to proof a point. This enables the usability analyst (moderator) to present the findings with visual, audio and statistical evidence,

Table 2 Descriptive statistics from eye tracker for user 1

User	Fixation and visit duration											
	AOI_1			AOI_2			AOI_3			AOI_4		
	N (C)	Mean (s)	Sum (s)	N (C)	Mean (s)	Sum (s)	N (C)	Mean (s)	Sum (s)	N (C)	Mean (s)	Sum (s)
Fixation	4	0.16	0.65	25	0.37	9.22	8	0.31	2.46	6	0.22	1.33
Visit	4	0.16	0.65	10	1.05	10.45	6	0.50	2.98	5	0.30	1.52

eventually increasing the persuasive power towards the design and development team to implement changes.

6.4 *Hofstede's Power Distance and RTA*

Among the seven users, it could be seen that user 1 has low power distance while the remaining users have high power distance. User 1 was from a native English-speaking country, while the remaining users were from Malaysia. Among the Malaysian users, two out of the six were quiet throughout the entire moderation. Although they were successful at completing the task (finding the date), they were not comfortable to TA because they had problems verbalising in English. This would make it difficult for the moderators to provide ratings on the interface without justification if there were only depending on RTA. But with the eye tracker, the moderators are able to provide the ratings with evidence as shown in the gaze opacity plot (Fig. 7) and with the time to first fixation metric.

Although users 2–4 managed to provide feedback, it was very succinct. This would not help the designers and developers to understand the degree of affective elements used in the design and how a user may react to it. The author finds this a common problem in usability studies conducted in Malaysia, whereby users are so focussed on completing the task and moving on to the next task. When the users are probed by the moderator on their feedback on the interface, they would provide a short or succinct answer, instead of providing a general comment on their UX and satisfaction level. There is still a strong belief that it is their fault when they are not able to complete a task, instead of criticising or praising a design of the interface.

User 6 managed to find the date. However, the CTA provided was in Malay language. This will require the moderator to translate to English. The eye tracker also enables the user and the moderator to perform an RTA. With the RTA, two interesting observations were made:

1. Even though user 6 mentioned during the RTA that the date was not found ("tak dapat jumpa tarikh"), during the playback, the moderator and user realized that the user was indeed staring at the date entry, but has mistakenly mentioned that the date was not found. The RTA feedback was corrected to "date was found" instead. In other words, eye tracking was able to reduce miscoding of information.

2. Unlike other users, user 6 was staring at the date and also staring away from the date. The user admitted that he was not in the right frame of mind to perform the test because he was rushing for another appointment after that. But the fact remained that the date entry was clearly visible by the user as shown by the fixations.

Despite advising the users that their feedback will remain anonymous, some users (user 5) were more comfortable on reporting some problems of the interface after the LBUT session. In these situations, the moderator was able to correlate the feedback provided with the eye-tracking visual cues to persuade the design and development team to improve the user interface.

7 Conclusion

From the PDI assessment performed, it was found that the existence of power distance in Malaysian is significant and this would make traditional usability testing problematic. Hence, the proposed LBUT is derived as shown in Fig. 2. To overcome the barrier, eye tracking is used in addition to TA during the LBUT. And the results from the case study do support the argument that despite the power distance and language barrier, the eye tracker is able to reveal key biometrics information. This is so because the LBUT method proposed with TA and eye tracking (Fig. 1) is able to reflect cognitive behaviour supported with visual cues to increase the persuasive power of the findings from the usability testing. In the future, it is recommended to use LBUT to enrich the UX of web and standalone interfaces, especially when power distance and language barrier is a constraint.

Acknowledgements The author would like to thank Mohamed Redzuan Abdullah from MIMOS Berhad who sponsored the MIMOS Usability/UX Lab that is equipped with the eye tracker. Many thanks go to Shi Tzuaan, Fahmi Mazlan and Zulkifle for facilitating the LBUT and Sujitra Balakrishnan for reviewing this chapter.

References

Abdollah, N., Soo, S.T., Sivaji, A., (2013). Eyes on OSH – Usability Testing with Eye-Tracking and User Preference Rating. Springer International Publishing. Book Section Advances in Visual Informatics. Lecture Notes in Computer Science Volume 8237, 2013, (pp 323–334).

Blandford, A., E, A., Hyde, J. K., Green, T. R., & Connell, I. (2008). Scoping Analytical Usability Evaluation Methods: A Case Study. *Human-Computer Interaction* (pp. 278–327).

Claudia, E., & Wilson, S. (2007). Identifying web usability problems from eye-tracking data. *British Computer Society* (pp. 119–128.)

Goh, K.N., Chen, Y.Y., Lai, F.W., Daud, S.C., Sivaji, A., Soo, S.T. (2013). A Comparison of Usability Testing Methods for an E-Commerce Website: A Case Study on a Malaysia Online Gift Shop. Information Technology: New Generations (ITNG), 2013 Tenth International Conference on , vol., no., pp.143,150, 15–17 April 2013

Hofstede, G., & Hofstede, G. J. (2005). Cultures and organizations: software of the mind (Revised and expanded 2nd ed.). New York: McGraw-Hill. ISBN 978-0-07-143959-6. OCLC 57069196.

Oshlyansky, L. (2007). Cultural Models in HCI: Hofstede, affordance and technology acceptance, Swansea University.

Sherman, P (2009), *UPA 2009 Salary Survey*. Usability Professional's Association. *http://usabilityprofessionals.org/usability_resources/surveys/2009salarysurvey_PUBLIC.pdf*. Accessed 15 Feb 2012.s

Sivaji, A., Mazlan, M. F., Shi–Tzuaan, S., Abdullah, A., & Downe, A. G. (2011). Importance of incorporating fundamental usability with social, trust & persuasive elements for E-Commerce website. *IEEEXplore Business, Engineering and Industrial Applications (ICBEIA)* (pp. 221–226).

Sivaji, A., Shi–Tzuaan, S. (2012). Website user experience (UX) testing tool development using open source software (OSS). Network of Ergonomics Societies Conference (SEANES), 2012 Southeast Asian, vol., no., pp. 1, 6, 9–12 July 2012.

Sivaji, A., Abdullah, M.R., Downe, A.G., Ahmad, W.F.W. (2013) Hybrid Usability Methodology: Integrating Heuristic Evaluation with Laboratory Testing across the Software Development Lifecycle. Information Technology: New Generations (ITNG), 2013 Tenth International Conference on , vol., no., pp.375,383, 15–17 April 2013.

Yammiyavar, P., Clemmensen, T., & Kumar, J. (2008). Influence of cultural background on nonverbal communication in a usability testing situation. *International Journal of Design*, 31–40.

Part IV
Eye Tracking and Education

Eye Tracking and the Learning System: An Overview

Bruce Allen Knight, Mike Horsley and Matt Eliot

1 Introduction

In everyday and learning tasks, the eyes have, firstly, the roles of locating and recognizing objects and then, secondly, directing the actions to make use of them (Land & Tatler 2009). The use of eye tracking can reveal important aspects about students' learning processes. Because eye tracking provides insights into the allocation of visual attention, it is very suited to study differences in learners' attentional processes. In this section of the book, the contributions focus on the visual processes that occur when participants are performing a task.

Eye-tracking research in the area of learning has been dominated by the use of eye tracking in investigating reading. During learning tasks, students use their eye movements to accommodate incoming stimuli, and if the information is in printed form, linguistic factors such as the frequency of words, the difficulty of the text and the background knowledge of the learner influence eye-movement patterns (Rayner 2009). Much research has been analytical, focussing on variables and components such as word frequency (White 2008), perceptual scans in reading, cognitive processes in reading (Rayner 2009), word predictability (Drieghe et al. 2005), number of meanings (Folk and Morris 2003), phonological properties of words (Ashby 2006) and semantic relations and word familiarity (Williams and Morris 2004). These valuable data have provided much useful information on how readers process text. The chapters in this part focus on the interaction between the learner and the stimulus in the context of task-oriented activities using eye-tracking methodology.

The use of eye-tracking methodologies is of increasing importance in many new technology-based research approaches in learning and education. The area of *Teaching Analytics*, for example, investigates the sort of data and information that

B. A. Knight (✉)
School of Education and the Arts, Central Queensland University, Mackay, Australia
e-mail: b.knight@cqu.edu.au

M. Horsley · M. Eliot
Learning & Teaching Education Research Centre, Central Queensland University, Noosaville, Australia

M. Horsley et al. (eds.), *Current Trends in Eye Tracking Research*,
DOI 10.1007/978-3-319-02868-2_22, © Springer International Publishing Switzerland 2014

teachers receive in dynamic learning environments (Reimann et al. 2013) and the Conferences of the Computer Supported Collaborative Environment held under the auspices of the International Society of the Learning Sciences (ISLS). Significant data are sourced from eye-tracking data analysis in collaboration with repertory grid analysis (Reimann et al. 2012).

Teaching and learning analytics generally offers new tools and resources, partially based on eye tracking, to allow teachers deeper insights into the learning process and to access new knowledge of how to provide feedback and guidance to learners. There are a number of learning analytics research groupings such as the Multimodal Learning Analytics Group at Stanford University (Worsley and Blikstein 2013; Blikstein 2013). Learning analytics researchers utilise eye-tracking methods to comprehensively and systematically describe, investigate and characterise the student learning process.

Studies of learners' cognitive and perceptual processes are taking research into a new phase of investigation, due mainly to the capability of eye-tracking software to produce data. These data can illustrate what learners are visually attending to (e.g. the Broadbent chapter which reports on what nursing students focus on when analysing an electrocardiogram (ECG)), the order in which they do it (e.g. school students in the Knight an Horsley chapter reading a passage and answering comprehension questions) and the length of time learners devote to a task (e.g. higher education students composing a response online as reported in the Persaud and Eliot chapter).

2 Research on the Mechanics of Reading Comprehension

Eye-tracking methodologies enable researchers to track a reader's eye movements, as well as their fixations (Rayner 1998) with saccadic movements between sentences and fixation times providing insights into comprehension activity. When completing a comprehension task, it is not easy for learners with learning and reading difficulties to determine which parts of the text are most relevant (and hence need to be attended to). By examining the eye movements of skilled and weak readers during a comprehension task, it is possible to compare the different strategies used by the different readers. These data can provide important insights into the attentional processes used by effective readers and then be applied to generate strategies to improve weak students' comprehension monitoring skills. For example, Van Gog et al. (2009) have indicated that a useful strategy to enhance novice (our equivalent of weak readers) attentional skills so that they perform more expertly (our equivalent of skilled readers) with better learning outcomes may be not only to model skilled readers' strategies and actions but also to display and discuss their eye movements.

Knight and Horsley analyse the mechanics of reading comprehension used by 15 skilled and weak readers completing a standardised comprehension test (National Assessment Program—Literacy and Numeracy (NAPLAN)) that is administered by all schools to determine students' reading and comprehension skills. *The* chapter outlines the use of eye tracking and electroencephalogram (EEG) methodologies that have been used to develop a model of the mechanics of reading. The model has added a set of macro-reading behaviours revealed during the development of the model to the micro-reading behaviours of reading individual words and phrases established by previous eye-tracking research into reading behaviour (e.g. Rayner 1998, 2009; Liversedge et al. 2011).

Early data report that skilled readers scanned quickly, continuously and consistently from the comprehension questions to the text, suggesting that what they believe to be a critical part of a text is aligned to comprehension questions. On the other hand, weak readers were reading linearly, displaying frustration and "giving up" on the task.

3 Eye-Tracking Research in Investigating Learning from Complex Visual Displays

Eye-tracking research has shown that the distribution of attention to a task is influenced by one's expertise (Charness et al. 2001). Van Gog, Paas and Van Merrienboer (2005) report that individual variation also occurs between individuals with smaller differences in expertise. Eye-tracking technology is used in the study reported here to investigate the cognitive processes used by nurses as they analyse the information provided in ECG displays of physiological data. Attention to the visual display is critical for patient care, and the research presented in this chapter by Broadbent explores the visual behaviour of novice (22 first-year students who often experience high cognitive load during new tasks), student nurses (18 final-year students) and 9 experienced nurses as they read ECGs.

Using a gaze model (Wolfe 1994) of novice and experts and applying it to reading visual displays, the author analyses the nurses' eye movements as they view four different ECG reports and answer multiple-choice questions on the medical condition that each display represents. The results report clear distinctions between the way in which the three groups approach the identification and assessment of areas of interest (AOIs) on ECGs. This research has broad application in nursing education and, in the current context, provides much useful data about what nursing students have learned in a real-life, non-threatening application of their skills.

4 Eye Tracking and the Development of Self Regulation

Eye-tracking research methodologies offer new approaches to investigating students' self-regulated learning behaviour. The need to examine the processes in the development of self-regulation to manage their own learning in a higher education online context is the focus of this chapter by Persaud and Eliot.

Hyona (2010, p. 173) suggests that "the eyes are guided both endogenously (i.e. to meet the learner's task-relevant goals) and exogenously (i.e. by perceptually salient stimulus characteristics)”. Bearing this in mind, Persaud and Eliot firstly report eye-movement data in the e-learning environment to identify what 11 first-year higher education students attend to and what they ignore whilst navigating through online learning tasks. Then to clarify participants’ visual behaviour during the activity, and to better understand students’ fundamental cognitive processes involved in composing a text, the researchers also used stimulated recall interview data to supplement and produce a more detailed description of the task behaviours.

Preliminary results report that working collaboratively with other students is influenced by how their work will be received by others and how students wanted to be perceived by others. Interestingly, the authors report that students’ eye movements during text composition (where they either looked at the monitor or looked at the keyboard) affected their cognitive and metacognitive processes as they composed text.

5 Eye Tracking on Websites

This chapter reports on usability testing by multilingual (English, Malay, Mandarin and Tamil) people as they navigate websites. Using eye-tracker data complemented with Think Aloud feedback, the authors report data on participants’ fixation count, fixation duration, time to first fixation and mouse click-related data as they interact with a web interface. The consistent results from the Think Aloud and eye-tracking data enabled the moderator to conclude that one website had a more efficient navigation strategy than other sites. This research has implications for how to enhance the design and navigation of websites as learners/consumers visualise and process displays.

6 Conclusion

Mayer (2010, p. 170) has stated that “A serious challenge for eye-tracking researchers is to find the sometimes-missing link between eye-fixation measures and learning outcome (or cognitive performance) measures”. Steps to meet this challenge have begun as the eye-tracking methodology used in each of the chapters in this part of the book provide a valuable insight into the cognitive and visual processes used in learning and to the improvement of cognitive performance. The studies, although few in number, have provided an overview of the potential of eye-tracking technology to impact learning and thus teaching in naturalistic settings.

For the final word on the future of eye-tracking research and the study of learning processes, we cite the work of Van Gog and Scheiter (2010, p. 98) who assert: “With technological developments making eye tracking more accessible (i.e. not only making equipment more affordable, but also making it easier to gather and analyse data), we believe that different applications of eye tracking will continue to increase in our field to study learning processes and improve the design of instruction”.

References

Ashby, J. (2006) Prosody in skilled silent reading: Evidence from eye movements. *Journal of Research in Reading, 29*, 318–333.

Blikstein, P. (2013). Multimodal learning analytics. In Dan Suthers & Katrien Verbert (Eds.), *Proceedings of the Third International Conference on Learning Analytics and Knowledge (LAK '13)*, ACM, New York.

Charness, N., Reingold, E., Pomplun, M., & Stampe, D. (2001). The perceptual aspect of skilled performance in chess: Evidence from eye movements. *Memory and. Cognition, 29,* 1146–1152.

Drieghe, D., Brysbaert, M., & Desmet, T. (2005). Parafoveal-on-foveal effects on eye movements in reading: Does an extra space make a difference? *Vision Research, 45,* 1693–1706.

Folk, J., & Morris, R. (2003). Effects of syntactic category assignment on lexical ambiguity resolution in reading: An eye movement analysis. *Memory & Cognition, 31,* 87–99.

Hyona, J. (2010). The use of eye movements in the study of multimedia learning. *Learning and Instruction, 20*(2), 172–176.

Land, M., & Tatler, B. (2009). *Looking and acting: Vision and eye movements in natural behaviour*. Oxford: Oxford University Press.

Liversedge, S., Gilchrist, I., & Everling, S. (2011). *The Oxford Handbook of Eye Movements*. Oxford: Oxford University Press.

Mayer, R. (2010). Unique contributions of eye-tracking research to the study of learning with graphics. *Learning and Instruction, 20*(2), 167–171.

Rayner, K. (1998). Eye movements in reading and information processing: 20 years of research. *Psychological Bulletin, 124,* 372–422.

Rayner, K. (2009). Eye movements and attention in reading, scene perception and visual search. *The Quarterly Journal of Experimental Psychology, 62,* 1457–1506.

Reimann, P., Hesse, F., Cerniak, G., Luckin, R., Vatrapu, R., Bul, S., Johnson, M., Halb, W., Utz, W., & Kossowski, M. (2012). Supporting teachers in capturing and analyzing learning data in the technology-rich classroom. International Conference of the Learning Sciences. July: Sydney.

Reimann, P., Jacobson, M., & Vatrapu, R. (2013). Handbook of Computer Supported Collaborative Learning.

Van Gog, T., Jarodzka, H., Scheiter, K., Gerjets, P., & Paas, F. (2009). Attention guidance during example study via the model's eye movements. *Computers in Human Behavior, 25,* 785–791.

Van Gog, T., Paas, F., & Van Merrienboer, J. (2005). Uncovering expertise-related differences in troubleshooting performance: Combining eye movement and concurrent verbal protocol data. *Applied Cognitive Psychology, 19,* 205–221.

Van Gog, T., & Scheiter, K. (2010). Eye tracking as a tool to study and enhance multimedia learning. *Learning and Instruction, 20*(2), 95–99.

White, S. (2008). Eye movement control during reading: Effects of word frequency and orthographic familiarity. *Journal of Experimental Psychology: Human Perception and Performance, 34,* 205–223.

Williams, R., & Morris, R. (2004). Eye movements, word familiarity, and vocabulary acquisition. *European Journal of Cognitive Psychology, 16,* 312–339.

Wolfe, J. M. (1994). Guided search 2.0: A revised model of visual search. *Psychonomic Bulletin & Review, 1*(2), 202–238.

Worsley, M., & Blikstein, P. (2013). Towards the development of multimodal action based assessment. In Dan Suthers, & Katrien Verbert (Eds.), *Proceedings of the Third International Conference on Learning Analytics and Knowledge (LAK '13)* 94–101 ACM, New York.

A New Approach to Cognitive Metrics: Analysing the Visual Mechanics of Comprehension using Eye-Tracking Data in Student Completion of High-Stakes Testing Evaluation

Bruce Allen Knight and Mike Horsley

1 Context

Reading is arguably the most critical skill for life success. Australia, like other Anglophone nations, has high proportions of children and adults with serious literacy weakness. It is common for up to 30 % of students at primary and secondary school to be experiencing significant reading weakness (Knight and Galletly 2011; Masters 2009; National Assessment Program Literacy and Numeracy (NAPLAN) 2010; Program for International Student Assessment (PISA) 2007, 2009).

In addition, almost half of all Australians over the age of 15 have literacy skills below a minimum level needed to manage the literacy requirements of a modern society (Australian Bureau of Statistics 2008).

Australian teachers are increasingly being held to account for their students' literacy and numeracy progress through the NAPLAN. To improve their students' literacy performance, teachers, schools, and educational systems have devised specific literacy teaching strategies and focussed these strategies toward improving NAPLAN scores. However, current strategies have not been informed by research about the interactive visual and cognitive mechanics and the processes of comprehension that students actively use as they complete the literacy comprehension component of the NAPLAN tests. The research presented here uses a new methodology that focusses on the visual and cognitive mechanisms used by students when responding to comprehension tasks that require interpretive and critical level responses. This new research has the capacity to underpin comprehension strategies for teacher use.

"It [comprehension] remains the key longitudinal and developmental goal of all school reading instruction" (Luke et al. 2011, p. 151). The research revolves around capacity building for teachers with a very clear purpose of relevance for

B. A. Knight (✉)
School of Education and the Arts, Central Queensland University, Mackay, Australia
e-mail: b.knight@cqu.edu.au

M. Horsley
Learning and Teaching Education Research Centre, Central Queensland University, Noosaville, Australia

M. Horsley et al. (eds.), *Current Trends in Eye Tracking Research*,
DOI 10.1007/978-3-319-02868-2_23, © Springer International Publishing Switzerland 2014

students with literacy difficulties. Currently, there is insufficient knowledge of how the skills and processes that students use to comprehend language interact. To better understand the precise and deliberate role of students' eye movements and emotions interacting with cognitive and metacognitive strategies in the processes involved in reading comprehension, the research will provide timely, reliable, and valid knowledge on the interaction of students' skills and processes required for successful comprehension. The outcomes will facilitate teachers and policy makers to plan and allocate resources to effectively assist students in achieving literacy outcomes.

2 Reading Comprehension

Reading comprehension is a complex cognitive process that engages both early literacy skills (including accurate word recognition, fluent access to word meaning, recognition of syntactic cues to sentence meaning) and sophisticated literacy (automatic and self-regulating) level processing of information (Knight and Galletly 2011). Sinatra and Pintrich (2003) describe the change that occurs as the "goal-directed and conscious initiation and regulation of cognitive, metacognitive, and motivational processes to bring about a change in knowledge" (p. 6). The reader makes inferences and creates connections between the text and the real world of their situated social and cultural practice so as to construct a mental representation and generate meaning from the ideas presented in the text (Kintsch 1998; Knight and Galletly 2011; Luke et al. 2011; McNamara and Magliano 2009; Snow 2010). During such cognitively demanding tasks, other factors that need to be considered in any analysis include the role of reader emotions, the nature of the task being undertaken and their motivation for completing the task.

Reading, thus, involves deciphering, comprehending and evaluating written texts to engage with the message it communicates. Gough and Tunmer's (1986) Component Model of Reading proposes that reading comprehension starts from two essentials which follow developmental patterns of attainment: Reading Accuracy (the recognition or pronunciation of the spoken words corresponding to written words including decoding and word recognition) and Language Comprehension, which comprises thinking and reasoning skills as well as vocabulary and literal and inferential comprehension.

The Component Model of Reading supports research evidence as to the nature and operation of cognitive and linguistic processes in reading and is embedded within the sociocultural milieu that typifies readers' lives. It acknowledges that research has shown that the fundamentals of word recognition processes and of language comprehension processes are different.

Although the Component Model of Reading describes the elements of reading comprehension, it does not consider how the processes operate and interact when students formatively appraise a text during comprehension. To consider how the processes interrelate, we focus on the Construction–Integration (CI) Model (Kintsch 1998) of comprehension.

The CI Model (Kintsch 1998) argues that discourse comprehension consists of a construction phase (a propositional network is constructed) and an integration phase where the network is integrated into the reader's memory to make inferences (propositions) that promote understanding. The CI Model assumes that multiple levels of memory representations (surface code using decoding and syntax cues, proposition base, and mental models to integrate information) are generated as part of the comprehension process (Graesser 2007). As inferences can be tracked through eye-tracking data where gaze patterns illustrate the processes used by readers to comprehend text and answer questions, this research will investigate the mechanics of the processes involved in reading comprehension.

3 Eye Movements During Reading

The fovea is the only part of the retina that is useful for reading, with visual accuracy being most favourable at the centre and decreasing toward the edge (Rayner 2009). The eyes move in small steps (saccades), with the perceptual abilities of the eyes dependent on the number of letters in a word. As gaze advances by only approximately seven to nine letters (Rayner 1998), individuals consciously process only a very small subset of the visual inputs. The eyes shift around with the outcome that "reading is nothing but the word-by-word mental restitution of a text through a series of snapshots" (Dehaene 2009, p. 17). Dehaene proposes the brain does not stop at a single-letter level but rather the visual system automatically regroups letters into higher-level graphemes, which are in turn automatically regrouped into syllables and whole words, suggesting that multiple levels of analysis coexist as meaningful reading and comprehension occurs.

Eye-tracking methodologies enable researchers to track a reader's eye movements, as well as their fixations as a text is read (Rayner 1998). Saccadic movements between sentences and fixation times provide insights into comprehension activity. Where a reader directs their attention is revealed by saccadic movements, how much attention is allocated is revealed by fixations, while regressions can indicate comprehension difficulties (Mikkila-Erdmann et al. 2008).

4 Eye-Tracker Literature

Reading then is a complicated process during which readers must control their eye movements to accommodate incoming printed information. Research has established that an interaction occurs between the text and the reader (Kintish 1988) and that language factors such as word frequency, text difficulty and the reader's background knowledge influence eye-movement patterns during reading (Rayner 2009). When reading complex English orthography (the range and irregularity of the spelling patterns used in English words), most students in English-speaking countries take several years to become proficient readers, and at least 6 years to reach adult proficiency, in comparison to readers from other countries which have a transparent orthography (Aro 2004; Knight and Galletly 2011).

Decades of reading studies using eye-tracking methodologies have focussed on experimental research. The narrow dimension of this research has led to an emphasis on the micro-aspects of reading individual words and phrases. Research on eye-movement control has reported that readers of alphabetic writing systems tend to fixate longer when a word appears less frequently (White 2008); with readers' eye movements also affected by word type (Carpenter and Just 1983); and word predictability (Frisson et al. 2005). Fixation time on a word can also be influenced by variables such as phonological properties (Ashby 2006) and word familiarity (Williams and Morris 2004).

As students gain reading accuracy skills and become proficient readers, their eye-movement behaviours undergo substantial changes. McConkie et al. (1991), for example, reported that the average fixation duration decreased from first grade to fifth grade, whilst the mean saccade length increased over the same period. It has also been reported that the number of fixations per word decreases continuously as well as a reduction in regression rates and an increase of word skipping as reading skills improve (Kaakinen et al. 2003).

Most studies have focussed on individual aspects of the reading components, rather than how students coordinate the mechanics to creating meaning. Some recent research has established that how much is processed during reading varies as a function of the skills of the reader, the difficulty of the text, and the purpose of the task (Rayner 2009). For example, the amount of preview benefit readers obtain (leading to fluency and comprehension) whilst engaged in the reading process varies as a function of the difficulty of the fixated word. If the fixated word is difficult to process, readers get little or no preview benefit from the word to the right of fixation (White et al. 2005). Most competition between words is solved automatically and unconsciously by skilled readers, while in some cases only the context permits understanding and/or pronunciation.

5 Methodology

The research explored skilled and weak readers' eye-movement patterns as they completed NAPLAN reading passages and answered comprehension questions that all students in Australian schools complete a number of times during their school life.

5.1 Aims

The research was designed to map the eye-movement behaviour of students, with data on the following variables being analysed:

1. What is happening when reading the text
2. Rereading habits

3. What students read first
4. Structure of reading scan
5. Collect data on students' emotions as they complete the tasks

5.2 Sample

Fifteen students in year 7 (aged 12–13) from a regional primary school in Queensland were involved in the study. Seven students were regarded as skilled readers (top achievement band) and eight were weak readers (bottom achievement band) based on their scores in the Australian NAPLAN test completed 2 years earlier. School tests and current teacher ratings supported the grouping of students into the skilled and weak reader categories.

5.3 Instrument

The NAPLAN tests assess skills in literacy and numeracy developed over time through the school curriculum. Knowledge and interpretation of language conventions in context are drawn upon in many reading comprehension questions. NAPLAN can involve students making decisions under complex, ambiguous constraints where the text presents a high load of information which needs to be continually processed and filtered. In year 7, reading texts usually includes a wide range of genres that may use scientific vocabulary and varied sentence structures requiring readers to understand the meaning of words as well as interpret the intention of a narrator, the motivation of a character, and an author's viewpoint.

The NAPLAN comprehension questions are designed such that students are required to use inferential and critical analysis skills to successfully complete the questions. In the experimental situation, students were provided with an online magazine test format that contained a range of texts that illustrated different writing styles. Students read the texts provided and then chose an answer from comprehension questions, which contained four alternative responses.

5.4 Apparatus

Eye movements are directly observable. Previously, the structure of experimental studies, together with unsophisticated eye tracker technology, has precluded studying the interactivity between eye movement and information processing patterns.

Sophisticated eye tracking is particularly useful to this study as it can provide direct, online, sensitive, and non-invasive indexes of processing. User experience (UX) research methodologies generate ways of collecting student strategy data using Tobii eye tracker hardware and software, with the allocations of gaze during an activity monitored and quantified. Saccades and fixations during the task provide information about the interactive processes occurring during the task.

6 Results

Early results show that skilled readers had increased reading times in comparison to weak readers. Early data report that skilled readers scanned continuously and consistently from the comprehension questions to the text, suggesting that when skilled readers come to what they believe to be a critical part of a text that aligns with answering comprehension questions, they return to previous sections of the text in an attempt to resolve any ambiguity that may have arisen from the text.

Figure 1 displays heat maps produced for a skilled and a weak reader on one of the tasks. In the heat map of the skilled reader, the map shows that the student has scanned the whole text on cockroaches (with paragraph 3 being the most intense) and also shows comprehensive reading of the comprehension questions. This student's reading behaviours included comprehensive scanning of the whole text to begin with and an alignment between comprehension questions and sections of the text. The most concentrated attention displayed in the heat map is looking at the alternative responses to answer the comprehension questions.

In contrast, the heat map of the weak reader as displayed in Fig. 1 is quite different. Neither the whole passage nor all of the alternative responses to the comprehension questions have been read. Most intensity is focussed on the early paragraphs and alternative responses in the comprehension questions. The reading behaviour of the individual suggested that the reading level was too difficult and the student "gave up". Other reading behaviours of the student included only linear reading left to right and top to bottom, no scanning of the whole text, and a focus on individual words.

7 Discussion

The early results have been used to develop an appreciation of the interactive processes of how learners "prime" text elements to gain an understanding of text and answer comprehension questions. These preliminary investigations of skilled and weak readers have been used to formulate a model which attempts to understand the mechanics used by students to comprehend text.

7.1 Mechanics of Reading

From our initial early research, we have conceptualised a framework for understanding how students interrogate text by measuring the components that represent the mechanics of reading comprehension. Underpinning the model are theories of the Component Model of Reading and the construction/integration model of comprehension discussed earlier.

Fig. 1 Heat maps of a skilled and weak reader

Fig. 2 Mechanics of reading

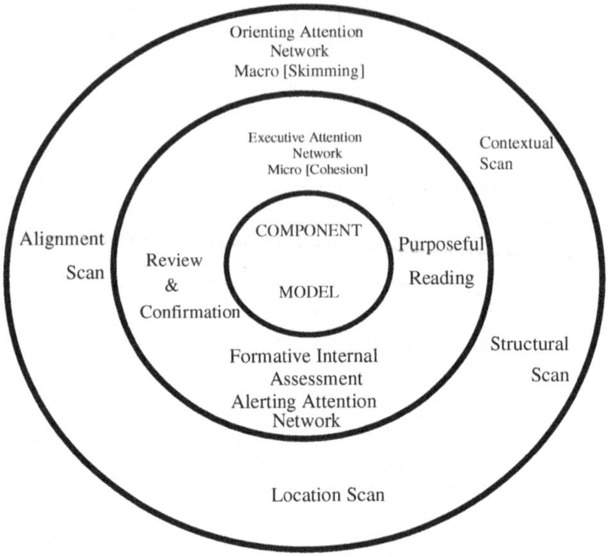

Figure 2 represents a diagrammatic representation that aims to scope reading comprehension by using a number of constructs that constitute comprehension processes needed for interpretive, critical, and creative levels of comprehension. Attention network theory (Posner and Petersen 1990) offered a theoretical framework to clarify the attentional processes that occur during a task. Posner and Petersen propose that there are three specific functionally and anatomically distinct attentional networks that learners use, and we have integrated them into the model by hypothesizing that readers' cognitive processes operate in the attention networks which act as mechanisms to comprehend text. Firstly, at the macro-level, an orienting attention network (Posner and Peterson 1990) operates when readers orient to the task by using structural scanning (headings and illustrations), location scanning (entire page and questions), and alignment scanning (interaction of questions and text that is operationalised by an individual in different texts and different formats). At the executive attention network level (Posner and Peterson 1990), micro-level mechanics are used by readers to decipher the text and guide purposeful reading. As this transpires, readers also actively coordinate comprehension questions and distracters by scanning (and re-scanning) and reading (and rereading) searching for alignment and confirmation of responses to the text. Posner and Peterson (1990) categorise this practice as functioning in an alerting attention network.

8 Conclusion

This research has demonstrated the usefulness of eye-tracking methodology in focussing on the visual and cognitive mechanisms used by students when responding to comprehension tasks. The results have been informally shared with practitioners who have supported the findings as relating to some existing teaching strategies and practices aimed at improving students' comprehension skills.

The model emerging from the research appears to provide early, reliable, and valid knowledge on the interaction of students' skills and processes required for successful comprehension. We acknowledge that more work is needed to enhance and refine the model and look forward to a large-scale study to test students' reading behaviours as they respond to comprehension tasks. It is anticipated that the outcomes will enable teachers and policy makers to plan and allocate resources to effectively assist students in achieving literacy outcomes.

References

Australian Bureau of Statistics. (2008) Adult literacy and life skills survey (ALLS 2006–2007). www.abs.gov.au/AUSSTATS/abs@.nsf/Lookup/4102.0Chapter6102008. Accessed 29 July 2010.

Ashby, J. (2006). Prosody in skilled silent reading: Evidence from eye movements. *Journal of Research in Reading, 29,* 318–333.

Aro, M. (2004). *Learning to read: The effect of orthography. Jyvaskyla.* Finland: University of Jyvaskyla.

Carpenter, P. A., & Just, M. A. (1983). What your eyes do while your mind is reading. In K. Rayner (Ed.), *Eye movements in reading: Perceptual and language processes* (pp. 275–307). New York: Academic Press.

Dehaene, S. (2009). *Reading in the Brain.* NY: Penguin Viking.

Frisson, S., Rayner, K., & Pickering, M. J. (2005). Effects of contextual predictability and transitional probability on eye movements during reading. *Journal of Experimental Psychology: Learning, Memory, and. Cognition, 31*(5), 862–877.

Gough, P. B., & Tunmer, W. E. (1986). Decoding, reading and reading disability. *Remedial and Special Education, 7*(1), 6–10.

Kaakinen, J., Hyönä, J, & Keenan, J.M. (2003). How prior knowledge, working memory capacity, and relevance of information affect eye-fixations in expository text. *Journal of Experimental Psychology: Learning, Memory, and Cognition, 29,* 447–457.

Kintsch, W. (1998). *Comprehension: A paradigm for cognition.* New York: Cambridge University Press.

Knight, B. A., & Galletly, S. (2011). Developing an informed and integrated teaching approach for students with reading-accuracy difficulties in the primary school. In D. Lynch & B. Knight (Eds.), *Issues in Contemporary Teaching* (Vol. 2, pp. 65–89). Brisbane: AACLM Press.

Luke, A., Dooley, A., & Woods, A. (2011). Comprehension and content: Planning literacy in low socio-economic and culturally diverse schools. *Australian Education Researcher, 38,* 149–166.

Luke, A., Woods, A., & Dooley, A. (2011). Comprehension as social and intellectual practice: Rebuilding curriculum in low socioeconomic and cultural minority schools. *Theory Into Practice, 2,* 157–164.

Masters, G. (2009). *Improving literacy, numeracy and science learning in Queensland primary schools: Preliminary advice.* Melbourne: ACER.

McConkie, G. W., Zola, D., Grimes, J., Kerr, P. W., Bryant, N. R., & Wolff, P. M. (1991). Children's eye movements during reading. In J. F. Stein (Ed.), *Vision and visual dyslexia* (pp. 251-262). London: Macmillan Press.

McNamara, D. S., & Magliano, J. P. (2009). Towards a comprehensive model of comprehension. In B. Ross (Ed.), *The psychology of learning and motivation* (pp. 297–284). New York: Elsevier Science.

Mikkilä-Erdmann, M., Penttinen, M., Anto, E., & Olkinuora, E. (2008). Constructing mental models during learning from science text. Eye tracking methodology meets conceptual change. Teoksessa D. Ifenthaler. In P. Pirnay-Dummer & J. M. Spector (Eds.), *Understanding models: Essays in honor of Norbert M. Seel* (pp. 63–79). New York: Springer.

National Assessment Program: Literacy and Numeracy (NAPLAN). http://www.naplan.edu.au/.

Posner, M., & Petersen, S. (1990). The attention system of the human brain. *Annual Review of Neuroscience, 13,* 25–42.

Program for International Student Assessment (PISA). (2007). Science competencies for tomorrow's world: Executive summary: OECD.

Program for International Student Assessment (PISA). (2009). Top of the class: High performers in science in PISA 2006: OECD.

Rayner, K. (1998). Eye movements in reading and information processing: 20 years of research. *Psychological Bulletin, 124,* 372–422.

Rayner, K. (2009). Eye movements and attention in reading, scene perception, and visual search. *The Quarterly Journal of Experimental Psychology, 62*(8), 1457–1506.

Sinatra, G., & Pintrich, P. (2003). The role of intentions in conceptual change learning. In G. Sinatra & P. Pintrich (Eds.), *Intentional conceptual change* (pp. 1–18). Mahwah, NJ: Lawrence Erlbaum Associates.

Snow, C. Reading Comprehension: *Reading for learning. International Encyclopaedia of Education* (3rd ed). http://www.sciencedirect.com/science/referenceworks/9780080448947. Accessed 11 July 2012.

White, S. J. (2008). Eye movement control during reading: Effects of word frequency and orthographic familiarity. *Journal of Experimental Psychology: Human Perception and Performance, 34,* 205–223.

White, S. J., Rayner, K., & Liversedge, S. P. (2005). Eye movements and the modulation of parafoveal processing by foveal processing difficulty: A re-examination. *Psychonomic Bulletin & Review, 12,* 891–896.

Williams, R. S., & Morris, R. K. (2004). Eye movements, word familiarity, and vocabulary acquisition. European Journal of. *Cognitive Psychology, 16*(1/2), 312–339.

Comparing Novice and Expert Nurses in Analysing Electrocardiographs (ECGs) Containing Critical Diagnostic Information: An Eye Tracking Study of the Development of Complex Nursing Visual Cognitive Skills

Marc Broadbent, Mike Horsley, Melanie Birks and Nayadin Persaud

1 Introduction to Clinical Practice Context

Nurses operate in dynamic health-care environments. They are often in positions where they are required to undertake rapid diagnosis, provide emergency health-care responses and make instant risk assessments. Often, these activities occur in real-time, high-risk health-care environments and emergency situations in conditions that can be characterised as time pressured, complex and ambiguous.

Nurses need to have high levels of understanding of a range of diagnostic patient information, presented in multiple visual formats such as X-ray, electrocardiograph (ECG) and other common diagnostic data presented in specialised diagnostic formats. Analysing ECGs and interpreting their diagnostic information is a highly complex cognitive activity. Significant amounts of physiological data, displayed in ECGs, have to be processed, understood and analysed from complex stimuli, even if these stimuli are presented in static format.

Examining an ECG visual display with a view to understanding the meanings represented in the display is a precursor to correctly interpreting the display and providing an accurate diagnosis of the medical condition presented in the ECG visual display. Nurse education, both in university and field placement contexts, places high priority on understanding and applying the meaning of diagnostic information conveyed in visual display formats. Understanding the condition of patients revealed by such displays is a critical factor in determining patient outcomes. A promising feature of eye-tracking technologies and methodologies investigating

M. Broadbent (✉)
The University of the Sunshine Coast, Sippy Downs, QLD, Australia
e-mail: mbroadbent@usc.edu.au

M. Birks
James Cook University, Townsville, QLD, Australia

M. Horsley · N. Persaud
Learning & Teaching Education Research Centre, Central Queensland University, Noosaville, Australia

M. Horsley et al. (eds.), *Current Trends in Eye Tracking Research*,
DOI 10.1007/978-3-319-02868-2_24, © Springer International Publishing Switzerland 2014

cognitive processes in diagnostic situations is to examine eye movements as nursing students analyse the information provided in real visual displays.

2 Conceptual Background: Understanding Complex Visual Displays

Eye-tracking methodologies provide a method for analysing the way that clinicians view, attend to and process complex visual displays. Gaze plot analysis provides a way of analysing how clinicians direct their view of complex visual displays, such as diagrams, charts, and representations, and their features. Gaze plot analysis allows researchers to understand the sequence of clinicians' visual search, their attention to components in the visual field and the way that an individual clinician's attention is structured. Such aspects of the visual field may then provide a platform for cognitive processing and understanding.

Research on visual attention has explored the processes that determine feature and object selection in the visual field. Current models of visual selection propose that attention is guided by two 'attentional' systems in the guided visual salience model developed by Wolfe (1994). This guided visual search model proposes that a 'bottom-up' visual saliency system guides visual attention to salient regions with high feature contrast, and that a 'top-down' feature-specific visual system based on prior learning and experience (of the visual and cognitive) targets certain features and seeks them out (Folk et al. 1992). In the model proposed by Wolfe (1994) and Navalpakkam and Itti (2007), information from these different 'attentional' systems interacts and is combined to guide attention, the precursor of focus, concentration and the building block of interpretation.

Clinical understanding requires more than attention to components in the visual field. Attention represents the initial step in making sense of complex displays that contain critical clinical information important in patient care. For nurses at different stages of their career, viewing complex diagnostic displays such as ECG and X-ray results, identifying the important features of these displays and interpreting their meaning in the context of patient care can be qualitatively different. This research seeks to analyse these differences using eye-tracking research methodologies. The research will explore the different ways that nurses at different stages of their learning journey attend to complex visual displays.

3 Novice and Expert Gaze Models Applied to Reading Complex Visual Displays

Smul et al. (2009) have proposed a novice expert eye gaze model. This model draws on the visual-salience approach developed by Wolfe (1994). In this model novice gaze patterns are determined by the saliency features of the visual field, termed bottom-up influences, because they reflect the visual field. Novice gaze patterns are believed to reflect the 'visual salience of the stimuli', such as its visual feature or saliency that draws the attention of the viewer. In contrast, experts' gaze patterns are dominated by top-down influences that determine the entire structure of gaze patterns. These top-down influences comprise three prior knowledge structures that determine gaze patterns and gaze pattern structures. These structures include:

a. Experience of the visual stimuli which influence gaze as a result of prior learning
b. Viewers' knowledge of the domain which affects the target of the gaze that exists in wider chunks of information
c. Specific goals of viewing as experts focus their gaze on identifiable components of what they are viewing (in sharp contrast to novices)

Pattern recognition based on experience, domain knowledge and vision goals contribute to wide differences in the gaze patterns of novices and experts.

In relation to the impact of bottom-up novice gaze models, Maljovic and Nakayama (1994, 1996) have suggested that in 'pop-out search' visual saliency, search is fostered when the target repeats. Maljovic and Nakayama (1994), supported by Becker (2008), also proposed a priming effect. In this proposal, 'attentional biases' to select previous target features create a bottom-up or automatic target search for pop-out features that is unaffected by knowledge about the upcoming target (Becker 2008a). 'Pop out does not occur automatically, but depends on repeating the target feature value.' This approach has significant implications for exploring how novice and expert nurses view ECGs because all ECG displays have some familiar features that provide repeated pop-out effects.

Becker, in a number of publications (2008, 2009, 2011), has developed a priming theory that proposes that bottom-up and top-down influences are, in part, due to priming with repeated pop-out effects guiding top-down visual input that feature specific visual effects.

Eye-tracking methodologies provide an opportunity to interpret cognitive processes by analysing eye gaze data.

4 The Research and Study Design

4.1 Description of Study Design

The research used an exploratory design in which participants were asked to view four different 12-lead ECGs. After entering demographic data, the participants were asked to view the four different ECGs at their own pace and answer multiple-choice questions on the medical condition that each ECG and their associated rhythm strips represented.

Each of the four different 12-lead ECGs represented different cardiac conditions. The 12-lead ECG is a critical diagnostic tool for all clinicians to understand and interpret. They form the basis for appropriate cardiac treatment. Each ECG contained a different diagnostic situation that was presented to the participants.

The different ECGs and the associated cardiac conditions they represented were:

1. *First-degree atrioventricular (AV) block*: This ECG portrayed a delay in the heart's electrical conduction. This delay in electrical conduction can progress into further electrical conduction issues and cause rhythm disturbances of the heart.
2. *Normal sinus rhythm*: This ECG represented the normal rhythm of a heart.
3. *Anterior myocardial infarction*: This ECG represented muscle (myocardial) damage to the anterior aspect of the heart. In this condition, a coronary artery has become occluded. This condition can cause life-threatening cardiac arrhythmia.
4. *Inferior myocardial infarction*: This ECG represented myocardial damage to the inferior aspect of the heart. In this condition, a coronary artery has become occluded. This condition can cause life-threatening cardiac arrhythmia.

These ECGs were sourced from freely accessible images on the Internet and their veracity and suitability for inclusion in the study were assessed by the principal investigator who has postgraduate qualifications in critical care nursing and 15 years experience in ECG interpretation. The tracings included standard ECGs encountered in clinical practice.

Two types of ECGs were selected for this research. The first described above as (1) and (2) contained diagnostic information that can be evidenced in all 12 leads across the ECG. As these ECGs represent normal rhythm (2) and a rhythm disturbance (1), the diagnosis will conventionally be made using the lead (Ld) II rhythm strip that runs across the width of the ECG and the bottom of the page.

The second type of ECGs used were those with diagnostic criteria found in a specific set of leads. As these ECGs represent damage to a specific area of the heart myocardium, the diagnosis is made through attention to the specific leads that reflect electrical activity of the affected part of the myocardium. However, there is also a requirement to view the Ld II rhythm strip to determine any rhythm disturbance associated with the myocardial damage.

4.2 Data Collection Method

Participants' eye movements were recorded on a Tobii 120 eye tracker, and the data were collected and analysed using Tobii Studio software. The Tobii 120 tracks eye movements at a resolution of 1,280 pixels at a controller refresh rate of 60–75 Hz. The Tobii 120 eye tracker allows 15 ° of head movement 60 cm from the screen and tracks corneal movements to identify gaze focus, fixation and saccade at 15 ms.

4.3 Procedure

Each participant was tested individually. The sequence of the test was:

a. Introduction to the nature of the test
b. Calibration of the Tobii 120 to the participant's gaze (eight participants wore glasses)

c. Testing which included demographic data questions built into the gaze test using Tobii Studio software
d. Gaze testing on 4-, 12-lead ECGs with a multiple-choice question for each ECG
e. Participants viewing their own gaze plots captured by the Tobii 120 Studio software and hardware

The study was conducted in Queensland between April and July 2011 at Central Queensland University, Noosa Campus.

4.4 Participant Sample

The 49 participants comprised three distinct groups:

- Twenty-two undergraduate nursing students recently enrolled in first year of the Bachelor of Nursing (BN). These were students with no prior nursing experiences or study of nursing. These students participated in the study while attending first-year BN study units. These units did not include ECG knowledge or skills.
- Eighteen pre-service nursing students in the final year of their 3-year BN degree. During the BN, these students undertook significant learning experience in ECG recognition, understanding and diagnostics. This included course components focussed on learning to read and interpret ECGs and clinical field experiences that may have included knowledge and skills in interpreting ECGs. In total, the third-year cohort had received information about ECG across three terms with the final unit of study, Critical Care Nursing, having a module on ECG interpretation including the normal ECG, ST segment elevation in infarction and Ld II rhythm strip interpretation.
- Nine expert nurses, clinicians with extensive nursing training and experience, who were employed in clinical settings in areas of nursing requiring continual analysis of ECG data such as intensive care and emergency departments. The experienced clinicians had training similar to the year 3 (Y3) cohort at an undergraduate level, clinical experience, and possibly postgraduate education including ECG interpretation.

5 Data Analysis

Interpretation of an ECG is a critical component for the accurate diagnosis of cardiac disorders, both disorders of rhythm and disorders associated with a reduction or cessation of blood flow to areas of the heart muscle (O'Brien et al. 2009). As noted previously, the ECGs selected for this study comprised two different kinds: those with diagnostic criteria found in a specific set of leads and those with diagnostic criteria that can be evidenced in all leads across the ECG.

The ECGs with diagnostic criteria in a specific set of leads demonstrated an inferior infarction (cell death) pattern (No. 4) and an anterior infarction pattern

(No. 3). The ECGs where the diagnostic criteria can be found in any lead are sinus rhythm (normal health rhythm; No. 2) and first-degree AV block (delayed conduction through the top chambers of the heart; No. 1).

The analysis proceeded in the following way:

Step 1—Gaze plot exploration. A range of gaze plots from participants from each cohort (novice, third-year BN and experienced clinicians) was examined to ascertain the sequence of gaze plots and the areas of highest frequency and visual attention.

Step 2—Area-of-interest analysis was undertaken. For each ECG, an area of interest (AOI) was selected that contained the critical diagnostic information. This AOI was different for each ECG.

Step 3—Area-of-interest data were generated. AOI data were generated to compare the three cohorts in relation to:

a. Time to first fixation on the AOI
b. First fixation duration
c. Fixation count
d. Visit duration

Step 4—Heat map analysis of AOI. A heat map analysis was undertaken to compare the three different groups in terms of the fixation count and frequency of visit to the critical area AOI within the entire ECG.

The following analyses of the data and the findings reflect the steps identified above.

6 Results from the ECGs with Diagnostic Criteria in Specific Leads

6.1 ECG 3—the Anterior Infarction ECG

6.1.1 Selection of AOI

The diagnostic criteria for an anterior infarction are the presence of Q-waves in leads V1–V4 and degrees of ST segment elevation in leads V1–V6. Therefore, the AOI was set in the leads demonstrating the most significant ST segment changes associated with the anterior infarction pattern on this ECG. V2 and V3 demonstrated high ST segment elevation and the development of Q-waves in V3.

6.1.2 Heat Map Analysis

The year 1 (Y1) heat map demonstrates a clustering of vision with the most intense gazes being V3, Ld II and the information on the bottom right-hand corner of the ECG that describes the electrical magnitude of the ECG. Another feature of this ECG is that the gaze plots are also generally associated with the changes of rhythm

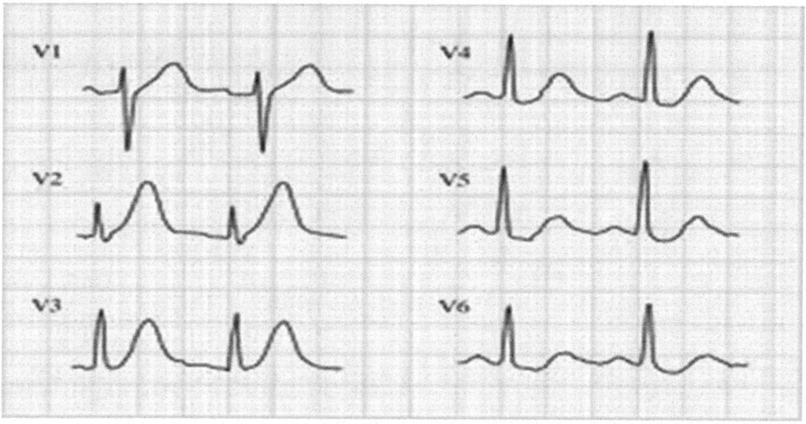

Fig. 1 Critical diagnosis elements of ECG 3

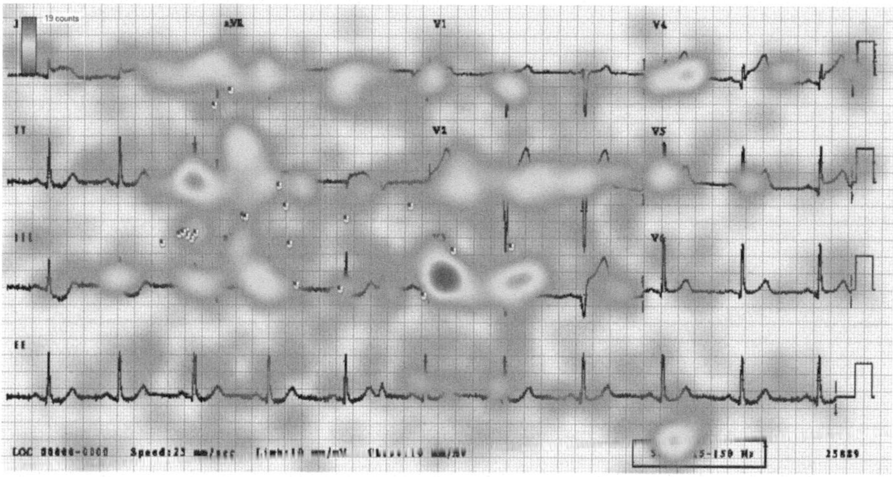

Fig. 2 Year 1 novice heat map anterior infarction

pattern in the various segments of the ECG. The Y1 cohort paid little attention to Ld II for diagnosis of rhythm. This heat map demonstrates the novice identifying the abnormal and also identifying changing patterns and other prominent features within the ECG that result from the bottom-up effects of visual saliency.

The Y3 heat map shows a cluster of vision specifically at the most abnormal leads on the ECG, Lds II and aVL, which are not associated with the diagnostic criteria for this ECG. They did however focus on the AOI to a lesser extent and demonstrated observation of Ld II associated with diagnosis on rhythm. The Y3 cohort demonstrates an awareness of the importance of examining Ld II to determine the

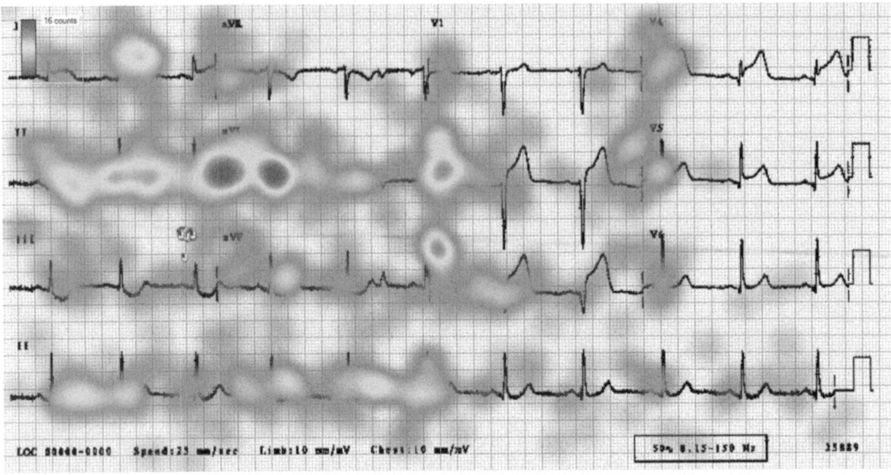

Fig. 3 Year 3 heat map anterior infarction

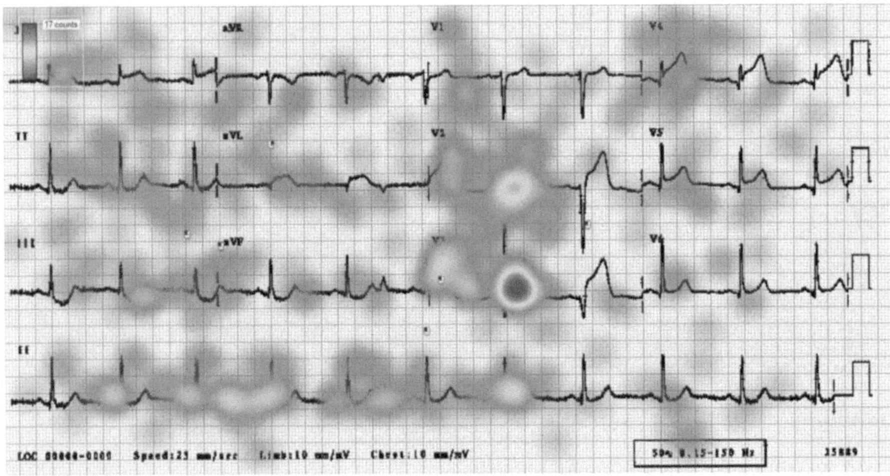

Fig. 4 Experienced clinician anterior infarction heat map

underlying rhythm. This cohort notices the AOI for this ECG but appears distracted by the other non-diagnostic ECG changes in Ld II and aVL.

The experienced cohort has two distinct areas of visual activity. The first being the AOI, and the second being a protracted gaze along Ld II for assessment of the underlying rhythm. The other feature of this heat map is that the entire area of the ECG has been examined, with a focus on the inferolateral leads that show reciprocal changes that are important but not specifically diagnostic. This heat map reflects what would be expected, identification of the significant changes and a focus on them, an examination of the entire ECG and a focus on Ld II for assessment of underlying rhythm.

6.1.3 Analysis of Time to First Duration, Fixation Count, Visit Duration

To build a working model of gaze structure and gaze pattern in each cohort, a comparison of time to first duration, fixation and visit duration for each cohort was undertaken using the Tobii 120 software package.

The AOI for this ECG encompassed the leads demonstrating the most significant ST segment changes associated with the anterior infarction pattern on this ECG. V2 and V3 demonstrated high ST segment elevation and the development of Q-waves in V3.

The Y1 and experienced cohort identified the AOI fairly quickly (Y1 mean = 2.25 s, Y3 = 3.07 s) while the Y3 group took double the time to fix on the AOI (6.23 s). Having identified the AOI, the experienced clinicians remained fixed on it for the first time for longer (0.31 s) compared to the Y1 (0.27 s) and Y3 cohort (0.29 s). The slightly longer first fix duration may indicate an awareness of significance of the finding. This significance is supported by the fact that analysis of the fixation count, the number of times participants fixated on the AOI, demonstrates that the Y1 and Y3 groups went back to the AOI less than half the time of the experienced group, with the Y3 group revisiting the AOI less than the other groups (Y1 = 15 times, Y3 = 11 times and Exp = 30.57 times). This trend is also evident in the total visit duration to the AOI. As well as revisiting the AOI less, the Y3 cohort spent less time observing the AOI (3.39 s) compared to the Y1 cohort (4.63s). The experienced clinician cohort spent more than double the amount of time examining the AOI (9.13 s).

These findings show that the Y1 cohort and the experienced cohort see the AOI in about the same amount of time, possibly because the Y1 cohort respond to the visual salience pop out as they see the changes in rhythm and the amplitude information as being different. They spend about the same amount of time looking at the AOI in the first instance as the other cohorts, revisit and observe the AOI slightly more than the Y3 cohort but significantly less than the experienced group.

The Y3 cohort take more than double the amount of time to identify the AOI than the other groups. The Y3 cohort spend about the same gaze time viewing the AOI but exhibit less visit duration, for less time than the other cohorts—this indicates they are not aware of the significance of the changes in the AOI. This finding is surprising as the Y3 cohort have completed three consecutive terms of study that includes instruction on ECG interpretation.

The experienced cohort identify the AOI relatively quickly and while they do remain fixed on the AOI only slightly longer than the other cohorts, they revisit it and spend longer looking at it than the other groups. This finding would indicate an awareness of the significance of the AOI and possibly demonstrates the cognitive processing required of diagnosis. The Y1 and Y3 cohort both demonstrate a lack of understanding of the significance of data in the AOI and suggest that there has been no net gain in knowledge from the ECG instruction received by the Y3 cohort.

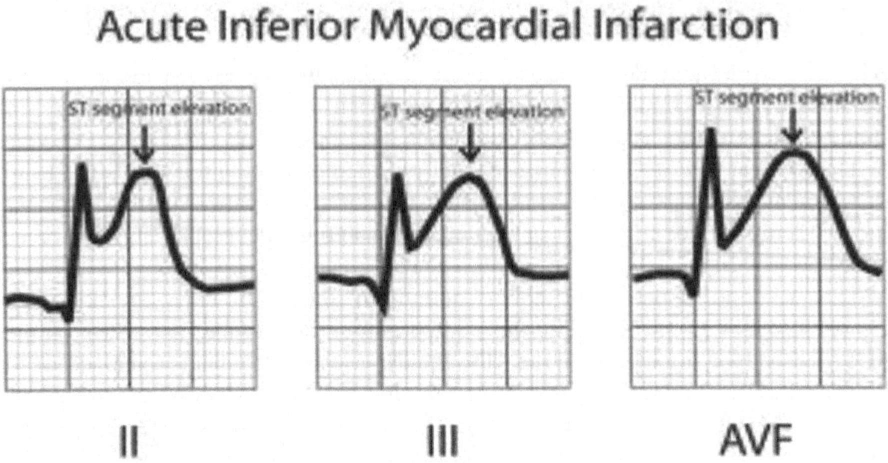

Fig. 5 Critical diagnosis elements of ECG 4

6.2 ECG 4: the Inferior Infarction ECG

6.2.1 Selection of AOI

The diagnostic criteria for an inferior infarction are the presence of Q-waves in Ld II, III and AVF and degrees of ST segment elevation in the same leads. Therefore, the AOI was set in those leads on this ECG.

6.2.2 Heat Map Analysis

The Y1 heat map indicates widespread viewing over the entire surface of the ECG. The areas of most and longest viewing duration were related to non-clinical information. This includes superfluous material on the tracing such as the letters aVF, V3, V2 and AVL and information that relates to the speed the ECG was recorded at and the voltage associated with rhythm amplitude. The viewing of clinically significant information is varied across the ECG and similar time is spent on the AOI as is spent on ECG changes across the entire ECG, including V1–V6 where there are both clinically significant changes and changes associated with the tracing of the ECG. The concentration on the extraneous information represents the visual saliency of pop out.

The Y3 heat map demonstrates a more ordered approach to viewing with clusters of activity in the AOI and above and below it. While the Y3 cohort focussed on the clinically significant components (ST segments), there was an equal amount of interest in the labels of the ECG components II, III, aVF suggesting students may have been drawing a correlation between the leads and the changes within them. This phenomenon is also observed in Ld I and on the rhythm strip across the bottom

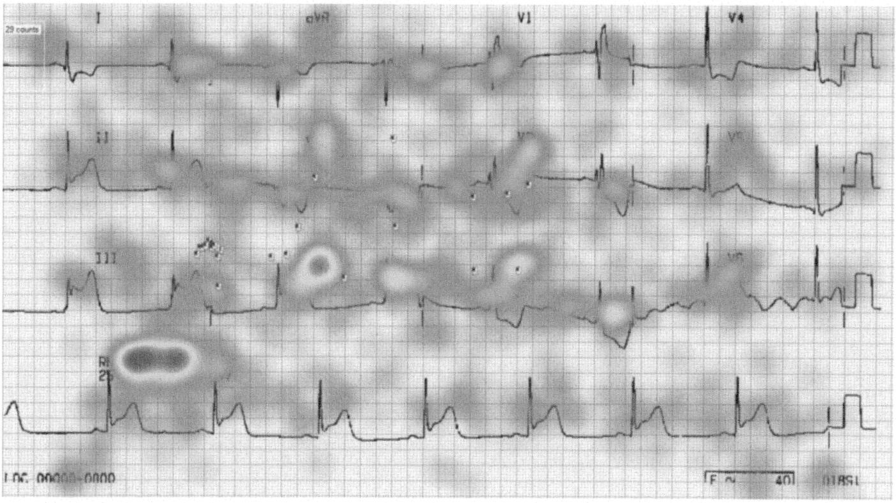

Fig. 6 Year 1 cohort heat map

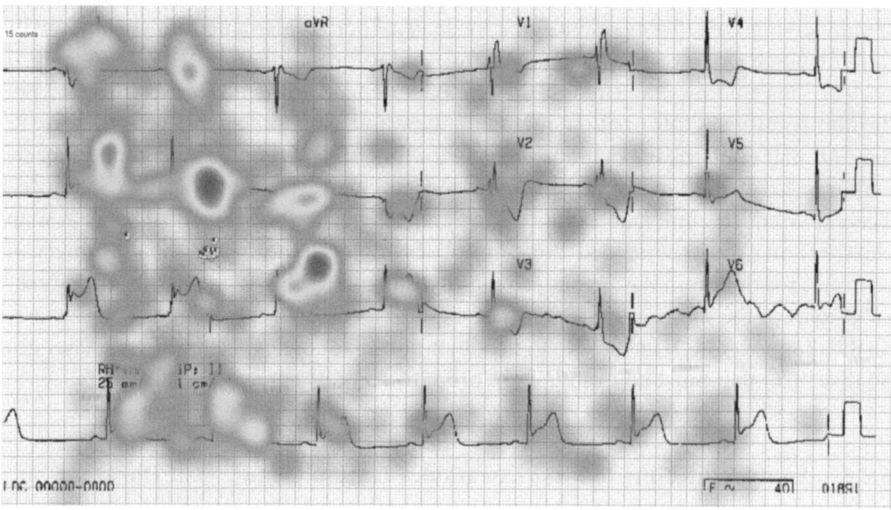

Fig. 7 Year 3 cohort heat map

of the ECG where their attention is also drawn to the information on the ECG that relates to the speed the ECG was recorded at and the voltage associated with rhythm amplitude as were the Y1 group. The Y3 cohort spent very little time viewing V1–V6 where evidence of other significant ECG changes are obvious.

The heat map of the experienced clinician cohort demonstrates prolonged viewing across the ECG. There is a concentration of activity in the AOI, which includes time

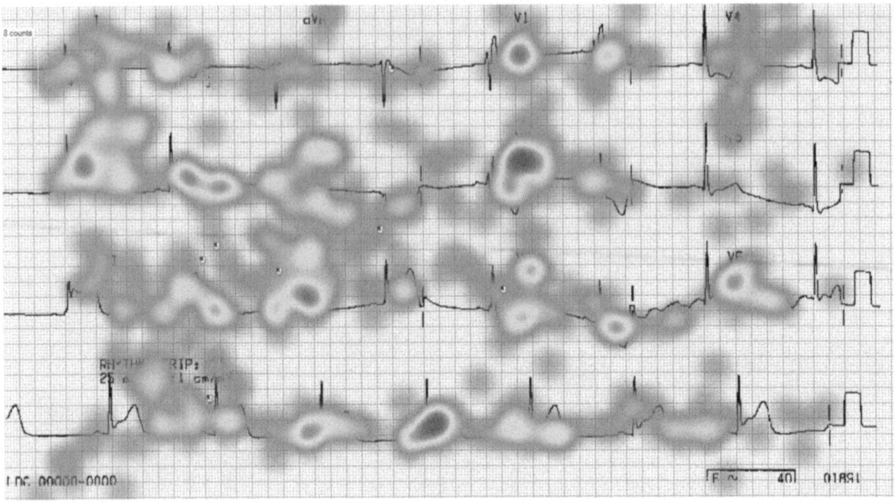

Fig. 8 Experienced clinician heat map

spent viewing the lettering identifying the leads. There is also a significant amount of time focussed on the other clinically significant information in leads V1–V6 and Ld I, aVL. This finding may suggest that the experienced clinicians are as interested in the non-significant information, including the wandering tracing in V6, as they are in the AOI that contains the diagnostic leads for this ECG. Of note is the greater observation of Ld II rhythm strip by the experienced cohort than the Y1 and Y3 cohorts. The experienced clinicians examined the entire length of the tracing. This heat map clearly shows the experienced clinicians observing both the area of the P–R interval, essential for determining the underlying rhythm and the ST segment elevation present as part of the inferior infarction. The experienced cohort place as much emphasis on the diagnostic AOI as they do to the other non-diagnostic, but significant, changes on this ECG that must be examined in order to correctly understand the ECG.

6.2.3 Analysis of Time to First Duration, Fixation Count, Visit Duration

The Y1 and experienced cohorts first viewed the AOI in much the same time (Y1 = 1.75 s, Exp = 1.82 s). However, the Y3 cohort viewed the AOI first but viewed the AOI in shorter periods of time—1.11 s. All three groups viewed the AOI for about the same amount of time on the first view (Y1 = 0.23 s, Y3 = 0.22 s, Exp 0.27 s). Similar to the results of the previous ECG, the experienced cohort viewed the AOI a larger number of times, 17.33, compared to the Y1 and Y3 cohorts, 11.88 and 12.35 times, respectively. Again, the experienced cohort spent longer viewing the AOI (5.09 s) compared to the Y1 cohort (3.90 s) and the Y3 cohort (3.90 s). This result may signify the experienced cohort recognising the significance of the AOI and focussing on these features of the AOI to confirm the diagnosis.

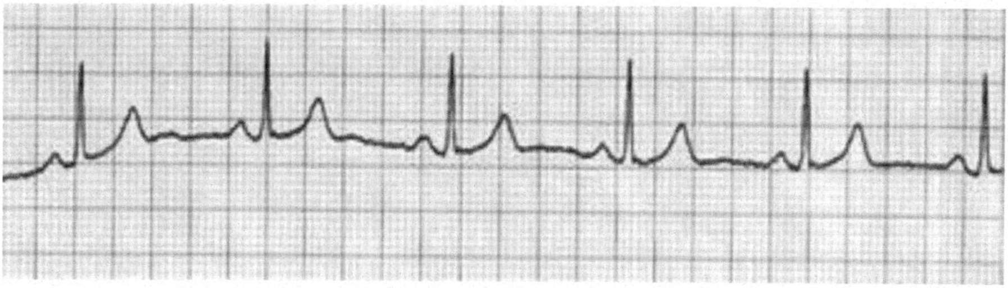

Fig. 9 Area of interest (AOI) of sinus rhythm strip

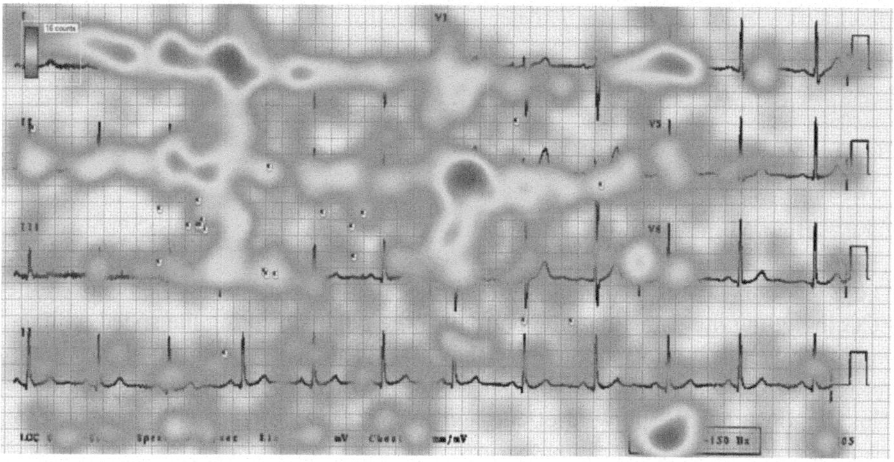

Fig. 10 Year 1 cohort map

7 Results from the ECGs with Diagnostic Criteria that can be Evidenced in All Leads Across the ECG

7.1 ECG: the Normal Sinus Rhythm ECG 2

7.1.1 Selection of AOI

The normal ECG is characterised by the complete lack of any abnormalities. Due to the highly complex nature of ECGs, and the ability to determine sinus rhythm in any part of the ECG, the AOI was set across the first half of the Ld II rhythm strip. This is the rhythm strip that is used to determine underlying rhythm in the absence of any abnormalities.

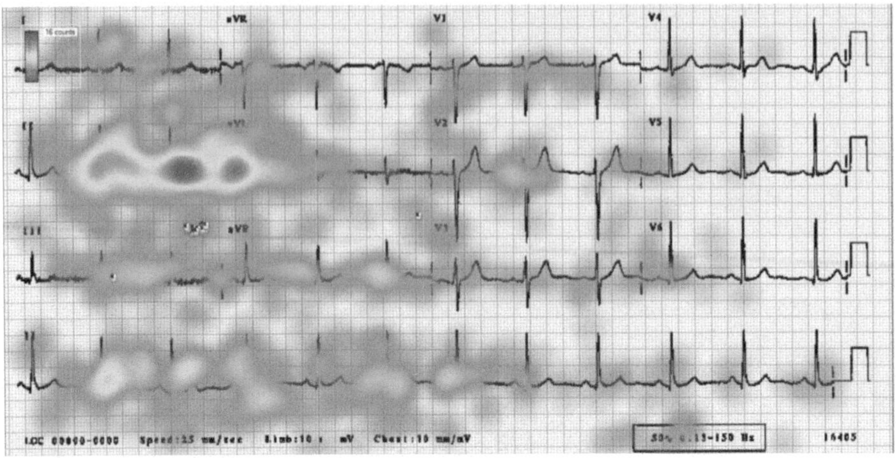

Fig. 11 Year 3 cohort heat map

7.1.2 Heat Map Analysis

The Y1 cohort heat map demonstrates widespread viewing of all elements of the ECG including the data along the bottom of the page indicating the speed, voltage and other technical information about the ECG. Prolonged visualisations are found across the ECG where the changes in the lead, and therefore the rhythm, are evident. This demonstrates the pop-out effect—where patterns are repeated. There is scant visual activity in the Ld II rhythm strip.

The Y3 heat map shows visual activity predominantly across the first half of the ECG and in the Ld II rhythm strip. There is a focus on leads II and aVL.

By contrast, the experienced cohort demonstrate widespread viewing of the ECG with a focus on Ld II and aVF, as did the Y3 cohort, yet the reason for this is uncertain. The two important features of this ECG are that there is extensive viewing of the Ld II rhythm strip, particularly in the first half where the main visualisation has occurred, and there are areas of extended gaze over the T-waves in V1, V2 and V3. This is likely due to the unusual elevation of the T-waves indicating that the experienced cohort is viewing these to rule out abnormalities.

The similar pattern of viewing between the Y3 cohort and the experienced cohort suggests the Y3 cohort is familiar with the importance of the Ld II rhythm strip.

7.1.3 Analysis of Time to First Duration, Fixation Count, Visit Duration

The Y1 cohort took more than twice as long (8.49 s) as the experienced cohort (3.66 s) and nearly twice as long as the third-year cohort (4.88 s) to view the AOI. All three groups viewed the AOI for about the same amount of time (Y1 and Y3=0.27 s, Exp=0.3 s). The Y3 and experienced cohort viewed the AOI a similar number of times (Y3=8.22, Exp=9.33) compared to the Y1 cohort who only viewed the AOI 4.86 times. Similarly, the Y3 and experienced cohorts spent longer

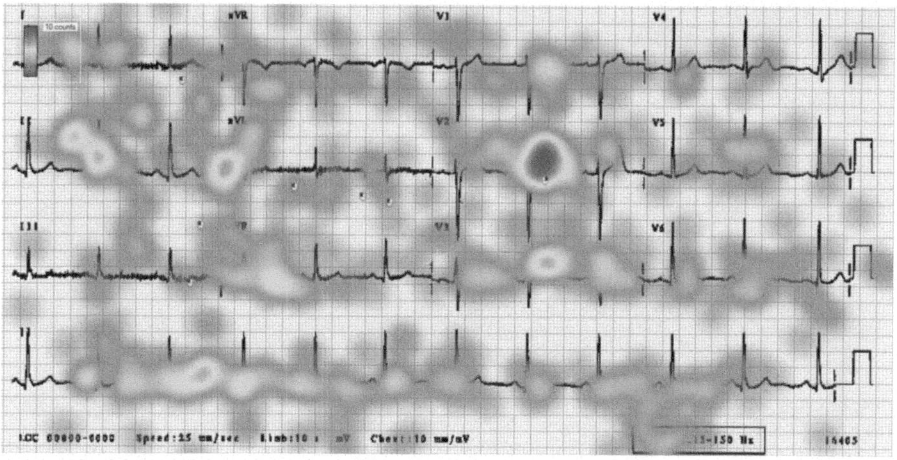

Fig. 12 Experienced clinician cohort heat map

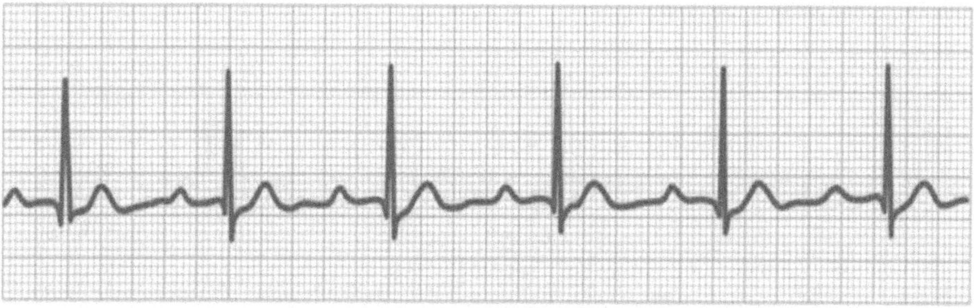

Fig. 13 Area of interest (AOI) of the first-degree AV block

viewing the AOI (Y3 = 0.95 s, Exp = 1.20 s) than the Y1 cohort (0.53 s). These data reinforce what has been determined for the heat map that the Y1 cohort are not aware of the importance of Ld II and while the Y3 cohort are, but that they are not viewing it for as long or as repeatedly as the experienced cohort. This finding has significant implications for instruction and education in relation to developing ECG clinical knowledge and skills.

7.2 ECG—the First-Degree AV Block ECG 1

7.2.1 Selection of AOI

This ECG is very similar to the normal sinus rhythm ECG in that diagnosis can be made in any lead; however, the focus should be on the Ld II rhythm strip. The diagnostic criterion for this ECG is a prolonged P–R interval. Therefore, the AOI was set across the first half of the Ld II rhythm strip.

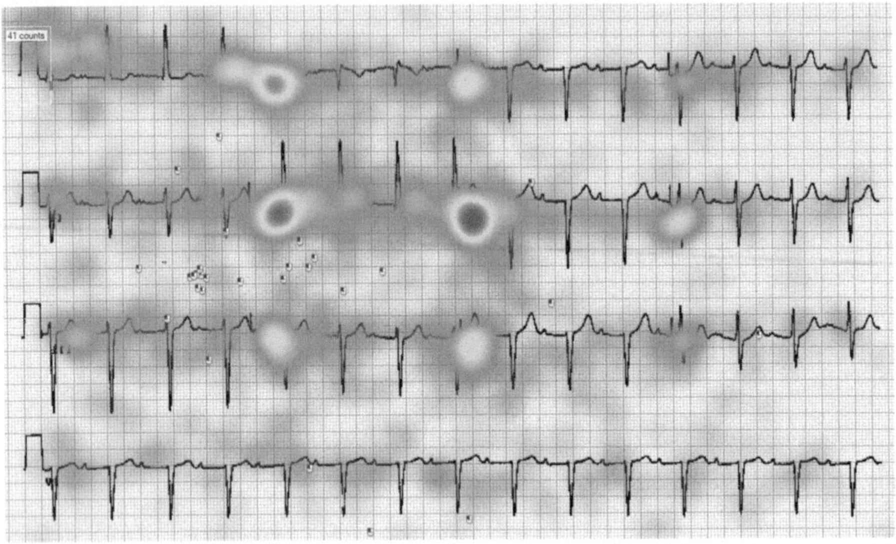

Fig. 14 Year 1 cohort heat map

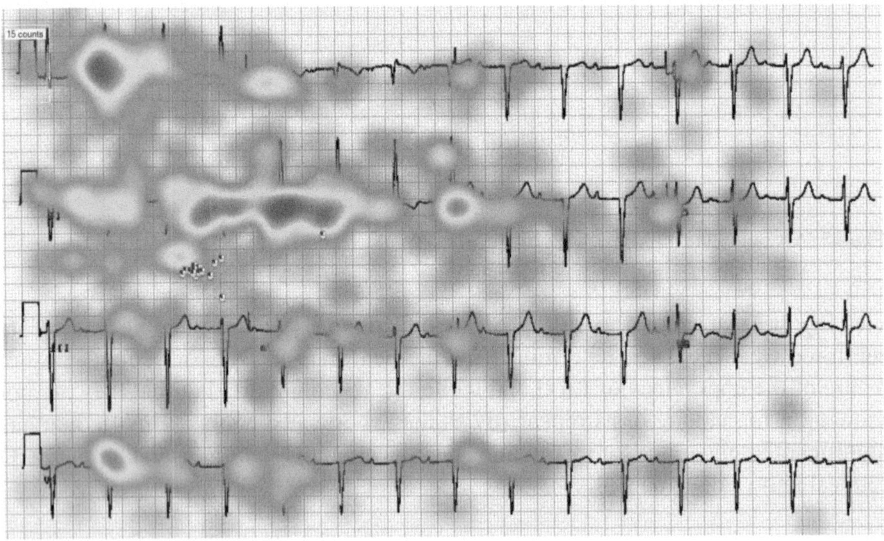

Fig. 15 Year 3 cohort heat map

7.2.2 Heat Map Analysis

Similar to the sinus rhythm ECG, the Y1 cohort heat map demonstrates prolonged visualisations across the ECG where the lead, and therefore the rhythm, changes. In particular, the Y1 cohort visualise the lettering at these points. This finding suggests

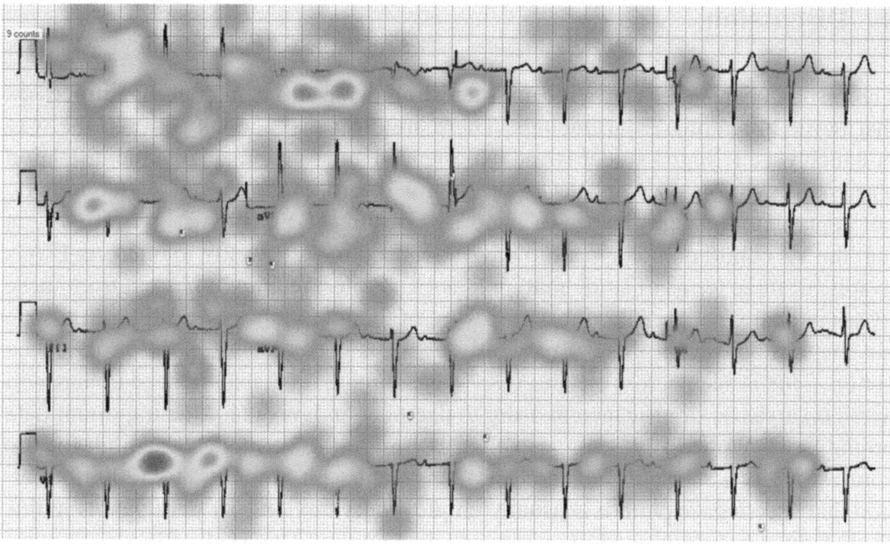

Fig. 16 Experienced clinician cohort

that the students are responding to visually salient aspects of the ECG display. There is little visual activity and focus in the critical Ld II rhythm strip.

There are similarities between the Y3 heat map for this ECG and the sinus rhythm ECG in that the visualisations are predominant across the first half of the ECG; particular attention is paid to Ld I and aVL and the Ld II rhythm strip is visualised.

The experienced cohort visualise the entire ECG. There are areas of visual concentration over the inverted T-waves in aVR. Again, the experienced cohort visualise the Ld II rhythm strip as would be expected. Of note are the prolonged visualisations over the extended P–R interval, i.e. the gaze is to the left of the R-wave, whereas in the Y3 cohort the gaze is to the right of the R-wave suggesting that while the Y3 cohort have examined the Ld II rhythm strip they are not aware of the abnormalities.

7.2.3 Analysis of Time to First Duration, Fixation Count, Visit Duration

The Y1 cohort took 16.41 s to first visualise the AOI, Y3 9.49 s and the experienced cohort 11.32 s. Having visualised the AOI, the experienced cohort observed the AOI for 0.62 s compared to the Y3 and Y1 cohorts who visualised the AOI for less than half that time (Y3 = 0.37 s, Y1 = 0.29 s). The Y3 and experienced cohort return to the AOI repeatedly (Y3 = 3.30, exp = 4.67 times) while the Y1 cohort only revisit the AOI 1.64 times. The overall time spent at the AOI is 1.49 s by the experienced cohort compared to 0.65 s by the Y1 cohort and 0.34 s by the Y3 cohort.

8 Discussion

The results of this study provide an interesting perspective on cognitive processes that occur in reading ECG tracings. There are clear distinctions between the way in which year 1, year 3 students and experienced nurses approach the identification and assessment of AOIs on ECGs. While it is clear that the Y1 and Y3 cohorts do not appreciate the significance of abnormalities in the ECG, what is surprising is that the Y3 cohort do not appear to have advanced in their ability to interpret ECGs in spite of three terms of study that include ECG interpretation. The data from these ECGs indicate that the Y3 cohort are aware of the importance of Ld II rhythm strip but they do not revisit this as frequently or focus on it as long as the experienced cohort. This is seen across all the AOIs suggesting that the experienced cohort will dwell on areas of significance in order to decipher and recognise the important data on the ECG. Wolfe (1994) suggests that 'bottom-up' gaze models use high feature contrast regions as a point of interest. Consistent with Smut et al.'s (2009) assertion that such a gaze pattern is usually employed by novices, Y1 participants in this study tended to be attracted to extraneous data on the ECG, such as letters or numbers, or areas where there is a distinctive pattern, such as where lead changes occur.

While the experienced cohort do not necessarily visualise the AOIs in more or less time, they do focus on them for longer periods and return to them more frequently than the other groups. This would be consistent with them having the knowledge to identify abnormalities and also the need to focus on these abnormalities to confirm details important for interpreting the diagnostic information provided. The experienced cohort, therefore, tend to target specific visual effects, suggesting a 'top-down' gaze model characteristic of experts (Wolfe 1994; Smut et al. 2009).

The Y3 cohort appear to be aware of the significance of viewing the Ld II rhythm strip yet are not aware of the clinical significance of the information there as evidenced by the findings of the first-degree AV block ECG. This outcome is reflected throughout the findings where the Y3 cohort appear to know what they need to find but are unsure of what the finding indicates. There is evidence, therefore, of 'attentional biases' (Maljovic and Nakayama 1994) in this cohort. Consistent with Becker's (2008) priming theory, third-year students appear to be primed by exposure to 'pop-out' features, possibly as a result of instruction on ECG interpretation, that enable identification of specific visual effects, yet they are unable to interpret their significance. It is evident that the Y3 cohort fall between novice and expert and behave accordingly in combining characteristics of both top-down and bottom-up systems of viewing the ECG tracing.

The findings of this study have implications for nursing practice, education and research, specifically in the design of instruction in reading complex diagnostic displays.

References

Ariasi, N., & Mason, L. (2011). Uncovering the effect of text structure in learning from a science text: An eye-tracking study. *Instructional Science, 39*(5), 581–601.

Becker, S. I. (2011). Determinants of dwell time in visual search: Similarity or perceptual difficulty? *PLoS ONE*, Vol. 6, No. 3.

Becker, S. I., Folk, C. L., & Remington, R. (2009). The role of relational information in contingent capture. *Journal of Experimental Psychology: Human Perception and Performance.*

Duchowski, A. (2002). A breadth-first survey of eye tracking applications. *Behavior Research Methods, Instruments, and Computers, 34*(4), 455–470.

Navalpakkam, V., & Itti, L. (2007). Search goal tunes visual features optimally. *Neuron, 53*(4), 605–617.

O'Brien, K., Cannarozzi, M., Torre, D., Mechaber, A., & During, S. (2009). Training and assessment of ECG interpretation skills: Results from the 2005 CDIM survey. *Teaching and Learning in Medicine, 21*(2), 111–115.

Wolfe, J. M. (1994). Guided search 2.0: A revised model of visual search. *Psychonomic Bulletin & Review, 1*(2), 202–238.

Wolfe, J. M. (1998). Visual search. In Pashler, H. (Ed.), *Attention.* London: University College London Press.

The Development and Refinement of Student Self-Regulatory Strategies in Online Learning Environments

Nayadin Persaud and Matt Eliot

1 Introduction

In the higher education setting, and more particularly higher education e-learning settings; students are increasingly required to manage their own learning. This management process incorporates self-regulatory skills. Students' skills include activating prior knowledge, setting goals, creating networks of support, seeking scaffolding, responding to feedback, reflecting, and creating an awareness of their own personal learning.

The need to examine and understand the processes in the development of self-regulation in higher education settings is essential primarily because self-regulation is critical to student learning. Significant theoretical research has focussed on self-regulated learning. However little is known about how higher education students develop and refine self-regulated learning strategies and behaviours in the e-learning context over time. Students' actual behaviors and the motivations that influence those behaviors are key elements in terms of understanding how students navigate their course work and how students self-manage/regulate learning in the online setting.

2 Self-Regulated Learning

Defining self-regulated learning (SRL) is not as clear as one would expect as different theorists have developed the concept within the context of varying epistemologies. For example, according to Furberg & Ludvigsen socio-cultural analysis seeks to understand learning as a process that is embedded in and interdependent on complex social and cultural contexts (2008). Also, socio-cultural approaches to explaining learning emphasize social, cultural, institutional and historical interaction as

N. Persaud (✉) · M. Eliot
Central Queensland University, Noosaville, Australia
e-mail: n.persaud@cqu.edu.au

M. Horsley et al. (eds.), *Current Trends in Eye Tracking Research*,
DOI 10.1007/978-3-319-02868-2_25, © Springer International Publishing Switzerland 2014

being critical for the way human thinking develops and learning occurs (Vygotsky, 1978).

SRL refers to the setting of goals and planning one's own learning by monitoring and refining learning strategies until the achievement of the goals. A commonly accepted socio-cognitive definition has been proposed by Azevedo, Johnson, Chauncey, & Burkett (2010) who suggest SRL is an active, constructive process whereby learners set meaningful learning goals and monitor and control their cognitive and metacognitive processes in the service of those goals. However, from the socio-cultural perspective, McCaslin and Burross offer a definition with a wider scope involving cultural tools and elements. McCaslin and Burross (2011) consider adaptive learning to be more representative of the self/social relationship. They argue a preference for the term adaptive learning as the term self-regulation may not recognize the degree in which social and cultural objects may co-regulate and influence control over behaviors. The researchers, while continuing to refer to the phenomena as SRL in this chapter, recognize the foundational aspects of the social and cultural artefacts and historical influences within the dynamic environment of online learning and their impact on all members of the learning community.

From a socio-cultural perspective social interaction is the source of self-regulatory control (Schunk and Zimmerman, 1997). Thus, while participating in online collaborative group work, students' sense of self are constantly being challenged and developed signifying that co-constructed interactions with the group, larger community and environment are mandatory and pervasive and necessary for learning.

Lastly, SRL processes have social origins and an amalgamation of individual and social phenomena appears to underlie how students develop self-regulation within the learning environment. In terms of online collaborative group work, there is a growing body of research exploring the nature of social interaction and how social interactions may impact the individual's development of self-regulatory processes. In publications surrounding the social aspects of SRL, Hadwin and Oshige (2011); Janssen, Erkens, Kirschner, & Kanselaar (2012); Järvelä, Järvenoja, & Näykki (2013) enrich the debate amongst researchers worldwide while striving to develop a common language and methods for exploring investigations into self-regulation, co-regulation and socially shared regulation in computer supported learning environments.

2.1 Goals and Motivation

Most theories that describe SRL highlight the importance of goals. Goals give purpose and meaning to individuals participating in an activity (Kaplan & Maehr 2007; Maehr & Nicholls, 1980). Research has shown that having goals is essential to SRL. Having goals and having the motivation to achieve goals interact in complex ways and it is through this interaction that regulation manifests.

Attention is one fundamental process in goal directed behavior. Goal directed behaviors require the focusing of attention on relevant stimuli and ignoring irrelevant detractors (Lavie, Hirst, de Fockert, & Viding, 2004). In order to better understand the relationship between attention and goal directed behavior as participants complete online learning tasks, it is necessary to examine where a participant gazes or the manner in which a participants' eyes move. Calvi, Porta, and Sacchi state "from eye position tracking, and indirect measures such as gaze position, fixation numbers, and fixation duration it is possible to draw information about attention, stress, relaxation, problem solving, successful learning, etc. (2008)." In a similar stance, Poole and Ball (2006) note fixations can also reveal the "amount of processing being applied to objects".

The eye tracker is essential to capture where participants look or gaze as they complete goal oriented tasks for online university course in which they are enrolled. However, other methods like stimulated recall interviews are necessary in order to evaluate eye movement data collected while a participant completes an online forum discussion task. A more detailed discussion of the methods is discussed later in this chapter.

2.2 Self-Regulated Learning and Hypermedia Environments

Overall, research into SRL has employed a range of mostly positivist research methodologies. Azevedo, Cromley, Moos, Greene & Winters (2011) investigated the role of scaffolding in facilitating student learning within an experimental hypermedia environment. The participants were randomly assigned to one of three scaffolding groups in which human tutors assisted with a) providing help with content and support for regulating of learning, b) help with only support for regulating of learning or c) no help at all. Findings from this research indicate that participants were more successful at learning a challenging science topic when they were given assistance with both content and assistance with the regulating of their learning.

In 2008, Winters, Greene, & Costich analyzed 33 empirical studies exploring self- regulated learning in computer based learning environments. In their findings, they highlight that particular task characteristics and learner goal orientation may be related to SRL. This is particularly important as many experimental studies control for specific factors by design. Often these studies are not designed to consider the participant's "goal" or motivations for participating in the study as it may not be meaningful to the study. Goals are essential for understanding regulation of learning; thus, researchers must consider the participant's goals, the actual process of regulating, and the self-regulating behaviors.

Using a case study approach, Whipp and Chiarelli (2004) focused on six students participating in a master's program who had experience with learning in the online format, Whipp and Chiarelli (2004) explored the manner in which SRL strategies are used and adapted as students participate in an online course for one term. Data was collected through student interviews, instructor interviews, and reflective journals completed by students. Whipp and Chiarelli found that students adapt traditional self-regulatory strategies to cope with learning in the online environment. Traditionally, SRL skills, like help-seeking, involved phone calls and personal contact with a course lecturer or peer. However, the online adaptation of help-seeking skills involve utilizing web based helpers, using student postings as models, and accessing technical expertise (Whipp and Chiarelli, 2004). Additionally, they found that elements of the course design prompted students to use specific self- regulatory strategies and lastly they suggest that students' goal orientation and interests facilitate what SRL strategies are employed.

2.3 Self-Regulated Learning and Eye Tracking Research Methodologies

Eye tracking research methodologies offer new approaches to investigating some aspects of student SRL behavior. Eye tracking research methodologies include the use of current eye tracking cameras and sophisticated analysis software in order to develop new ways to explore cognitive and social processes. Eye tracking methods allow researchers to obtain data regarding where a participant allocates attention, length of focused attention, and gaze paths of eye movement behavior. More specifically, eye movement data in the e-learning environments allow researchers to understand what students pay attention to and what they ignore while navigating through online learning tasks within learning management systems.

A number of studies have used eye tracking methods to explore specific SRL behaviors and strategies that appear in hypermedia settings such as self-monitoring while reading (Hyönä, Lorch Jr, & Kaakinen, 2002); pausing when composing text (Johansson, Wengelin, Johansson and Holmqvust (2010) and decision making (Muldner, Christopherson, Atkinson & Burleson, 2009). Additionally, Wiklund-Engblom (2010) explored SRL in a specific e-learning course. Her study examined how novice e-learners manage their own learning in an online environment and how their learning might be better facilitated. The previous studies often used data sources such as eye tracking metrics, key stroke analysis software, observations of eye gaze data, and retrospective cueing. For example, Kostons, van Gog, & Paas (2009) investigated if and how retrospective cueing from the eye movement data could be used to stimulate participants' recall of learning behaviors. One notable finding from their study suggested that eye tracking data could serve as a metacognitive tool assisting students in reflective self-assessment of their learning. Lastly, van Gog, Kester, Nievelstein, Giesbers, & Paas (2009) also state that eye movement data may be useful for studying changes in cognitive structures at an individual level over time.

2.4 Text Composition and Eye Tracking

In terms of text composition, we found that the students could be grouped into two categories that were identified as "monitor gazers" and "keyboard gazers". These terms: "monitor gazer" and "keyboard gazers" have emerged from the research of Johansson and colleagues. Johansson et al. (2010) investigated differences regarding the production of text between participants who mainly focus on the monitor and participants who mainly focus on the keyboard. Gaze data, specifically fixations and fixation durations, are important for this study as they reveal location and length of participants' focus/attention as they work toward a goal.

Johansson et al. (2010) investigated the visual behaviors of writers with differing levels of typing skills. In their study, they used an eye tracker to gather data from 28 university students as they composed text using a keyboard and a computer monitor. They described their participants as "keyboard gazers" and "monitor gazers" as their primary objective was to investigate where students looked as they composed text. On the continuum of typing ability, they reported that the "keyboard gazers" are less skilled typists than the "monitor gazers". Johansson and colleagues assumed differences in working memory in regards to tying ability in their findings. They suggested that monitor gazers can devote a larger share of their working memory to higher level processing like planning and editing while they are in the process of composing text. This research is relevant for the current study as Johansson and colleagues investigated connections between SRL (including working memory and attention) and online composition.

In this study, the researchers sought to investigate participant behavior in a wider range of online learning contexts. Therefore, the primary use of the eye tracking technology in this study was as a stimulus for participants' reflections on their own eye movement behavior.

3 Rationale for Study Design

There is a great need to investigate SRL processes and behaviors in the online university setting using multiple methods. Anson and Schwegler (2012) confirm eye tracking methodologies may reveal new data about the psychological, visual, social and educational dimensions of both writing and reading in the online university setting. Whipp and Chiarelli (2004) emphasize the need for additional studies which investigate the motivational beliefs of students and how those beliefs influence student learning in online environments. Additionally, Whipp and Chiarelli call for more longitudinal studies investigating the complex nature of SRL, and specifically focusing on if, how, and why students use specific SRL strategies as they complete other higher education coursework. Broadly, Schraw (2010) advocates that researchers employ a wide variety of technology based methods to investigate self-regulatory and cognitive learning processes while suggesting that greater use be made of multiple data collection and outcome measurements such as eye tracking and stimulated recall approaches.

As a result, this study utilizes a longitudinal approach to investigating student behavior in e-learning environments. The eye tracker data provides a capture of visual attention as a student completes an online forum task and the stimulated recall interview gathers data about motivational influences on SRL strategy use.

A few assumptions about the use of the eye tracking data have been taken into consideration in the design of the study. The use of the eye tracker in this study is based on the assumption that there is a relationship between fixations, gaze patterns, and cognitive and metacognitive functions. As this research uses the eye tracker's

Table 1 Tobii Technology Assumptions and Researcher's Design Response

Eye Tracking Assumptions taken from Tobii (2010) white paper	Researcher's Design Response
Some fixations do not translate into conscious cognitive process. Tobii suggests following up eye tracking test with interview to assess participants 'motivations and expectations.'	Researcher did not inquire about every single fixation. A stimulated recall interview was done immediately after the eye tracking event that inquired about motivations and rationale behind some specific eye movement's behaviors.
Fixations can be interpreted in different ways depending on the context and objective of the study. Tobii suggests that a clear understanding of the objective of the study and clear planning of the tests are important...	Participants who volunteered for this study were given instructions to complete broadly defined forum related learning tasks. There were a variety of tasks that participants completed as part of their coursework. For example, some participants had required reading or online videos to watch before completing their forum task while others simply posted a general question or short response to a relevant forum for their coursework. As the tasks varied from student to student the researcher adjusted the prompts and questions to correlate with the task demands of the student to gain more insight into motivation and metacognitive skills usage.
Individuals move their eyes to relevant features of a scene. Some features are primarily detected by the peripheral area of our visual field and sometimes we filter features that may be irrelevant to us. For example, we may filter banners or advertisements when using a website. Tobii suggests a follow up interview or think aloud protocol to address the reasons for why a certain aspect of a visual scene may have been ignored.	The researcher accounted for participants' visual filtering as the Learning Management System layout may have provided visual features that were completely ignored or overlooked. During the recall interviews, the participant was questioned primarily about what they focused their attention on and why. The process of gathering the information to complete the task was more relevant and significant to the research questions.

video capture as a foundation of the stimulated recall interview; it is necessary to take into account each assumption and how the methods were modified in order to elicit clear and meaningful data for this study.

Table 1 provides Tobii Technology's white paper (2010) on these assumptions and also highlights some of the considerations that were made during the data collection and analysis phases of this research project.

As noted in the above table, the follow up interview when working with eye tracking data adds value and rigor to the method. The stimulated recall interview prompts included questions about a variety of learning behaviors that were exhibited by the participant during the completion of the task as visually captured by the eye tracker.

4 Research Questions

This chapter reports on an ongoing investigation of how and why students employ certain self-regulatory strategies as they progress toward their learning goals. The research focuses on text composition as a window to investigate SRL processes.

Specifically, the researchers are investigating how students' self-monitor while reading and composing text through observations of participants' visual attention during online learning tasks.

This study utilizes eye tracking methods in combination with qualitative video stimulated recall procedures to explore the primary research questions of:

1. What does eye gaze data reveal about how students allocate attention as they complete online goal-oriented tasks?
2. How do students' relative concerns for presentation of self in the online environment interact with their self-regulated learning behaviours?

The following sections describe methods, data analysis, findings to date, and implications for further research.

5 Methods

A mixed methods approach using both qualitative and quantitative data was used to answer the research questions (Hessie-Biber 2010). Qualitative methods include pre-interviews and stimulated recall methods. Quantitative methods included surveys and analysis stemming from data collected by an eye tracker. This chapter focuses on the qualitative and eye- tracking data collected for a subset of study participants.

5.1 Participants

Ethical clearance for the study was approved by the researchers' institution. All participants completed an informed consent process and were aware of the longitudinal nature of this research. Informed consent forms included a section for the participants to agree to their image being recorded. Participants were invited to a brief orientation in the eye tracking lab and were given the opportunity to ask questions about the study and the eye tracking procedures before they formally signed the consent form.

The first round of data collection occurred in August 2012 with 21 participants from a variety of academic disciplines including nursing and education. The students agreed to return to the lab during the second year of their degree program. The sample included 17 females and 4 males who were all in the first year of their study. The average age of all participants in the study was 27.66.

This chapter examines the data collected from a subset of the participants in the first data collection period. Specifically, the subset includes 11 students who completed an online discussion forum task which required them to compose text. The discussion forum was a part of the enrolled course.

The 11 participants (P1, P2, P3, P4, P6, P7, P12, P15, P17, P18, and P19) were completing their university degree in either a fully online mode of study or a blended mode of study. Two participants, P12 and P15 participated in course work that was exclusively delivered online. The rest of the participants (nine) engaged in a blended mode of study. This blended mode allowed for participants to primarily complete their course work online and participants attended residential schools held on university campuses once during the term. Attendance at residential schools was mandatory for these students.

During the initial data collection phase, all 11 participants completed a learning task in an online discussion forum. Participants were instructed to choose a forum discussion task that was relevant to their coursework. Four participants created a posting in a large group forum where their lecturer had prepared a specific topic for the discussion. Five participants created a posting in a small group forum where their lecturer had created a space for small group discussions related to a group assignment and in which no specific topic was set. Lastly, two participants created a posting in a forum that was part of a large group forum where the lecturer had not set a topic. Overall, nine participants posted their responses by selecting the "post to forum" icon. Two participants chose not to post their responses and both of these responses were in the type of forum that the lecturer had prepared a specific topic for discussion.

5.2 Eye Tracker

The study used a PC integrated with a Tobii X120 eye tracker with a 17″ display having a maximum resolution of 1,280 × 1,024 pixels. Eye tracking software, Tobii Studio, recorded eye movements and screen dynamics. The eye tracker has a tracking or frame rate of 50 Hz, and looks like a normal computer display with a small camera mounted atop the monitor. The eye tracking apparatus is located below the monitor. The camera recorded the participant' natural movements as they navigated through the online learning task. The video captured the participant sitting in front of the computer, his or her head movements, and facial expressions. The researcher used the video replay from the eye tracking data to capture overt behaviors to stimulate discussions of cognitive and socio-cultural motivations underpinning participants' eye movement.

5.3 Observation

The researcher recorded observations about participants' behaviour while they completed their tasks online. The documentation of these observations served two purposes. First, the behaviour exhibited may have been a SRL behaviour that the researcher inquired about during the post-interview. Second, the researcher documented any words spoken or verbal sounds the participant may make while completing the task.

5.4 *Video-Stimulated Recall*

The purpose of stimulated recall (interviews in this study) is to investigate cognitive processes that are not usually evident by other methodologies (Gass and Mackey 2000). The video capture was used to examine participants' thoughts, decisions, and reasons for exhibited behaviour. The video-stimulated recall interview focused on the participants' experience and behaviour during the e-learning task. The participant was asked to describe their motivations and goals for particular behaviors while watching a video which revealed gaze patterns of the e-learning task they completed. Edwards-Leis (2007) explains that the use of video provides a visual stimulus because it documents and captures participants' actions and can further trigger memory cues of their participation in a recorded event.

This method of using eye tracking and verbal cued interviews as examined in e-learning environments was included in a 2010 study exploring students experience in e-learning environments. Wiklund–Engblom (2010) incorporated video stimulated recall interviews while exploring the learning process of students as they engaged in an e-learning course. She highlights the value of using observation and video-stimulated recall because the combination of methods provide a detailed overview of the participants' experience and learning process during the completion of the e-learning coursework. In addition, Wiklund-Engblom noted that the eye tracker supported the stimulated-recall interview allowing the discussions with the participant to be more focused on the eye gaze behavior displayed.

There are advantages and disadvantages when using stimulated recall methods or cued retrospective interviews. Van Gog, Paas, van Merriënboer, & Witte (2005) highlight some of these strengths and limitations of cued retrospective interviews. A strength of this method is that it may lead to the active reconstruction of thoughts resultant from the participant reviewing their eye movement behavior instead of solely depending on memory. A limitation may be that the participant responses and thoughts related to those behaviors may include irrelevant information.

Video stimulated recall methods are beneficial to the validity of the re-search as the participants are asked to describe what they experienced during the e-learning task. Eger, Ball, Stevens, & Dodd (2007) highlight some of the benefits of using a retrospective report noting minimal cognitive load is placed on the participant, silence [while completing the learning task]; increased time for them to compose their thoughts and direct access of where they were looking on screen [during the learning task]. These benefits of this method described by both Wiklund-Engblom (2010) and Eger et al. (2007) and Van Gog et al. (2005) have all been taken into careful consideration during the research design phase of this research project.

5.5 *Data Analysis*

Two separate interviews were conducted with each participant during their time in the eye tracking facility. The first interview was a pre-interview. The pre-interview was designed to elicit information regarding: 1) participants' ideas relating to their long term and short term goals in regards to their coursework and 2) their description of the task that they wanted to complete while they were in the eye tracking facility. The researcher utilized a pre-interview script with open ended prompts for each participant. This script provided consistency between participants as the structure of the procedures and the questions were the same. Participants were allowed to express their goals and ideas about achievement through the open ended format of the questions. The researcher used additional prompts or questions in order to clarify or gain a better sense of participants' ideas as described. The pre-interview was audio recorded using a digital voice recorder and transcribed.

The second interview, the stimulated recall interview (SRI) was conducted immediately after the participant completed their online learning task. (See section 5.4 for a detailed description of the stimulated recall interview).

The pre-interviews and stimulated recall interviews were analysed using a thematic analysis (Boyatzis, 1998). Themes are essentially patterns within data sets. Themes can be useful for organizing and describing the possible observations as well as highlight or interpret aspects of a phenomenon (Boyatzis 1998). The interviews were then transcribed and entered into data analysis software, NViVo.

Thematic analysis comprises three separate stages: Stage I, deciding on sampling and design issues; Stage II, developing themes and a code; and Stage III, validating and using the code (Boyatzis 1998). Using a multi-coder process ensured the validity of the codes. After the 21 interviews were transcribed and coded, the researcher had identified over 100 codes. These 100 codes were further aggregated into broader themes. A few of these themes are included in the findings below and include assessment related behavior in terms of motivation, cognitive overload and the learning management system, and impression management.

6 Findings

6.1 Overview

In terms of the qualitative data sources, the social environment of the learning management system was depicted by participants as both a support and a challenge in terms of self-regulation. As this is a longitudinal study, it is also important to keep in mind that the participants in the study were first year university students. As mentioned earlier the eye tracker was primarily used to capture participant gaze as they completed online learning tasks. The researcher chose to debrief eye movement behaviors that supported or were in contrast to what the participant revealed during the pre-interview based on the participants' responses during the pre-interview about goals and motivations. The findings from the thematic analysis indicate that: 1) the gaze of a participant as they navigated through an online learning task varied depending on their goals and motivation; 2) participants were cognitively overloaded by the learning management system; and 3) impression management was revealed as a primary concern when composing forum responses in the online setting.

6.2 Assessment as Motivation

A wide variety of motivations drive the goal directed behaviors participant exhibited as they navigated online coursework. The dominant influences for forum selection and decisions to complete and post into forums are percentage grades, marks, or assessment criteria. Students feel compelled to contribute in forums for the sheer number of points they can accumulate that may influence their final mark in a course. This is, of course, a typical and unsurprising response for the majority of students whether or not they are participating in online course settings or in traditional face to face course settings. However, through the use of the eye tracker software, the researchers can gather data and observe and identify behaviors such as using other students' postings as models, rereading criteria sheets, and how students use lecturer prompts to help them formulate responses in forums. The development and usage of these self-regulatory and co-regulatory strategies are essential to understanding how participants interact with and cope with online course work and the significance of acquiring and using these skills in their future studies.

As the SRL literature highlights, goals serve as a reference point for performance (Koch and Nafziger, 2011). Students are constantly monitoring and adapting their behaviors as they set and interpret expectations, complete short term goals and work toward longer term goals. In this study, as participants worked toward completing a forum task goal they expressed a variety of motivations for behaviors. These task goals were heavily influenced by the way the lecturer set and communicated the task, the way in which the student interpreted the task, and the way in which the participant interacted with artefacts like criteria sheets and course materials.

Participant behavior is influenced by their concern with assessment marks and their longer term goal of achieving a good grade or mark for the course. One factor that greatly influenced whether or not they posted their forum response was whether it contributed to their marks. For example, participant 15 states, "So a small percentage of the mark is required to be participating in the forums. So I choose to do it because I want the marks but it also leads into our discussion in my follow-up tutorial on Thursday". Participant 15 clearly wants the participation points, however when further questioned about why she specifically participates in this forum she elaborates with:

> *P15: I enjoy the analysis of the story and I enjoy looking at the questions and reflecting on the lecture and forming an answer, so the discussion question helps me to make sense of the story, otherwise I could just not even read the story.*

Participant 15 is most likely motivated by participation marks or points that lead toward a perceived better grade. However, she remarks that were it not for the discussion question provided by the lecturer, she may "not even read the story". Her motivation for "performing" and in this case completing an "analysis" which is something she "enjoys" doing, helps her to "make sense of the story". At the same time the discussion question is another motivating factor for the reading assignment to be completed. Thus there are multiple factors of motivations that often appear in the complex online learning environment. While prompts from the lecturers serve as motivation to some participants, other participants use the forum discussion board to regulate others.

Five of the participants, who were participating in small group discussions, all posted their responses, because their forum postings contributed toward a participation grade. In these stimulated recall interviews, participants reveal how they co-regulate the behaviors of other group members. Co-regulation occurs when individuals set and hold standards and goals for each other in relation to progress and contributions to the group task (Järvelä and Hadwin, 2013). The task that these participants engaged in was designed to be a collaborative group assignment. It was necessary to use the forums to communicate with each other. The forum setting became a space for the social interaction that was necessary to facilitate movement toward their shared goal. Hence, in the example below, Participant 2 uses the forum to co-regulate the behavior of the other members of her group.

> *P2: I've been looking at it, yeah. I've got an assessment criteria here that's got the exact wording of what we need to make sure is included in our essay, and I know from doing other essays, it's easy to miss a line, you know, a few words that are key to the finishing of the topic. So when I have read it back, I notice that in all of our chats we hadn't covered that at all, so I just wanted to make sure that today that we did some research on those.*

Participant 2 has reviewed previous posts and has made a judgment on the direction in which her group needs to move. Probing further into how Participant 2 will know that her message will convey what she means, in terms of the criteria sheet, she explains,

> *P2: Because I've literally picked up words that we've missed and I've made sure I've put them in that discussion topic and I've said that, "So far in our chats, we haven't covered these subjects, so would everyone like to research those today just to make sure that we do get them in there?"*

Participant 2 has been explicit in her forum post about criteria standards being met as she values her grades. Clearly, she is trying to achieve the highest mark possible for this assignment and attempting to monitor her group's direction with the task. She is monitoring the content and the progress of the final product by being explicit about the specific steps that need to be undertaken by the group. Rogat and Linnenbrink-Garcia (2011) describe the behavior of Participant 2 as content monitoring. In this view, group members make specific recommendations for the content of the piece they are working on. In this example, Participant 2, may be perceived as a more capable peer as she is monitoring the assignment against the criteria sheet and being assertive in the discussion forum. This interaction is advantageous to her group as they can now become more aware of missing parts of the assignment.

6.3 Cognitive Overload – Eye Movements

Socio-cultural theory posits that motivation and goals depend on interaction with people and the tools that the culture provides to help form their own view of the world. The previous statement is reflective of the participants' interaction and feelings about the learning management system. Participants revealed, through their statements, their discontent with working with the learning management system. Participants experienced cognitive load challenges due to interface design issues. The complexity of the online learning environment was challenging for participants as they communicated they felt "overloaded" by emails from other students as well as frequently getting "lost" in the hyperlinked environment. As this was their first year in university and the learning management system (Moodle) was fairly new to most of the participants, they struggled to cope with the software as well. In this study, the way that the interface or in-formation is presented and structured inhibited the process of learning in a logical and straightforward manner.

For example, Participant 6 had adapted the way she used email in order to avoid going on Moodle.

> *P6: Yeah, I guess because I – I email everyone, like I use email quite a lot, so for me, like the first place I would go to when I come online would be my emails because I figure it's the first thing most people think about, you know, when they're trying to get a hold of someone or wanting to get in contact with someone so, "Okay, I'll just email them," so yeah, I think most people use email so, like – and I do too, so I probably use it a lot more than Moodle, than I do the forum post because I don't usually want to go all the way into Moodle and then you have to go in and in, you know, whereas you just go into – you just log in and then you just go into your emails.*
>
> *Researcher: Okay, so you find the email, you described it as better than Moodle because you go in and in and in.*
>
> *P6: Yeah. Because when I log in, like emails are just there, I just open it, it opens and then I've got what I need, it's all there; whereas if I go into Moodle, you know, there's my other classes and then there's posts from other students about other things and then I get a little distracted and I'll read what they wrote and, you know, I'll read what they wrote and then, you know, there's posts from the lecturers and stuff who make announcements and then I'm like, "Oh, okay, I'll read what they wrote." So then I end up, you know, reading everything on Moodle, whereas – and then I forget the whole purpose of logging in was to check for one response, so sometimes I feel I just need to check my emails first before Moodle.*

Participant 6 elaborates on how Moodle "distracts" her from staying on task as she is faced with a variety of social influences. She notes that she feels obligated to read the "posts from other students" and "posts from lecturers" which results in forgetting "the purpose of logging in".

Participants' learning seemed to be hindered with large amounts of text and feeling lost in the hyperlinks within course content. They expressed frustration about each course having a layout that hinders their access to course materials and useful information. Even though, they matured and grew accustomed to the system during their first year; it will be interesting to note which coping strategies they have developed in terms of the learning management system during the second year of their respective programs.

There is strong evidence that the process of using a learning management system as a way of delivering information must be revisited and perhaps rethought.

Information that is constantly changing and content that may be changing as part of the online learning environment must be taken into account by all members of the institution. From the socio-cultural perspective, this aspect of the learning environment is quite an important space as it is the dominant space where all of the cultural artefacts for their program are found. Understanding the impact and being mindful of this space must be a priority for all institution in order to assist students to become more successful learners.

6.4 *Impression Management*

Participant concerns with how they might be perceived in the online environment strongly influences their choices and behaviors (i.e., not wanting to be the first to post and looking at others postings before starting their post). Echoing Goffman's work on presentation of self (1959), participants' attempts at impression management often guided their engagement with online information and limited their self-expression.

Impression management refers to the process whereby individuals attempt to control the impressions others form of them (Leary and Kowalski, 1997). Goffman (1959) asserts that individuals construct the way they present themselves to an audience either intentionally or unintentionally. The online platform allows learners to present themselves in a variety of ways through the images they project of themselves. For example, in an online learning management system, a student may behave in a way that is different from the way that they would behave in a social networking site. Goffman views people as "actors" who engage in "performances" before "audiences". In Table 2 below, a sub-sample of responses from the stimulated recall interview is presented. The statements reveal participants' concerns regarding the manner in which they present themselves in online discussion forums.

During the stimulated recall interviews, the researcher debriefed participants on particular eye moment behaviors. These eye movement behaviors provided insights into visual patterns showing how the student completed the forum task. For example, eye movement data tracked where and how long they looked at particular items, what and how many different screens they looked at, and how they visually gathered information to complete the forum task. As the eye tracking software provides an enormous set of data points per minute in terms of fixation and saccades, the researcher asked about specific eye movements.

In the table below, the eye movement behavior column characterizes the movements of the eye tracking gaze data utilized by the researcher used as a prompt to inquire about motivations and actions. The researcher, to elicit motivations and metacognitive skills used, inquired about the gaze behavior by asking the participant to: 1) explain what they were experiencing at specific moments; 2) to describe the actions of their eye movement and 3) why they used those particular strategies. The third column reveals the participant responses and motivations behind their eye movements and why they chose those specific behaviors as they completed the task. Lastly, the final column reveals the SRL themes that have emerged from the verbal feedback during the stimulated recall interviews with the participants.

Table 2 SRL strategies and impression management

Participant	Eye Movement Behavior	Verbalizations about Impression Management	SRL Strategies
P2	Characterized by the software capturing a scanpath super imposed over text composed by the participant. This particular scanpath goes from left to right with a number of fixations and then a backtracking scanpath to the beginning of the paragraph.	*So now I'm rereading it just to make sure it still sounds that I've – that I've managed to get my request across, but that it also still reads back friendly, upbeat, not too instructional. So I'm just rechecking it just to make sure that if I was receiving this email that I would feel like I know what the person wants but that they're also saying, "Hi, and how's everything going.*	Self-monitoring by re-reading, self-reflecting
P4	The characteristics involved in this specific eye movement occurred when the participant moved and clicked the cursor in the text box and the eye movement behavior showed a large fixation that grew in size, indicating that the participant seemingly held their gaze at that specific spot for a few seconds.	*... I always just watch what I'm saying and I want to say it the right way without offending anyone or anything like that, so I just really think about what I'm typing before I start typing...So it's easy for them to understand. I don't sound like an idiot, so I'm trying – I suppose I'm just trying to type in a different way than what I would speak it.*	Planning by pausing and thinking about how to communicate with the audience
P12	There were no specific eye movements that paired with this statement as the participant was generally reflecting on how they evaluated their process for completing the task.	*I'm trying to – I always try to make it – make myself seem smart. So I'm trying to formulate a way of saying it concisely and saying it like intelligently...*	Self-monitoring by evaluating
P18	Characterized by the software capturing a scanpath super imposed over text composed by the participant. This particular scanpath goes from left to right with a number of fixations and then a backtracking scanpath to the beginning of the paragraph.	*Well I want people that read it to be able to understand what I'm saying and I want it to sound good so I'm trying to put it in the best words that I can think of... I don't want people to judge me and go, "Oh, that sounds really dumb.*	Self-monitoring by re-reading and self-reflecting

Impression management influences students' behavior as they make a plan for constructing text. Participant 2 wants her post to come across as "friendly" and "upbeat", thus she rereads her material trying to imagine how she would feel if she were receiving the same message. She self-monitors by re-reading the text she has constructed because she is concerned with the image she is trying to convey to the audience. Similarly, Participant 12 wants to "seem smart" and Participant 4 does not want to "sound like an idiot", so they both spend a significant amount of time thinking about what they will write. There is a range of self-regulatory processes (self-monitoring, planning, and reflecting) between the participants. However, the motivations for the behaviors are similar. They address the need for participants to present themselves in a favorable way to their online classmates and professor. A social setting, like the online learning management setting, causes a heightened awareness of self, as students rely heavily on their text and construction of dialogue to represent what they know, their values and beliefs. This concern with impression management is apparent in and supported by the self-report data about the SRL processes participants engage in when constructing text in online forums.

7 Conclusions

The complex and dynamic online learning environment places a variety of demands on students. In order to become successful learners, students must manage their own learning especially when studying by distance or blended modes. Areas of concern resulting from this investigation include: 1) implications of eye gaze data regarding how students allocate attention as they work toward a meaningful online goal oriented tasks and 2) how participants present themselves in the online environment and how this presentation influences their self-regulatory behaviors.

7.1. Assessment as Motivation

University students are mostly assessment driven. Students' goals are framed by past performance and their preferred grades or marks. During the stimulated recall interviews, participants expressed that their visual behavior was impacted by the need to work toward meeting forum task goals and a common theme was about the potential to impact an overall grade or mark in the course. In the example provided in Table 2 of this chapter from Participant 2, the explanation of the motivation behind her eye movement behaviors align with socio-cultural themes highlighted earlier by Schunk and Zimmerman (1997). Thus, Participant 2's awareness of the social interaction of her forum post is influencing her behavior or self-regulatory control. Specifically, she is monitoring by re-reading and reflecting on her forum post to ensure she conveys a friendly tone. By using the criteria sheet, the participant is self- and co-regulating her own and her group's behaviors by attempting to construct a forum post to direct focus and attention to very specific aspects of the task. This behavior has the potential to impact on Participant 2's longer term goal of acquiring a "high" grade or mark for the course overall. However, in terms of co-regulation and motivation, Walker (2010) highlights that sometimes the process of co-regulation may both support and impede learning and motivation.

Broadly, online university students manage and regulate their behavior on perceived assessment criteria as well as the social environment in which they are participating. As they strive to understand cultural artefacts (i.e. criteria sheets) in their course programs, students are also struggling to understand the social nuances that occur in social settings like forum discussion boards. Artefacts like criteria sheets are strong foundational tools within these settings. Participants' in this study were highly aware and motivated by the assessment criteria sheet and used this tool to help identify their goals, self- and co regulate their behaviors and attempt to co-regulate the behaviors of others.

7.2 Cognitive Overload – Eye Movements

For this group of participants, their unfamiliarity with online learning environments placed a strong constraint on some of their self-regulatory behaviors. Specifically, they expressed frustration with the differing interface designs across multiple courses. These differences hindered their ability to navigate through course content in a logical and straight forward manner. Additionally, participants expressed frustration with the amount of content their lecturers placed in their weekly tasks. However, due frustrations, participants adapted their behavior to avoid challenging aspects of the learning environment.

The overall implications of the feedback provided by the participants suggest that university stakeholders need to be aware of the repercussions of placing excessive content on the learning management system, differing interface designs, and the subsequent impact of "information overload".

7.3 Impression Management

After the first round of data collection and analysis, the researcher found that the need for impression management influenced SRL processes. Specifically, when planning to compose text, participants were highly aware of how they set the tone of their response. Planning involved more than conceptualizing what they wanted to say and the focus shifted to how they wanted to say it. Their behaviors were motivated by the need to manage how their message was perceived by their peers and their lecturer. Participants also self-monitored by re-reading to check to ensure that their words sounded "friendly".

There are implications for course coordinators and stakeholders who support students, as there may be a mistaken assumption that students are confident in their impression management. Participants may have active social media experience and participate in Facebook, Twitter and other platforms, and may feel very comfortable posting ideas to these forms of social media. However, they may struggle to present themselves as scholars when they are placed online learning platforms in an academic setting.

Understanding the connections and implications of students' impression management and self-regulation in an online setting must be a priority for university stakeholders, particularly course coordinators, online tutors and lecturers. As this is data from the first data collection point, it is possible that students may not report impression management concerns in the second data collection point as they adapt and grow accustomed to the online learning environment. The stakeholders who have access to this online learning environment must understand how students' identities are continuously being shaped by the experiences of online learning and their participation in this academic and social environment.

References

Anson, C. M., & Schwegler, R. A. (2012). Tracking the Mind's Eye: A New Technology for Researching Twenty-First-Century Writing and Reading Processes. College Composition and Communication, 64(1), 151–171.

Azevedo, R., Cromley, J. G., Moos, D. C., Greene, J. A., & Winters, F. I. (2011). Adaptive content and process scaffolding: A key to facilitating students' self-regulated learning with hypermedia. Psychological Test and Assessment Modeling, 53(1), 106–140.

Azevedo, R., Johnson, A., Chauncey, A., & Burkett, C. (2010). Self-regulated learning with MetaTutor: Advancing the science of learning with MetaCognitive tools. In M.S. Khine, I.M. Saleh (Eds.), New Science of Learning: Computers, Cognition, and Collaboration in Education, (pp. 225–247). New York: Springer.

Boyatzis, R. (1998). Transforming qualitative information: Thematic analysis and code development. London: Sage.

Calvi, C., Porta, M., & Sacchi, D. (2008, July). E5Learning, an e-learning environment based on eye tracking. In Advanced Learning Technologies, 2008. ICALT'08. Eighth IEEE International Conference on (pp. 376–380). IEEE.

Edwards-Leis, C. (2007). Variations to stimulated recall protocols to enhance student reflection: I did, I saw, I remembered. In Proceedings of the 2006 International Education Research Conference In: 2006 AARE Conference, 26-30 November 2006, Adelaide, Australia. Retrieved from http://www.aare.edu.au/06pap/edw06717.pdf.

Eger, N., Ball, L. J., Stevens, R., & Dodd, J. (2007, September). Cueing retrospective verbal reports in usability testing through eye-movement replay. In Proceedings of the 21st British HCI Group Annual Conference on People and Computers: HCI... but not as we know it-Volume 1 (pp. 129–137). British Computer Society.

Furberg, A., & Ludvigsen, S. (2008). Students' Meaning-making of socio-scientific issues in computer mediated settings: Exploring learning through interaction trajectories. International Journal of Science Education, 30(13), 1775–1799.

Gass, S.M., and Mackey, A. (2000). Stimulated recall methodology in second language research, Mahwah, NJ: Routeledge.

Goffman, E. (1959). The presentation of self in everyday life. New York: Doubleday Anchor.

Greene, J. A., & Azevedo, R. (2010). Introduction: The measurement of learners' self-regulated cognitive and metacognitive processes while using computer-based learning environments. Educational Psychologist, 45(4), 203–209.

Hadwin, A.F., & Oshige, M. (2011). Self-regulation, coregulation, and socially shared regulation: Exploring perspectives of social in self-regulated learning theory. Teachers College Record, 113(6), 240–264.

Hadwin, A. F., Oshige, M., Gress, C. L., & Winne, P. H. (2010). Innovative ways for using gStudy to orchestrate and research social aspects of self-regulated learning. Computers in Human behavior, 26(5), 794–805.

Hesse-Biber, S. (2010). Mixed methods research: Merging theory with practice. New York: Guilford.

Hyönä, J., Lorch Jr, R. F., & Kaakinen, J. K. (2002). Individual differences in reading to summarize expository text: Evidence from eye fixation patterns.; Journal of Educational Psychology, 94(1), 44.

Kaplan, A., & Maehr, M. L. (2007). The contributions and prospects of goal orientation theory. Educational Psychology Review, 19(2), 141–184.

Koch, A. K., & Nafziger, J. (2011). Self-regulation through Goal Setting. The Scandinavian Journal of Economics, 113(1), 212–227.

Kostons, D., van Gog, T., & Paas, F. (2010). Self-assessment and task selection in learner-controlled instruction: Differences between effective and ineffective learners. Computers & Education, 54(4), 932–940.

Janssen, J., Erkens, G., Kirschner, P. A., & Kanselaar, G. (2012). Task-related and social regulation during online collaborative learning. Metacognition and Learning, 7(1), 25–43.

Järvelä, S., & Hadwin, A. F. (2013). New frontiers: Regulating learning in CSCL. Educational Psychologist, 48(1), 25–39.

Järvelä, S., Järvenoja, H., & Näykki, P. (2013). Analyzing regulation of motivation as an individual and social process. In S.Volet and M. Vauras (Eds.), Interpersonal regulation of learning and motivation: Methodological advances (pp. 170 – 188). New York:Routledge.

Johansson, R., Wengelin, Å., Johansson, V., & Holmqvist, K. (2010). Looking at the keyboard or the monitor: relationship with text production processes. Reading and writing, 23(7), 835–851.

Lavie, N., Hirst, A., de Fockert, J. W., & Viding, E. (2004). Load theory of selective attention and cognitive control. Journal of Experimental Psychology: General, 133(3), 339.

Leary, M. R., & Kowalski, R. M. (1990). Impression management: A literature review and two-component model. Psychological bulletin, 107(1), 34.

Maehr, M., & Nicholls, J. (1980). Culture and achievement motivation: A second look. In N. Warren (Ed.), Studies in cross-cultural psychology (Vol. 2, pp. 221–267). New York: Academic Press.

McCaslin, M., & Burross, H. L. (2011). Research on individual differences within a sociocultural perspective: Coregulation and adaptive learning. Teachers College Record, 113(6).

Muldner, K., Christopherson, R., Atkinson, R., & Burleson, W. (2009). Investigating the utility of eye-tracking information on affect and reasoning for user modeling. In User Modeling, Adaptation, and Personalization (pp. 138–149). Springer:Berlin.

Poole, A., & Ball, L. J. (2006). Eye tracking in human-computer interaction and usability research. In C. Ghaoui (Ed.), Encyclopedia of human computer interaction (pp. 211–219). Pennsylvania: Idea Group.

Rogat, T. K., & Linnenbrink-Garcia, L. (2011). Socially Shared Regulation in Collaborative Groups: An Analysis of the Interplay Between Quality of Social Regulation and Group Processes. Cognition and Instruction, 29(4), 375–415.

Schraw, G. (2010). Measuring self-regulation in computer-based learning environments. Educational psychologist, 45(4), 258–266.

Schunk, D. H., & Zimmerman, B. J. (1997). Social origins of self-regulatory competence. Educational psychologist, 32(4), 195.

Tobii Techonology(2010). Tobii Eye Tracking: An introduction to eye tracking and Tobii Eye Trackers [White paper]. Available online: http://www.tobii.com/en/eye-tracking-research/global/library/white-papers/tobii-eye-tracking-white-paper/

van Gog, T., Kester, L., Nievelstein, F., Giesbers, B., & Paas, F. (2009). Uncovering cognitive processes: Different techniques that can contribute to cognitive load research and instruction. Computers in Human Behavior, 25(2), 325–331.

van Gog, T., Paas, F., van Merriënboer, J. J., & Witte, P. (2005). Uncovering the Problem-Solving Process: Cued Retrospective Reporting Versus Concurrent and Retrospective Reporting. Journal of Experimental Psychology: Applied, 11(4), 237.

van Gog, T., & Scheiter, K. (2010). Eye tracking as a tool to study and enhance multimedia learning. Learning and Instruction, 20(2), 95–99.

Vygotsky, L. S. (1978). Mind in society: The psychology of higher mental functions: Cambridge: Harvard University Press.

Walker R. A. (2010). Sociocultural Issues in Motivation. In: P. Peterson., E. Baker, B. McGaw (Eds.) International Encyclopedia of Education. Vol. 6, (pp. 712–717). Oxford: Elsevier.

Whipp, J. L., & Chiarelli, S. (2004). Self-regulation in a web-based course: A case study. Educational Technology Research and Development, 52(4), 5–21.
Wiklund-Engblom, A. (2010). Exploring Conative Constructs and Self-Regulation of E-Learners: A Mixed Methods Approach. In Proceedings of the 27th ASCILITE 2010 Conference, Sydney, Australia, Dec 5-7. Retrieved from http://www.ascilite.org.au/conferences/sydney10/Ascilite%20conference%20proceedings%202010/ Wiklund_Engblom-full.pdf
Winters, F. I., Greene, J. A., & Costich, C. M. (2008). Self-regulation of learning within computer-based learning environments: A critical analysis. Educational Psychology Review, 20(4), 429–444.

Erratum

Chapter 15

Private and Public: Eye Movement and Eye Tracking in Marketing

En Li, James Breeze, Mike Horsley and Donnel A. Briely

M. Horsley et al. (eds.), *Current Trends in Eye Tracking Research*,
DOI 10.1007/978-3-319-02868-2_15, © Springer International Publishing
Switzerland 2014

DOI 10.1007/978-3-319-02868-2_15

In the original publication the last name of co-author chapter 15, is misspelt.
It should be: Donnel A. Briley

M. Horsley et al. (eds.), *Current Trends in Eye Tracking Research*,
DOI 10.1007/978-3-319-02868-2_26, © Springer International Publishing Switzerland 2014

Index

M. Horsley et al. (eds.), *Current Trends in Eye Tracking Research,*
DOI 10.1007/978-3-319-02868-2, © Springer International Publishing Switzerland 2014